Price AE- 10. -

→ Castiglione - .377.
  No 2. - .- 342
→ Moment Dist. _ 25f

THE ANALYSIS OF STRUCTURES

THE ANALYSIS OF STRUCTURES

BASED ON THE MINIMAL PRINCIPLES AND THE PRINCIPLE OF VIRTUAL DISPLACEMENTS

by

NICHOLAS JOHN HOFF

HEAD OF DEPARTMENT OF AERONAUTICAL ENGINEERING AND APPLIED MECHANICS
POLYTECHNIC INSTITUTE OF BROOKLYN
BROOKLYN, NEW YORK

JOHN WILEY & SONS, INC.

New York · London · Sydney

Library of Congress Catalog Card Number: 56–6503

PRINTED IN THE UNITED STATES OF AMERICA

TO THE MEMORY OF J. PAUL HOFF

P R E F A C E

Many students of engineering consider the analysis of structures a rather boring subject. They do not understand the justification of the assumptions made, and they get tired of memorizing scores of unrelated formulas, rules, and procedures. It is the hope of the author that in this volume he has proved that these objections to the theory of structures are not justified because the study of structures can be an exciting undertaking. The various methods used by structural engineers are all logically interconnected if one looks upon them from a unifying standpoint. Not a single formula need be memorized if the fundamental principles are understood; every method presented in this book is an almost obvious consequence of the basic principles and can be derived with a minimum amount of mathematical manipulation. The various devices suggested for increasing the accuracy of the calculations and reducing the work involved may appear as tricks to the uninitiated. But to the reader of this book, these devices should become obvious applications of the basic principles.

The first purpose of this volume is therefore the presentation of structural analysis as a logical and unified theory based on a small number of first assumptions. To achieve it, the principle of virtual displacements is chosen as the starting point. The second purpose is the development of the minimal principles of structural theory; they have been used so generally in the engineering literature of the last few years that no modern practicing engineer can afford to ignore them. Finally an effort is made to give the reader a thorough understanding of buckling phenomena.

The book is a consolidation of lecture notes prepared by the author for a first-year graduate course given to students of aeronautical, civil, and mechanical engineering and of applied mechanics. Although it is assumed that the reader has had an undergraduate training in engi-

vii

neering, the presentation is simple enough to allow a practicing engineer, who is a little rusty on mathematics and mechanics, to read the book by himself, without advice from a teacher. Whenever mathematical methods beyond the elementary principles of the calculus are used, they are explained in sufficient detail.

The fundamental information is presented in Articles 1.1 to 1.4 of Part 1, Articles 2.1 and 2.2 of Part 2, Article 3.1 of Part 3, and Article 4.1 of Part 4. These articles should be studied in the order given. The rest of the book contains applications to various structures, and principles of a less fundamental nature; it can be read by a person with some experience in structural analysis in almost any order. To facilitate a random study of the various problems treated, references to preceding articles have been kept to a minimum. Of course, the beginner will find it easier to follow the text in the order presented.

Footnotes are used sparingly because they tend to disrupt the continuity of presentation. Remarks usually given in footnotes are collected in an appendix which also serves as a bibliography.

It is hoped that the reader, who approaches the subject for the first time, will find it just as interesting as the author who has had over twenty-five years of experience in studying and applying it, and fifteen years of teaching it.

I am indebted to many of my friends with whom I have had occasion to discuss the subject. Among them I want to mention Professors Frederick V. Pohle and Philip G. Hodge Jr., of the Polytechnic Institute of Brooklyn. I am particularly grateful to my wife, Mrs. Vivian Church Hoff, for her continuous help during the years in which the material for the book was collected.

NICHOLAS J. HOFF

Brooklyn, February 1956

C O N T E N T S

ix

P A R T   O N E

The Principle of
Virtual Displacements

1.1 Importance of the Principle

The principle of virtual displacements is one of the simplest principles of mechanics and can even be set forth as an axiom.   Here it will be developed from the principle of the parallelogram of forces with the aid of the concept of work as defined in mechanics.

The principle is probably the most powerful tool of structural analysis. Though it can be derived without complex mathematics and applied to a group of problems with great ease, its application to many other problems necessitates care and a clear understanding of the fundamental concepts.

In this book the various theorems and methods of structural analysis are derived directly or indirectly from the principle of virtual displacements.   Its generalization is the first minimal principle discussed in Part 2.   In Part 3 the problem of the stability of a system is reduced to the establishment of equilibrium in a neighboring state, and thus it is made amenable to an analysis by the first minimal principle.   Finally in Part 4 the complementary energy principle is proved by means of the principle of virtual displacements.

The entire book is based, therefore, on the principle of virtual displacements.   For this reason the principle is derived and discussed in some detail in the next few articles.

1.2 Statement of the Principle for a Mass Point

The condition of equilibrium of a planar system of forces acting on a mass point† is usually stated in one of the following ways:

1. The sum of the vertical components of the forces and the sum of the horizontal components of the forces must vanish.

2. The forces must form a closed polygon.

3. The vector sum of the forces must be equal to zero.

† The mass point is the idealized concept of a small particle with a vanishingly small volume but a finite mass.

1

Each of these statements is a necessary and sufficient condition of equilibrium. *Sufficient* means that, if any one of the conditions is fulfilled, the mass point must be in equilibrium. On the other hand, if the mass point is in equilibrium, all the three conditions must be satisfied. This is indicated by the adjective *necessary*.

Since the equilibrium condition can be expressed in three different forms, it may be possible to find a fourth way of stating it. Indeed, the principle of virtual displacements is just another formulation of the condition of equilibrium.

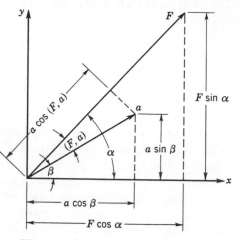

Fig. 1.2.1   Force and displacement.

This formulation involves the concept of work. The word *work* is commonly used in everyday language; in the terminology of mechanics it is only vaguely related to human labor and is given an explicit meaning only through its definition:

Work is the product of a force and the displacement of its point of application in the direction of the force. In the form of an equation

$$W = Fa \cos (F, a) \qquad\qquad 1.2.1$$

where W is the work, F the force, a the displacement of the point of application of the force, and (F, a) the angle subtended by the directions of F and a (see Fig. 1.2.1).

This definition may seem unnecessarily elaborate because of the intuitive feeling that a mass point must move in the direction in which the force is acting. This feeling, however, is fallacious. Newton's second law states only that the acceleration imparted to a mass point has the same direction as the force, but it says nothing about the velocity. Of course, the direction of the velocity is the same as that of the acceleration if the mass point is originally at rest, but this is true only if it is at rest when the force is applied.

A simple example can illustrate this point. If a stone is thrown horizontally, the only force acting on it during the motion is its weight if the aerodynamic forces can be disregarded. Although the weight acts vertically, the direction of motion is almost horizontal at the beginning, and the path becomes inclined downward only gradually. If time is counted from the moment when the missile leaves the thrower's hand, the initial conditions are zero vertical and a given finite horizontal velocity. Because of the latter, force and velocity have different directions. The work done by the weight is the weight times the vertical downward displacement, and not the weight times the total length of the path of the stone.

The need for a clear-cut definition of work is apparent when the work done by a single force is of interest although a number of forces are acting. For instance, one might wish to calculate the work quantities when a man pushes a sled up an inclined, snow-covered but sanded road where there is considerable friction.   The work done by the man is equal to the force exerted by him times the distance traveled in the direction of the force, while the work done at the same time by the weight of the sled is the weight times the difference in elevation.   The latter quantity is negative, since the weight acts downward and the vertical component of the displacement is directed upward.

Equation 1.2.1 can be rewritten with the aid of the notation of Fig. 1.2.1 in the following manner:

$$W = Fa \cos (F, a) = Fa \cos (\alpha - \beta)$$

$$= Fa (\cos \alpha \cos \beta + \sin \alpha \sin \beta)$$

$$= (F \cos \alpha)(a \cos \beta) + (F \sin \alpha)(a \sin \beta)$$

$$= F_x a_x + F_y a_y \qquad\qquad (a)$$

where F_x, F_y and a_x, a_y are the x and y components of the force F and the displacement a, respectively.   In a similar manner it can be shown that in three-dimensional space the work done by a force during a displacement is the sum of the work done by the three rectangular (Cartesian) components of the force:

$$W = F_x a_x + F_y a_y + F_z a_z \qquad\qquad 1.2.2$$

where F_x, F_y, and F_z are the three rectangular components of the force, and a_x, a_y, and a_z the three rectangular components of the displacement.

If a second force P is acting simultaneously with F in the planar system of Fig. 1.2.1, the work W_P done by it during the displacement a is

$$W_P = P_x a_x + P_y a_y$$

The total work W_{tot} done by F and P when they are applied simultaneously is therefore

$$W_{tot} = F_x a_x + F_y a_y + P_x a_x + P_y a_y$$
$$= (F_x + P_x)a_x + (F_y + P_y)a_y$$
$$= R_x a_x + R_y a_y$$

where R_x and R_y are the rectangular components of the resultant R of the forces F and P. Because of Eq. a, this equation can be written in the form

$$W_{tot} = Ra \cos (R, a) \tag{b}$$

Equation b states that the work done by forces F and P when they act simultaneously on a mass point during a displacement a of the mass point is equal to the work done by the resultant R of the forces F and P.

In general, the sum of the work done by a number of forces acting on a mass point during a displacement of the mass point is equal to the work done by the resultant of these forces during the same displacement. Therefore, if the forces are in equilibrium and consequently their resultant is zero, the sum of the work done by all the forces during any displacement of the mass point must be equal to zero.

It is therefore possible to assume arbitrarily a displacement for the mass point and to calculate the work done by all the forces acting on the mass point. If the forces constitute an equilibrium system, the sum of the work must vanish. This must be true irrespectively, whether the forces are caused by electric or magnetic attraction, by gravity, or by elastic distortions. It is important to realize that the assumed displacement has nothing to do with the real motion of the mass point; it is purely fictitious and serves solely as an artifice in determining whether the condition of equilibrium is satisfied. For this reason the magnitude and the direction of the forces acting on the mass point must be considered fixed, even though they would change during a real displacement. They cannot be affected by a purely fictitious displacement which only serves in the calculation of fictitious work.

Often the displacements are assumed very small in order to avoid the difficulty caused by changes in the forces. For reasons to be discussed later, infinitesimal virtual displacements have advantages in the investigation of finite elastic or rigid bodies. When the equilibrium of a system consisting of a mass point suspended from a spring is tested by the principle of virtual displacements, the displacement may be assumed indefinitely small.   The change of the elastic force in the spring during an indefinitely small displacement can certainly be neglected as compared to the original force in the spring.   It is not necessary, however, to assume the displacements of the mass point as infinitesimal.   Finite displacements can serve the purpose just as well, as long as they are thought of as fictitious displacements which must be multiplied by the forces corresponding to the real state of the spring, not to the state that would prevail after an actual finite displacement.

Though it should be clear from this discussion that the sum of the work done by all the forces acting on a mass point must vanish for any fictitious displacement if the forces constitute an equilibrium system, this does not imply that the converse is also true.   If the sum of the work is calculated for a fictitious displacement and is found to be zero, the mass point need not be in equilibrium.   An example will clarify this statement.

In Fig. 1.2.2 a vertical upward force of 2 lb and a vertical downward force of 1 lb are acting on the mass point.   Under these forces the mass point obviously cannot be in equilibrium.   Indeed, if a displacement of 1 in. upward is assumed, the upward force does 2 in.-lb and the downward force −1 in.-lb work.   Their sum is +1 in.-lb, not equal to zero which indicates that the system of forces is not in equilibrium.   On the other hand, if a horizontal displacement of 1 in. is assumed, the work done by each of the forces is equal to zero, and consequently their sum also vanishes, since the displacement is perpendicular to the direction of the forces.   In this case the sum of the work vanishes though the forces are not in equilibrium.   By investigating only the work done during this second displacement, one may draw false conclusions regarding the equilibrium.   Therefore, the condition of equilibrium of a mass point should be stated in the following final form:

Theorem† 1   A mass point is in equilibrium if the sum of the work done by all the forces acting on it is equal to zero for any fictitious displacement.

This is again a necessary and sufficient condition of equilibrium.   It is the statement of the principle of virtual displacements for a mass point.   Of course, it may seem unnecessary and useless to replace the simpler conditions of equilibrium given at the beginning of this article by this more complicated one.   However, it should be remembered that the three

† A theorem is a statement of a principle that can be proved.

elementary conditions can be used alternately to advantage in different problems.   It will be shown in later articles that frequently a generalized version of Theorem 1 permits the calculation of problems whose solution cannot even be attempted by means of the simpler conditions of equilibrium.

1.3   Extension of the Principle to Systems of Points

The principle of virtual displacements can be extended to apply to systems of mass points.   Figure 1.3.1 shows a system consisting of seven points.   The forces acting between them are assumed to be produced by electric attraction and repulsion.   The point at the center has a positive

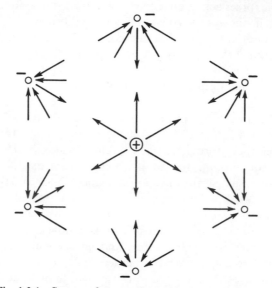

Fig. 1.3.1   System of mass points with electrostatic forces.

charge, and the other points have smaller and equal negative charges.   In such a system the forces are known to be proportional to the charges and inversely proportional to the square of the distances.   Their directions are represented by the arrows drawn in the figure.   With suitable charges the seven points constitute an equilibrium system.

The principle of virtual displacements can easily be applied to the investigation of the equilibrium of any one of the points.   The point in question must be displaced in some direction, and the work done by all the forces acting on the point must be calculated.   Since the displacement only serves as an artifice for the proposed investigation and is not an actual displacement, the forces must be taken as they are when the points are in the positions shown in Fig. 1.3.1.   If the point is in equilibrium under the specified forces, the sum of the work done by the forces during the fictitious displacement is zero.

The same reasoning applies to every point.   If the principle of virtual displacements is written in the form of an equation for each point, and the seven equations are added up, the resulting equation can be interpreted in the following manner:

A system of mass points is in equilibrium if the total   **Theorem 2**
work done by all the forces acting on all the points of the
system is equal to zero for every arbitrary displacement.

In the derivation of this theorem the nature of the forces acting on the points is of no consequence.   The reasoning would be the same for any other type of forces.   Therefore, Theorem 2 must also be valid for the

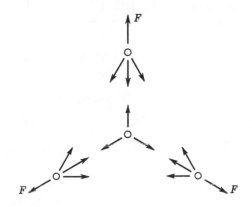

Fig. 1.3.2   System of mass points with elastic and external forces.

system of Fig. 1.3.2, which consists of four mass points connected by weightless springs.   In this system the springs are stretched under the action of the external forces.   In the state shown in the figure, the elastic forces, which are proportional to the elongations of the springs, are assumed to be in equilibrium with the external forces.   The application of the principle of virtual displacements to the point at the center does not differ from that in the previous example.   With the remaining three points the only novel feature is that one of the forces is an external force. This fact, however, cannot influence the application of the principle since the nature of the forces is immaterial.   If the equations containing the statement of the principle for each point are summed up again, the resulting equation has the same form as in the earlier example.   The total work can now be broken up into two parts, one corresponding to the work done by the external forces, and the other to that done by the internal forces.

Then the principle can be stated as follows:

> The total work done by the external forces plus the total **Lemma† 1**
> work done by the internal forces is equal to zero for any
> arbitrary displacement if the system of mass points is in
> equilibrium.

The converse of this statement is:

> A system of mass points is in equilibrium if the total work **Theorem 3**
> done by all the external forces plus the total work done by
> all the internal forces is equal to zero for any arbitrary
> displacement.

If the number of points is increased indefinitely, the statement of the principle of virtual displacements remains unchanged. It must therefore hold for any gas or liquid, and for any rigid, elastic, or plastic body. It should be remembered that, in its present derivation, only the principle of the parallelogram of forces and the concept of work have been used. Consequently the validity of the principle of virtual displacements does not depend on the principle of superposition and on Hooke's law.

1.4 Virtual Displacements

The next question is how many displacements need be investigated before the equilibrium of a system is established conclusively. Of course, it has already been mentioned that every possible displacement of the system of mass points must be considered, and, if the sum of the work is not equal to zero in even a single case, the system cannot be in equilibrium. Therefore the only remaining task is to determine how many displacements are possible for any particular system.

With a single mass point in a plane, every possibility of displacement is exhausted by an arbitrary displacement in the x direction followed by an arbitrary displacement in the y direction. If the work done by all the forces acting on the mass point is zero for the displacement

$$1. \quad x = a, \qquad y = 0 \qquad\qquad 1.4.1$$

and for the displacement

$$2. \quad x = 0, \qquad y = b \qquad\qquad 1.4.2$$

where a and b are arbitrary positive or negative (non-zero) numbers, it must also be zero for any combination of the two, that is, for any arbitrary displacement in the plane. This follows directly from Eq. 1.2.2 which establishes the rule for the calculation of the work using the components of the displacement.

---

† A lemma is a preliminary proposition used in proving another proposition or theorem.

It is not necessary, however, to stipulate the two displacements in such a manner.   One may just as well consider two displacements defined by

$$3. \quad x = c, \qquad y = d$$

and   1.4.3

$$4. \quad x = e, \qquad y = f$$

provided that

$$d/c \neq f/e \qquad\qquad 1.4.4$$

In inequality 1.4.4, d/c is the slope of the line connecting the original position of the mass point with its final position as defined by displacement 3; consequently it is a measure of the direction of displacement 3. Similarly f/e is a measure of the direction of displacement 4.   As long as these two directions are different, any displacement in the plane can be considered as a combination of displacements 3 and 4.   Consequently equilibrium is certain if the work done by all the forces is zero for displacement 3 as well as for displacement 4.   Any set of two independent displacements suffices in the investigation of the equilibrium of a mass point that can only move in a plane.   A mass point in a plane is said to have two degrees of freedom of motion.

In the more general case, when the mass point can move freely in space, it has three degrees of freedom of motion, and three independent directions of displacement must be investigated.   With the system of mass points shown in Fig. 1.3.2, each point has two degrees of freedom of motion if the displacements are restricted to a plane.   The total number of degrees of freedom of the system is therefore $4 \times 2 = 8$.   This means that the work done by all the forces acting on the system must be calculated for eight independent displacements.   As an independent displacement one may choose a non-vanishing displacement component of a single point, with all the other displacement components equal to zero.   But frequently it is more convenient to give simultaneously all the four points non-vanishing displacements.   Then again the establishment of zero work for eight independent combinations of the individual displacement components is proof of the equilibrium of the system of forces acting on the four mass points.

It is evident that the work involved in checking the equilibrium of forces by the principle of virtual displacements must become onerous when the number of mass points, and thus the number of degrees of freedom of motion, increases.   Fortunately it is often unnecessary to investigate all the possible displacement patterns.   This is true whenever constraints are imposed on the movement of the points.

As an example, a system of mass points will be considered on which internal and external forces are acting.   The forces may be caused by mass attraction, extensions of massless springs, electric attraction and repulsion, or by any other effect.   Let it be assumed that the ith and pth mass points are connected by a perfectly inextensible bar which nevertheless has no rigidity in bending, shearing, or twisting.   The bar represents a simple constraint: It prevents changes in the distance between points i and p.   If there are altogether n mass points and they are located in a plane, equilibrium of the system is established if it is found that the total work done

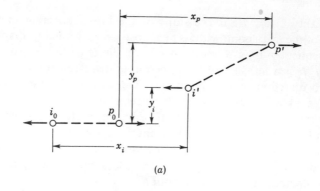

(a)

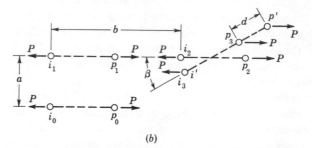

(b)

Fig. 1.4.1   Displacements of points i and p. (The subscript 0 indicates the initial position and the prime the final position.)

by all the forces is equal to zero during $2n$ independent combinations of the elementary displacement components.   Among these $2n$ displacement patterns there will be at least one in which the distance between points i and p is changed because $2n$ independent displacement patterns suffice for the representation of any arbitrary displacement.

Since the bar between points i and p has only extensional rigidity, it can transmit forces to points i and p only in the direction ip.   Equilibrium of the bar requires that the force exerted by it on point i be equal and opposite to that exerted by it on point p.   The magnitude of this unknown force will be denoted by P; its positive direction is shown in Fig. 1.4.1.

In the investigation of the equilibrium of the system of mass points by means of the principle of virtual displacements, all the mass points must be given arbitrary displacements. In the presence of internal constraints it is advantageous to consider the displacements as very small quantities. The initial positions of points i and p are indicated in Fig. 1.4.1 by the subscript 0, and their final positions by the prime. All displacements are shown greatly enlarged, because otherwise the final positions could not be distinguished from the initial ones in the drawing. In Fig. 1.4.1a the two components of the displacement of point i are x_i and y_i, and those of point p are x_p and y_p. Figure 1.4.1b shows how the same final positions can also be reached through four displacement patterns consisting of a vertical translation a, a horizontal translation b, a rotation β, and a stretching d.

In the calculation of the work done by all the forces acting on all the mass points of the system, it is convenient to divide the work into two parts. The first part consists of the work done by the two forces P exerted by the inextensible bar, and the second part contains the work of all the other forces Q in the system. The forces P do work during infinitesimal displacements only if the distance between points i and p is changed; this can be easily verified with the aid of Fig. 1.4.1b.

When points i and p are translated through the same distance in the direction of the rigid bar connecting the two, the positive work done by the force at one end is equal in absolute value to the negative work done by the force at the other end. When the two points are displaced through the same distance perpendicularly to the direction of the connecting line, neither force does any work. The same is true of a rotation of the connecting line about its midpoint if the angle β is very small, because then the forces P move perpendicularly to their line of action, but the work would not vanish if β were finite.[†] When finally the distance between points i and p is increased by d, the work done by the two forces is Pd. Consequently the principle of virtual displacements can be written in the following form when the displacement pattern of the n mass points involves a change in the distance ip:

$$\Sigma Qc + Pd = 0 \qquad\qquad 1.4.5$$

Here Q stands for any of the forces acting on the mass points of the system except force P, c is the displacement of the point of application of the force Q in the direction of Q, d is the change in the distance between points i and p, and the summation must be extended over all the forces Q.

If all the forces Q are known, Eq. 1.4.5 can be used to calculate P. In other words, whatever the magnitude of the forces Q, it is always

† With a finite rotation, the principle of virtual displacements would still be valid, but Eq. 1.4.5 would become inconveniently complicated.

possible to find a value for P that satisfies the condition of equilibrium expressed in Eq. 1.4.5. Because of the assumed infinite load-carrying capacity of bar ip in tension and compression, the bar can transmit a force P of any magnitude, and for this reason the actual magnitude of P is of no interest. Consequently Eq. 1.4.5, that is, the condition of equilibrium obtained by assuming a change in the length of an inextensional bar, is not needed for the establishment of the equilibrium of the forces acting on the system of n mass points.

Intuitively, a rigid body can be considered as a system of an indefinitely large number of mass points with such constraints imposed on the system that the distances between the mass points remain unchanged. In the investigation of the equilibrium of a rigid body it suffices to take into account displacement patterns that do not violate the constraints. These displacement patterns are known as rigid-body motions, that is, displacements in which the shape of the body remains unchanged. It is, of course, permissible to consider displacement patterns in which the distances between points of the body do change, but the resulting equations only yield the forces acting between the points inside the rigid body. These forces are of no interest. When the analyst needs the internal forces, he must give up the idealized concept of a perfectly rigid body and admit the possibility of relative motion between the mass points. In that stage of the investigation he considers the body elastic, plastic, or viscous, depending on the material, its temperature, and the magnitude of the internal forces.

In general, whatever the nature of the constraints imposed on a system of mass points, the necessary and sufficient condition of equilibrium can be stated in the following way:

> A system of mass points is in equilibrium if the work done   **Lemma 2**
> by all the forces acting on the system vanishes for any
> displacement pattern that does not violate the constraints.

Displacements that are compatible with the constraints will be designated as virtual displacements. A great simplification of the analysis results from the fact that only virtual displacements need be considered when the equilibrium of a system is established, and displacement patterns which are incompatible with the constraints can be disregarded. Thus the number of cases that must be investigated is greatly reduced.

With this definition of the adjective *virtual*, the principle of virtual displacements can now be stated in the following final form for a system of mass points:

> A system of mass points is in equilibrium if the work done   **Theorem 4**
> by all the forces acting on the mass points of the system is
> zero for every virtual displacement.

It might be mentioned that, from the very beginning of the derivation of the principle, use has been made implicitly of the idea that virtual displacements alone need be considered in the investigation of the equilibrium of forces. This was the situation when plane systems were discussed. The assumption of a plane system involves the restriction of the displacements of the mass points to one plane. The fact that displacements out of the plane are disregarded is equivalent to assuming that sufficient reactions are always available to establish the equilibrium of the system in a direction perpendicular to the plane.

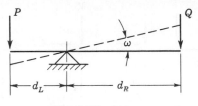

Fig. 1.5.1   Lever.

1.5   Application to Rigid Bodies

As a first example of the equilibrium of a rigid body the lever shown in Fig. 1.5.1 is investigated. It must be in equilibrium if the work done by all the forces acting on it is zero for any virtual displacement. Because the lever is considered a rigid body, changing its shape would infringe upon the constraints imposed upon its particles. Translations of the lever as a rigid body would also violate the constraining condition since a point of the lever is attached to the fulcrum. The only motion that remains is a rotation about the fulcrum, and all the conditions of equilibrium are satisfied if the work done by all the forces is zero during such a rotation.

A rotation through the angle ω is indicated in Fig. 1.5.1. If ω is assumed small, the work done during the rotation by force P is $P\omega d_L$, and that by force Q is $-Q\omega d_R$. The principle of virtual displacements requires for equilibrium that

$$P\omega d_L - Q\omega d_R = 0$$

that is,

$$P = Q(d_R/d_L) \qquad\qquad 1.5.1$$

which is the well-known law of the lever.

It should be noted that the reaction force at the fulcrum does not figure in the calculations. It is one of the advantages of the principle of virtual displacements that reactions need not be taken into account if they do no work during displacements compatible with the conditions of the system, that is, during virtual displacements.

In an investigation of the equilibrium of the rigid beam shown in Fig. 1.5.2, it is found that there are no virtual displacements. The shape of the beam cannot change because the beam is a rigid body, and rigid-body displacements are prevented by the supports. The requirements expressed in the principle of virtual displacements are automatically satisfied, and the body is in equilibrium under any forces that may be applied to it. Nevertheless the principle can be useful for the calculation of the reactions at the supports if displacements incompatible with the constraints are undertaken.

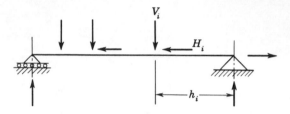

Fig. 1.5.2   Simply supported beam.

The most general rigid-body motion in the plane consists of two translations and one rotation. If these are taken as a vertical translation v, a horizontal translation h, and a rotation ω about the right-end support of the beam of Fig. 1.5.2, the requirement of a vanishing total work can be expressed by the equation

$$v \sum_{i=1}^{n} V_i + h \sum_{i=1}^{n} H_i + \omega \sum_{i=1}^{n} V_i h_i = 0 \qquad 1.5.2$$

where V_i, H_i, and h_i are the vertical component, the horizontal component, and the distance from the right-hand support to the ith force, and n is the number of forces acting on the beam, the reactions included. Equation 1.5.2 is an identity in v, h, and ω; that is, it must be satisfied for any values of v, h, and ω. If h and ω are assumed to be zero, while v is not zero, Eq. 1.5.2 can be satisfied only if the first sum vanishes. In a similar manner it can be shown that each one of the sums must vanish by itself. Hence, identity 1.5.2 is equivalent to the following three equations:

$$\sum_{i=1}^{n} V_i = 0, \qquad \sum_{i=1}^{n} H_i = 0, \qquad \sum_{i=1}^{n} M_i = 0 \qquad 1.5.3$$

where

$$M_i = V_i h_i \qquad 1.5.4$$

is the moment of the ith force with respect to the right-hand support. Equations 1.5.3 are the well-known conditions of equilibrium of a rigid body in the plane. They can be used for the calculation of the three reaction components if the applied forces are known.   Equations 1.5.3 also show that three reaction components can be determined from the equilibrium conditions of a plane system of forces applied to a rigid body. If the rigid body is supported in such a manner that only three reaction components can act on it, as is the case with the beam of Fig. 1.5.2, the support is statically determinate.   If the number of possible reaction components is less than three, the body can move; if greater than three,

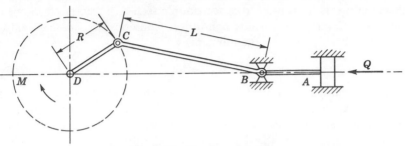

Fig. 1.5.3   Crank mechanism.

the support is redundant, and the reaction components can be determined only if the elastic deformations of the body are taken into account.   When the body is assumed to be perfectly rigid, the reactions cannot be calculated if there are more than three possible reaction components in the plane.

The principle of virtual displacements can be employed to advantage in the investigation of the equilibrium of forces and moments in mechanisms. For such purposes a mechanism can be defined as an assembly of rigid bodies with one degree of freedom of motion.   Thus a single virtual displacement suffices for the establishment of the equilibrium of the system of forces acting on a mechanism.

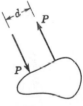

Fig. 1.5.4
Couple.

Figure 1.5.3 shows a crank mechanism in which the moment M acting on the crankshaft has to be calculated when the force Q on the piston is known.   Inertia forces are to be neglected.   Before the problem proper can be attacked, it is necessary to determine how the work done by a moment can be calculated.   A concentrated moment is the limiting case of a couple $Pd = M$ (see Fig. 1.5.4) when P increases indefinitely, d approaches zero, and the moment of the couple remains constant.   The most general displacement of the rigid body to which the couple is applied consists of two translations and a rotation if only plane problems are considered. The translations can be taken parallel and perpendicular, respectively, to the direction of the forces P, and the point of application of one of the forces may be chosen as the axis of rotation. The work done by the couple is obviously zero for a translation perpendicular to P, since the displacement is perpendicular to both forces, and for a translation parallel to P, since the work done by one of the forces is equal in magnitude and opposite in sense to that done by the other force. During a small rotation ω about the point of attack of one of the forces, the other force is displaced a distance ωd in its own direction.† Hence, the work done by the couple is $P\omega d$. When the distance d is decreased and the force P is increased in the transition from the couple to the concentrated moment, the product $Pd = M$ remains constant. Consequently, in the limiting case the work W done by the moment is

$$W = M\omega \hspace{4cm} 1.5.5$$

When the couple is applied to a body which is free to move in three-dimensional space, there are three more possible displacements, namely, a translation perpendicular to the plane of the moment and two rotations in planes perpendicular to the plane of the moment. During these displacements the work done by the forces P is zero.

It can be concluded therefore that a concentrated moment does work only if the body to which it is applied undergoes a rotation in the plane of the moment, and the work can be calculated then from Eq. 1.5.5.

With the crank mechanism shown in Fig. 1.5.3 the only possible displacement is a displacement d of the piston in the sense of Q, and a simultaneous rotation θ of the crank opposite the sense of M. As the elements of the mechanism are rigid bodies, no internal work is involved, and only the work done by the external forces need be calculated. The external forces are Q and M. (The latter, a moment, is considered a generalized force.) The reactions at B and D can do no work if friction and elastic or plastic deformations are ruled out. Equilibrium is certain if

$$Q\,\delta d - M\,\delta\theta = 0$$

where δd is the increment in the displacement d of point B and $\delta\theta$ is the increment in the rotation θ of the crank CD. This equation can be solved for M:

$$M = Q(\delta d/\delta\theta) \hspace{4cm} 1.5.6$$

The moment M can be calculated from the force Q if the ratio $\delta d/\delta\theta$ is known. With the aid of the concept of the instantaneous center of

---

† The rotation must be assumed as small in a virtual displacement because the direction of the forces must remain unchanged. The rule given for the work of the two forces is correct even for finite rotations if the direction of the forces relative to the body is maintained unchanged.

rotation this geometric problem can be solved easily for any position of the crank.   In Fig. 1.5.5 point B of the connecting rod is guided on a horizontal path.   Its motion can be considered for an infinitesimal length of time to result from a rotation about an instantaneous center situated somewhere on the vertical above B.   Simultaneously point C of the same connecting rod must move perpendicularly to the crank.   For an infinitesimal length of time therefore the motion of C can be represented by a rigid-body rotation about an instantaneous center situated somewhere

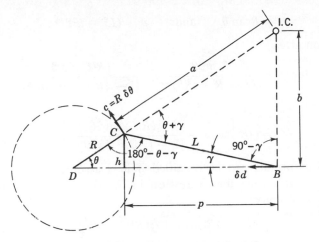

Fig. 1.5.5   Instantaneous center of rotation.

to the right on the continuation of the center line of the crank.   The two statements can be true simultaneously only if the instantaneous center of the connecting rod is the point of intersection of the two lines mentioned.   Then the ratio of the displacement δd of point B to the displacement $R\,\delta\theta$ of point C is the same as the ratio of the distances of the two points from the instantaneous center:

$$\delta d / R\,\delta\theta = b/a$$

or

$$\delta d / \delta\theta = R(b/a) \qquad\qquad 1.5.7$$

Substitution of the right-hand member of Eq. 1.5.7 in Eq. 1.5.6 gives

$$M = QR(b/a) \qquad\qquad 1.5.8$$

Since the ratio b/a can be easily determined graphically for any angle θ, the problem can be considered solved.

If an analytic expression is preferred to the graphical solution, no difficulty is encountered in expressing a and b in terms of R, L, and θ. The law of sines gives (see Fig. 1.5.5)

$$\frac{b}{a} = \frac{\sin(\theta + \gamma)}{\sin(90° - \gamma)} = \frac{\sin\theta\cos\gamma + \cos\theta\sin\gamma}{\cos\gamma} = \sin\theta + \cos\theta\tan\gamma$$

On the other hand,

$$\tan\gamma = h/p$$

and

$$h = R\sin\theta \quad\text{and}\quad p = (L^2 - h^2)^{1/2}$$

Substitution gives

$$\tan\gamma = \frac{R\sin\theta}{(L^2 - R^2\sin^2\theta)^{1/2}} = \frac{(R/L)\sin\theta}{[1 - (R/L)^2\sin^2\theta]^{1/2}}$$

Consequently

$$M = QR\left[\sin\theta + \frac{\frac{1}{2}(R/L)\sin 2\theta}{[1 - (R/L)^2\sin^2\theta]^{1/2}}\right] \qquad 1.5.9$$

A simplified formula can be obtained if it is assumed that L is large compared to R. Then the expression in the denominator is approximately equal to unity:

$$M = QR[\sin\theta + \tfrac{1}{2}(R/L)\sin 2\theta] \qquad 1.5.10$$

Finally, a rough approximation can be had if the connecting rod is assumed to be indefinitely long, and then

$$M = QR\sin\theta \cdot \qquad 1.5.11$$

1.6   The Forces in the Bars of Frameworks

From the standpoint of analysis the simplest elastic body is the ideal framework. It is a structure consisting of straight bars connected by ideal frictionless joints. The framework is known as a truss, or plane framework, when all its bars lie in a plane; when this condition is not met, the structure is a space framework. All external forces acting on an ideal framework are applied at the joints, and the weight of each bar is assumed to be concentrated at the two end points of the bar, that is, at the joints. It follows then from the conditions of equilibrium of any bar that the two forces transmitted to it from the joints at its ends must act along the axis of the bar, and must be equal in magnitude and opposite in sense. The bars are consequently either in pure tension or in pure compression.

It is convenient to begin the analysis of the framework with a rigid-body displacement pattern. This would be a virtual displacement even if the bars were rigid and rigidly connected at the joints. The most general rigid-body displacement in the plane consists of two translations and one rotation. During these motions only the external forces do work, and the statement of the principle of virtual displacements leads to the same conditions of equilibrium of the external forces as were obtained in the preceding section.

In the most general pattern of displacements each joint is displaced and each bar takes on some arbitrarily deformed shape. The work done by the external forces is easily calculated, but a little more care is necessary in determining the work done by the internal forces which act between the particles, or the mass points, that constitute the bar.

The virtual displacements of a bar can be resolved in two rigid-body translations, a rigid-body rotation, and a change in the shape and the length of the bar. An argument paralleling that presented in connection with the inextensible bar of Fig. 1.4.1 would prove that the two tensile forces P, which in this discussion are assumed positive as shown in the figure when they act from the joints on the bar, do no work during the rigid-body motions if the virtual displacements are very small. When the shape and the length of the bar are changed, these external forces (they are external as far as the bar is concerned) do work, and the work is equal to the product of the force P by the increase in the distance between the two end points of the bar. But according to Lemma 1 the internal work plus the external work is equal to zero for any system that is in equilibrium; consequently the work done by all the internal forces in the bar must be equal to $(-Pd)$, regardless of the deformed shape assumed by the bar.

This conclusion is of great importance, since it provides the rule for the calculation of the internal work in an ideal framework. It also proves that the internal work depends only on the displacements of the joints and is entirely independent of the changes in shape assumed for the individual bars. (This is not so astonishing if it is remembered that, according to the assumptions, the bars can resist neither shearing forces nor bending moments.) It follows then that displacement patterns involving changes in the shape of the bars can be disregarded when the principle of virtual displacements is used to determine the conditions of equilibrium of the ideal framework. Consequently the effective number of degrees of freedom of a plane framework having j joints is $2j$, as the most general virtual displacement can be described by stipulating two independent arbitrary displacements (for instance, in the directions of the x axis and of the y axis) for each one of the j joints.

These $2j$ very small but otherwise arbitrary displacements can always be resolved into three displacement patterns in which the distances between the joints are preserved, and $2j - 3$ further displacement patterns in which the distances, in general, are changed. This should be clear from Figs. 1.6.1 and 1.6.2. In the former the four joints A, B, C, D are given arbitrary displacements both in the x and in the y directions. The original

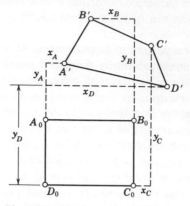

Fig. 1.6.1   Displacements of joints.

positions of the joints are marked by the subscript 0, and the displaced positions by the $'$. Figure 1.6.2 shows how these final positions can be reached over entirely different paths. The four joints are first translated

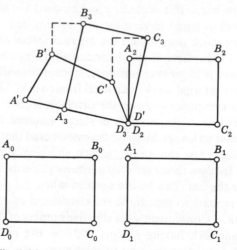

Fig. 1.6.2   Resolution of displacements of joints.

in the x direction to positions marked by the subscript 1. A translation in the y direction carries the joints into position 2, and a subsequent rotation into position 3. The magnitudes of the translations and the rotation are so chosen, that when they are completed, one of the joints, D, has reached its ultimate position, and one of the bars, AD, has attained its ultimate direction.   A subsequent stretching of bar AD succeeds in bringing a second joint, D, into its final position.   Further x and y displacements of all the other joints carry each joint into the position indicated by the prime.

The procedure remains the same for any number of joints.   The three rigid-body displacements and one stretching can always carry two joints into their final positions, and two more displacements are needed, in general, for every additional joint.   The number of steps necessary for the completion of the displacement is $2j$ in this procedure, as in the original one indicated in Fig. 1.6.1.

This proves that the $2j$ steps of the second procedure are fully equivalent to and can replace the $2j$ steps of the original procedure.   On the other hand, it was mentioned at the beginning of this article that the three rigid-body displacements lead to the well-known three conditions of equilibrium of the plane external-force system.   Thus $2j - 3$ virtual displacements remain for the investigation of the equilibrium of the internal forces, and they can be used to establish $2j - 3$ conditions of equilibrium.

The unknowns in the framework are the forces in the bars.   If their number is n, and the equation holds

$$n = 2j - 3 \qquad\qquad 1.6.1$$

the number of unknowns is equal to the number of available equations. In this case the forces in the bars, in general, can be calculated.   Frameworks for which Eq. 1.6.1 is valid are called statically determinate.   When n is smaller than $2j - 3$, the framework is incomplete.   It has at least one degree of freedom of motion and can hardly be considered a structure; it is rather a mechanism.   When n is greater than $2j - 3$, the framework is stiff, but the equilibrium conditions alone do not suffice for the calculation of the forces in the bars.   A method for calculating the forces in such so-called redundant frameworks will be given in Art. 1.8.

It should be mentioned here that the rule given in Eq. 1.6.1 is a necessary but not a sufficient condition of static determinateness.   The bars must also be arranged in a suitable manner.   The proof of this statement is simple: Remove one diagonal from a statically determinate bridge truss, and add it as a second cross-bracing element to the structure in a different bay.   In this process the total number of elements remains the same, and Eq. 1.6.1 holds after the change if it held before.   Nevertheless the second bay is now redundant while the first one is unbraced and the structure is reduced to a mechanism.

These considerations can readily be extended to three-dimensional space frameworks.   A space framework having j joints has $3j$ degrees of freedom of motion, since each of its joints may be given arbitrary x, y, and z displacements.   The same arbitrary final positions of each joint can also be reached in another manner.   The space framework can be translated first in the directions of the coordinate axes until the final position of one of its joints is reached.   Three subsequent rotations about axes passing through this joint and parallel to the axes of coordinates will succeed in giving a bar attached to this joint its final direction, and at the same time in bringing a second bar attached to the joint into the plane defined by the final positions of the two bars.   During these six motions the distances between the joints remain unchanged.   The final position of the second end point of the first bar can be reached by stretching the first bar, and the final position of the second end point of the second bar through two non-collinear displacements in the final plane of the two bars.   Consequently nine displacements are needed in this procedure to reach the final positions of three joints.   Obviously three additional displacements for each additional joint suffice to reach any arbitrary configuration of the space framework.

This proves that the $3j$ displacement patterns of the second procedure are equivalent to and can replace the $3j$ individual displacements of the first procedure.   If the principle of virtual displacements is applied to the six rigid-body displacement patterns of the second procedure, the well-known conditions of equilibrium of a three-dimensional system of forces result:

> If a system of forces in three-dimensional space is in   **Theorem 5**
> equilibrium, the sum of the components of the forces in
> any of three non-coplanar directions and the sum of the
> moments of the forces about any of three non-coplanar
> axes must be equal to zero.

There remain therefore $3j - 6$ arbitrary displacements from which $3j - 6$ independent equations can be derived for the calculation of the unknown internal forces, namely, the tensions in the bars of the space framework.   If the number of bars is n, and the equation holds

$$n = 3j - 6 \qquad\qquad 1.6.2$$

the $3j - 6$ equations can be solved, in general, for the $3j - 6$ unknowns. The space framework is statically determinate in such a case.   If the number of bars is smaller, the framework is incomplete; if greater, it is redundant.   Again Eq. 1.6.2 is a necessary but not a sufficient condition.

In the foregoing considerations the principle of virtual displacements was applied to derive the formulas for the number of bars needed in a statically determinate framework.   The principle can also provide a rule for the calculation of the forces in the bars.

For this purpose let one of the joints of the space framework be displaced through a distance a in the direction of the x axis.   The work done by all the forces of the system is obviously equal to the product of a by the sum of the x components of the tensions in the bars attached to the joint and of the external forces applied to the joint.   This work must be zero for equilibrium.   But, as a is not zero, the sum of the x components must vanish.   If virtual displacements are taken in the y and z directions, two more conditions of equilibrium are obtained.   Altogether

$$\Sigma X = 0, \qquad \Sigma Y = 0, \qquad \Sigma Z = 0 \qquad\qquad 1.6.3$$

Calculation of the tensions in a space framework from Eqs. 1.6.3 is known as the method of joints.   In a plane framework the same method is used, but the last of the three equations is not written.

Two remarks may be of interest in connection with these considerations. The first concerns the expression $(-Pd)$ obtained for the internal work. If P is a tensile force in and d a virtual elongation of a bar, the internal work is negative.   Although the negative sign may be surprising at first sight, it is correct, since an elongation is imposed on a bar in spite of the internal forces and not by them.

The second remark is just a reminder that in the present article no assumptions were made regarding the material of the bars of the framework.   The conclusions reached are valid therefore, whether the material is elastic, plastic, or viscous.   Hooke's law and the principle of superposition need not hold.

1.7   Displacements of Joints

When a statically determinate framework is subjected to known loads, the forces in its members can be calculated as outlined in the preceding article.   Under these forces the members change their length and the elongations can be determined without difficulty if the stress-strain law of the material is known.   With the commonly used structural materials, the elongations are very small in the range of stress allowed by the safety codes;   for instance, with steel, Young's modulus and the maximum working stress can be taken as 30×10^6 and 30×10^3 psi, respectively, from which it follows that the maximum strain is 10^{-3}.   This means that no bar in the framework can be stretched more than about one thousandth of its initial length.

The shape of a statically determinate framework is completely determined by the lengths of its members and by their arrangement.   If a bar is removed from an unloaded statically determinate framework and replaced by one of different length, the shape of the framework is changed without the introduction of any stress. A draftsman can determine the shape of the new framework by laying off the bars with their proper lengths. In the same manner he can draw the shape of a framework whose members are stretched because of the loads, and by comparing the new shape with the initial one he should be able to determine the displacements of the joints.

The trouble is that the stretched lengths of the bars differ so little from the initial lengths that they cannot be distinguished from each other in an engineering drawing. Nevertheless a framework bridge can easily deflect several inches. The straightforward geometric approach is therefore not accurate enough for solving the purely geometric problem of obtaining the displacements of the joints of a framework from the known changes in the lengths of its members.

For this reason it is advisable to abandon this geometric approach in favor of a method based on mechanics and to calculate the displacements from the principle of virtual displacements. The application of the principle to the problem of deflections is known as the dummy load method. In it a unit force, denoted as the dummy load, is assumed to act at the joint whose displacement is sought. The direction of the dummy load must be the same as that of the component of the displacement to be determined. As the first step of the analysis, the forces F, caused in the bars by the dummy load, are calculated.

The internal forces, that is, the tensions F in the bars, are in equilibrium with the external forces which consist of the dummy load and its reactions. Consequently the work done by all these forces must be zero for any virtual displacement. The artifice employed is to consider the real displacement d of the joint and the real elongations of the bars as a generalized virtual displacement. This is certainly permissible because any displacement pattern not violating the constraints can be chosen as a virtual displacement pattern. Of course, d is not known, but the changes in the lengths of the bars corresponding to it, that is, the real elongations e of the bars, are available.

The interesting situation then arises that the real displacements of the system under the prescribed loads are considered the virtual displacements while the equilibrium system of forces employed, namely, the dummy load system, is entirely fictitious and completely independent of the real loads. The only reason for introducing such concepts is, of course, that they provide the simplest method for computing the displacements of the joints.

The next step in the analysis is the calculation of the work done by the dummy load system during the deformations of the structure caused by the real loads. This work can be subdivided into two parts, the work done by the internal forces and that done by the external forces. The former, W_i, is the negative of the products of the forces caused by the dummy load and the real extensions e of the bar, summed up over all the bars contained in the framework:

$$W_i = -\Sigma Fe \qquad 1.7.1$$

The work done by the external forces, W_e, is the dummy load times the real displacement d of the joint to which the dummy load is applied. If the magnitude of the dummy load is unity, the external work is numerically equal to the displacement sought:

$$W_e = d \qquad 1.7.2$$

This is true provided that the supports of the framework are rigid and the

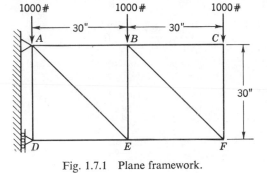

Fig. 1.7.1 Plane framework.

reactions do no work during the displacements. Because of Lemma 1,

$$W_e + W_i = 0$$

and thus

$$d = \Sigma Fe \qquad 1.7.3$$

where the summation must include every bar in the framework.

The deflection d can be calculated from Eq. 1.7.3 if the extensions e of the bars in the real load system (not in the dummy load system) are known. In most applications the extensions are assumed to be related to the loads through Hooke's law. For the purpose the present calculations Hooke's law can be stated as

$$e = PL/EA \qquad 1.7.4$$

where P is the tension in the bar caused by the real (not the dummy) load system, L is the length, A the cross-sectional area of the bar, and E is Young's modulus of its material.

The computation of the displacement can best be carried out in tabular form; details of the work are shown in the example of the framework in Fig. 1.7.1. (It is assumed that suitable lateral supports are available to prevent the buckling of compression members.) The cross-sectional area of each bar is taken as 0.1 sq in., and the material as steel with a Young's modulus of 30×10^6 psi. The quantity to be determined is the vertical displacement of joint F.

The tensions P caused by the actual loads can be found, for example, by the method of joints. They are listed in column 2 of Table 1.7.1. The elongations e of the bars are obtained by multiplying these forces by $R = L/EA$. Column 3 lists the values of R, and column 4 those of e.

Table 1.7.1. Computation of Displacement of Joint F

[1]	[2]	[3]	[4]	[5]	[6]
Bar	P, lb	R, in./lb	e, in.	F, lb	Fe, in.-lb
AB	1000	10^{-5}	10^{-2}	1	10^{-2}
BC	0	10^{-5}	0	0	0
DE	-3000	10^{-5}	-3×10^{-2}	-2	6×10^{-2}
EF	-1000	10^{-5}	-10^{-2}	-1	10^{-2}
AD	0	10^{-5}	0	0	0
BE	-2000	10^{-5}	-2×10^{-2}	-1	2×10^{-2}
CF	-1000	10^{-5}	-10^{-2}	0	0
AE	2828	1.41×10^{-5}	4×10^{-2}	1.414	5.656×10^{-2}
BF	1414	1.41×10^{-5}	2×10^{-2}	1.414	2.828×10^{-2}

$$\Sigma Fe = 18.484 \times 10^{-2}$$

Next a dummy load of 1 lb is assumed to act vertically downward at joint F. The tensions caused by it are denoted by F and are listed in column 5. The product Fe is entered in column 6 for each bar, and the individual entries are summed up. Because of Eq. 1.7.3 the vertical deflection d of joint F is

$$d = \Sigma Fe = 0.185 \text{ in.}$$

if three digits are retained as physically significant.

The salient feature of this calculation is the combination of the elongations e computed for the real force system P with the fictitious force system F of the dummy load for the purpose of determining the virtual work. The calculation can be carried out in exactly the same manner if the material of the bars does not follow Hooke's law, except that then Eq. 1.7.4 must be replaced by an equation representing the appropriate relation between force and elongation.

The procedure can also be used for space frameworks because no argument in the derivation of the formulas is restricted to plane structures.

A suitable choice of the dummy load, or of a group of dummy loads, can reduce materially the work of computation in many problems. As an example, let the rotation of bar CF be calculated for the framework and loading of Fig. 1.7.1. In this case the suitable group of dummy loads consists of a unit force acting horizontally to the right at joint C, and a unit force acting horizontally to the left at joint F. An argument similar to that advanced in the derivation of Eq. 1.5.5 would prove that the work done by this dummy load system is equal to the moment of the couple times the rotation (in radians) of bar CF. The computations are carried out in Table 1.7.2, which is a continuation of Table 1.7.1.

Table 1.7.2. Computation of Rotation of Bar CF

[1]	[2]	[3]	[4]
Bar	e, in.	F, lb	Fe, in.-lb
AB	10^{-2}	1	10^{-2}
BC	0	1	0
DE	-3×10^{-2}	-1	3×10^{-2}
EF	-10^{-2}	-1	10^{-2}
AD	0	0	0
BE	-2×10^{-2}	0	0
CF	-10^{-2}	0	0
AE	4×10^{-2}	0	0
BF	2×10^{-2}	0	0

$$\Sigma Fe = 5 \times 10^{-2}$$

Column 2 of the new table is identical with column 4 of the old table, since obviously the elongations of the bars in the actual load system are not influenced by the choice of the dummy loads. The tensions F caused by the dummy loads, on the other hand, are different in the two tables. From Table 1.7.2,

$$\Sigma Fe = 0.050 \text{ in.-lb} = -W_i$$

The moment of the dummy load group is given by

$$M = 1 \times 30 = 30 \text{ in.-lb}$$

The work W_e done by the dummy load group is

$$W_e = M\beta = 30\beta$$

where β is the rotation of bar CF. Hence, Lemma 1 gives

$$30\beta - 0.050 = 0$$

The solution for β is

$$\beta = 0.00166 \text{ radian} = 0.0957°$$

1.8   The Redundant Framework

The calculation of the forces in a redundant framework is based on the principle of superposition.†   For the purpose of this article the principle can be stated in the following manner:

> If a system of forces P_1 acting on an elastic body causes a   **Theorem 6**
> stress σ_1 and a displacement d_1 at some particular point
> of the elastic body, and if a second system of forces P_2
> acting alone (without P_1) causes a stress σ_2 and a displace-
> ment d_2 at the same point, then under the simultaneous
> action of systems P_1 and P_2 the stress is $\sigma_1 + \sigma_2$ and the
> displacement is $d_1 + d_2$.

The principle of superposition depends for its validity on Hooke's law. This can be shown by computing the deformations of a tension test specimen in a testing machine.   Let P_1 be a tensile force of 1000 lb, and the elongation of the specimen caused by this force $d_1 = 10^{-3}$ in.   Let P_2 also be a tensile force of 1000 lb.   Obviously the elongation caused by P_2 alone must be $d_2 = 10^{-3}$ in.   According to the principle of superposition, the elongation under the simultaneous action of loads P_1 and P_2 is $d_1 + d_2 = 2 \times 10^{-3}$ in.   On the other hand, $P_1 + P_2 = 2P_1 = 2000$ lb; that is, doubling the load results in doubling the displacement.   Consequently the elongation is proportional to the load, which is Hooke's law.   If the tensile specimen is a rubber cord, Hooke's law is not valid for large elastic deformations, and the elongation corresponding to an increase of the load from 1000 to 2000 lb is much greater than the elongation caused by an increase of the load from zero to 1000 lb.   For the rubber cord therefore the principle of superposition cannot hold.

It is not quite so obvious that in certain structures the principle of superposition may not be valid even though the material follows Hooke's law.   Among others the beam column is such a structure; it is defined as a beam on which compressive axial loads act simultaneously with the transverse loads.   Figure 1.8.1 shows a beam column on two supports.

Let the dotted line represent the deflected shape of the beam under the transverse load W when the end loads P are absent.   This deflected shape $y_1 = f(x)$ can be calculated by any routine method from the bending moment $M_0 = (W/2)x$ in the right-hand half of the beam; the moment in the left-hand half can be given by a similar formula.   When subsequently the end loads P are applied, they act on a variable lever arm y_1 and cause additional bending moments $M_1 = Py_1$.   The additional bending moments cause additional deflections y_2, which in turn give rise to additional bending

---

† A redundant framework for which the principle of superposition is not valid because of the non-linear behavior of its material is discussed in Art. 4.2.

moments M_2. This process of reasoning and the corresponding calculation by step-by-step approximations can be continued until in the nth step the additional bending moments M_n are found small enough to be neglected.†

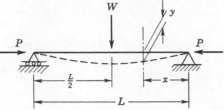

Fig. 1.8.1   Beam column.

Instead of finding the results by step-by-step approximations a differential equation can be derived and solved, as is done in Art. 2.4.   The solution indicates that, if W and P are increased to $2W$ and $2P$, the deflections become larger than twice the original deflections.   Consequently the principle of superposition does not hold for the beam column, even though its material follows Hooke's law.

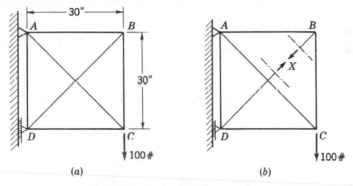

Fig. 1.8.2   Redundant framework.

Similar behavior can be anticipated with thin-walled structures under compressive, shearing, and more complex types of loading when the displacements caused by the loads are large enough to alter materially the loading of the structure.   In framework theory, on the other hand, it is generally assumed that this is not so and the principle of superposition is valid.   Forces in the bars of space and plane frameworks calculated on these assumptions have been found to agree well with results of experiments.

The method of calculating the tensions in a redundant framework will now be discussed with the aid of Fig. 1.8.2.   The framework itself is

† This is not true for an unstable structure for which the process diverges.   The convergence criterion of stability is discussed in Art. 3.8.

shown in Fig. 1.8.2a. It is redundant because the four joints of the truss are connected with six bars instead of the five required by Eq. 1.6.1 for static determinateness. The framework is supported in a statically determinate manner since the frictionless pivot at A can transmit any horizontal and vertical reactions and the frictionless slide at D any horizontal reaction.

The fictitious section through diagonal BD shown in Fig. 1.8.2b transforms the framework into a statically determinate one. With diagonal BD out of action, the tensions F_0 caused in the remaining five bars by the 100-lb load can be calculated without difficulty. They are listed in column 3 of Table 1.8.1. Column 2 contains the values of $R = L/EA$

Table 1.8.1. Forces in Redundant Framework

[1]	[2]	[3]	[4]	[5]	[6]	[7]	[8]
Bar	$R,$ in./lb	$F_0,$ lb	$F_1,$ lb/lb	$F_0F_1R,$ in.	$F_1^2R,$ in./lb	$XF_1,$ lb	$F_{tot},$ lb
AB	10^{-5}	0	-0.707	0	0.5×10^{-5}	39.7	39.7
BC	10^{-5}	0	-0.707	0	0.5×10^{-5}	39.7	39.7
CD	10^{-5}	-100	-0.707	70.7×10^{-5}	0.5×10^{-5}	39.7	-60.3
AD	10^{-5}	0	-0.707	0	0.5×10^{-5}	39.7	39.7
AC	1.41×10^{-5}	141	1	200×10^{-5}	1.41×10^{-5}	-56.2	84.8
BD	1.41×10^{-5}	0	1	0	1.41×10^{-5}	-56.2	-56.2
			Total	270.7×10^{-5}	4.82×10^{-5}		

computed for steel having a modulus $E = 30 \times 10^6$ psi and for a cross-sectional area of 0.1 sq in. In reality, however, there is no section through diagonal BD, and the fictitious section transmits an unknown force X (positive when tensile). It is easy to see what the tensions must be in the five active bars of the framework under the force X. The values obtained for $X = 1$ lb are listed in column 4 whose heading is F_1. The tension $F_{1 \cdot BD}$ in the cut diagonal BD is obviously 1 lb. With $F_{0 \cdot MN}$ the force caused in a typical bar MN of the cut system by the external load, and $F_{1 \cdot MN}$ the force caused in the same bar by a unit load applied in the fictitious section, the total force $F_{tot \cdot MN}$ in bar MN of the complete (non-cut) system must be

$$F_{tot \cdot MN} = F_{0 \cdot MN} + XF_{1 \cdot MN} \qquad 1.8.1$$

From Eq. 1.8.1 the total force in the typical bar MN of the actual framework can be computed if X is known. The magnitude and the sign of X can be obtained from the condition that the two sides of the fictitious section through diagonal BD cannot separate. No relative displacement between the two sides is possible because the section is not real but fictitious.

A relative displacement between two sides of a section can be calculated by the dummy load method. If a group of two equal and opposite unit loads is assumed in the fictitious section in place of the forces X in Fig. 1.8.2b, the dummy load group and the tensions caused by it in the bars of the framework certainly constitute an equilibrium system. Because of the principle of virtual displacements the work done by the two forces of the group and by the tensions caused by them in the bars must add up to zero for any arbitrary displacement of the joints. On the other hand, the two equal and opposite dummy loads $X = 1$ do no work when the two sides of the fictitious section are translated simultaneously through the same distance in the direction of the bar or perpendicularly to it. They do no work either when the bar is rotated. They do work only if one side of the section is translated a distance δ further in the direction of the bar than is the other side of the section; that is, the two sides separate or overlap. The work is then the product of dummy load and relative displacement, which is δ when the magnitude of the dummy load is unity. If the dummy loads are considered the external loads of the system, the external work W_e can be written as

$$W_e = \delta \qquad\qquad 1.8.2$$

The elongation of a bar under the actual force F_{tot} in the redundant framework is $F_{tot}R$. The force caused by the two dummy loads was calculated earlier and is listed in column 4 of Table 1.8.1. (The dummy loading is identical with the loading $X = 1$, which was used in connection with the calculation of the tensions.) In accordance with Eq. 1.7.1 the tensions F_1 corresponding to the dummy loading do the following internal work during the displacements of the joints caused by the actual loads F_{tot}:

$$W_i = -\Sigma F_{1 \cdot MN} F_{tot \cdot MN} R_{MN} = -\Sigma F_{1 \cdot MN}(F_{0 \cdot MN} + X F_{1 \cdot MN})R_{MN} \qquad 1.8.3$$

In this equation the summation must be extended over every bar of the framework. The statement of the principle of virtual displacements, that is,

$$W_e + W_i = 0$$

together with Eqs. 1.8.2 and 1.8.3 can be solved to give explicitly the relative displacement δ of the two sides of the fictitious section:

$$\delta = \Sigma F_{1 \cdot MN}(F_{0 \cdot MN} + X F_{1 \cdot MN})R_{MN} \qquad 1.8.4$$

Because no relative displacement can take place between the sides of a fictitious, not a real, section, the following condition must be satisfied:

$$\Sigma F_{1 \cdot MN}(F_{0 \cdot MN} + X F_{1 \cdot MN})R_{MN} = 0$$

Since X is a constant, the equation can also be written in the form

$$\Sigma F_{0 \cdot MN} F_{1 \cdot MN} R_{MN} + X \Sigma F^2_{1 \cdot MN} R_{MN} = 0 \qquad 1.8.5$$

When solved for X, Eq. 1.8.5 yields

$$X = -\frac{\Sigma F_{0 \cdot MN} F_{1 \cdot MN} R_{MN}}{\Sigma F^2_{1 \cdot MN} R_{MN}} \qquad 1.8.6$$

Here, as before, the subscript MN refers to the typical bar, and the summation must be extended over every bar of the framework.

In Table 1.8.1, column 5 contains the products $F_0 F_1 R$, and column 6 the products $F_1^2 R$. Summation over all the bars of the framework gives

$$\Sigma F_0 F_1 R = 270.7 \times 10^{-5} \text{ in.}$$

and

$$\Sigma F_1^2 R = 4.82 \times 10^{-5} \text{ in. per lb}$$

Consequently

$$X = -270.7/4.82 = -56.2 \text{ lb}$$

The products XF_1 are listed in column 7 of Table 1.8.1. The total tensions F_{tot} are computed from Eq. 1.8.1 and are listed in column 8.

The values in the last column show that the forces are more evenly distributed in the bars when diagonal BD is present than when it is absent. Nevertheless diagonal AC carries about 1.5 times the force acting in diagonal BD. The difference in the load carried by the two diagonals is due to the deformations of bars AD and BC through which the load in diagonal BD must pass, while the load in diagonal AC is transmitted directly from the point of application of the external load to the support at A. This conclusion can be easily checked. When bars AD and BC are perfectly rigid, the value of R for them is zero. Then the products $F_0 F_1 R$ and $F_1^2 R$ also vanish for AD and BC. The new values of the sums are

$$\Sigma F_0 F_1 R = 270.7 \times 10^{-5} \text{ in.}, \qquad \Sigma F_1^2 R = 3.82 \times 10^{-5} \text{ in. per lb}$$

and the unknown force X becomes

$$X = -270.7/3.82 = -70.7 \text{ lb}$$

The tensions in the bars of the modified framework are computed in Table 1.8.2. As anticipated, the two diagonals carry equal loads in the framework when bars AD and BC are perfectly rigid.

Table 1.8.2.   Forces in Modified Framework

Bar	AB	BC	CD	AD	AC	BD
F_0 lb	0	0	−100	0	141.4	0
XF_1 lb	50	50	50	50	−70.7	−70.7
F_{tot} lb	50	50	−50	50	70.7	−70.7

If diagonal BD is actually cut in the framework of Fig. 1.8.2, relative displacements of the two sides of the section become possible, but a force X can no longer be transmitted through the section. In Eq. 1.8.4 the actual force is then $F_{0 \cdot MN}$, and the elongation of the typical bar MN is $F_{0 \cdot MN} R_{MN}$. Hence the relative displacement δ_0 of the two sides of the real section is

$$\delta_0 = \Sigma F_{0 \cdot MN} F_{1 \cdot MN} R_{MN} \qquad 1.8.7$$

where the summation must be extended over all the bars of the framework. On the other hand, if the 100-lb external load is removed and only two equal and opposite forces $X = 1$ lb are applied to the two sides of the actual section, the force in the typical bar is $F_{1 \cdot MN}$. The elongation of this bar is $F_{1 \cdot MN} R_{MN}$, and the relative displacement δ_1 of the two sides of the real section is, according to Eq. 1.8.4,

$$\delta_1 = \Sigma F^2_{1 \cdot MN} R_{MN} \qquad 1.8.8$$

Equations 1.8.7 and 1.8.8 give a physical meaning to the quantities in Eq. 1.8.5 from which the unknown force X in the redundant framework was calculated. With their aid Eq. 1.8.5 can be written in the modified form

$$\delta_0 + X \delta_1 = 0 \qquad 1.8.9$$

This equation can be stated in the following manner:

The relative displacement of the two sides of the fictitious section caused by the given external loading, plus X times the relative displacement caused by the loading $X = 1$ lb, must be zero.

Consequently of all statically possible values of X that value actually prevails in the section which closes the gap developed by the external loads.

The procedure developed for the singly redundant truss can be easily generalized to hold for any arbitrary redundant plane or space framework. When the number of redundant bars is n, the following steps must be undertaken in the analysis:

1. The framework must be transformed into a statically determinate one by means of a fictitious section through each of n bars.

2. The tensions F_0 in the bars of the statically determinate framework caused by the given external loads must be calculated.

3. In reality an unknown force X_1 is acting in the first fictitious section. With the assumption that $X_1 = 1$ lb, the tensions F_1 in the bars of the cut framework can be calculated.

4. In reality unknown forces $X_2, X_3, \cdots X_n$ are acting in the 2nd, 3rd, $\cdots$ and the nth sections, respectively. If their value is assumed to be 1 lb, the tensions $F_2, F_3, \cdots, F_n$ can be calculated in turn in the cut framework.

5. The actual force in the typical ith bar is

$$F_{\text{tot}\cdot i} = F_{0i} + X_1 F_{1i} + X_2 F_{2i} + \cdots + X_n F_{ni} \qquad 1.8.10$$

6. The magnitude of the unknown forces X can be determined from the requirement that the relative displacements of the two sides of each of the n fictitious sections must be zero. The assumption of dummy loads $X_1 = 1$ lb, $X_2 = 1$ lb, $\cdots X_n = 1$ lb leads to the following n simultaneous linear equations:

$$\delta_{10} + X_1 \delta_{11} + X_2 \delta_{12} + \cdots + X_n \delta_{1n} = 0$$
$$\delta_{20} + X_1 \delta_{21} + X_2 \delta_{22} + \cdots + X_n \delta_{2n} = 0$$
$$\cdot \qquad \cdot \qquad \cdot \qquad \qquad \cdot \qquad \cdot$$
$$\cdot \qquad \cdot \qquad \cdot \qquad \qquad \cdot \qquad \cdot \qquad 1.8.11$$
$$\cdot \qquad \cdot \qquad \cdot \qquad \qquad \cdot \qquad \cdot$$
$$\delta_{n0} + X_1 \delta_{n1} + X_2 \delta_{n2} + \cdots + X_n \delta_{nn} = 0$$

where

$$\delta_{pq} = \Sigma F_{pi} F_{qi} R_i \qquad 1.8.12$$

In this equation i refers to the typical bar, and the summation must be extended over all the bars of the framework.

7. After the calculation of the sums indicated in Eq. 1.8.12, the n linear equations 1.8.11 can be solved for the n unknowns $X_1, X_2, \cdots, X_n$. Substitution of their values in Eq. 1.8.10 gives the final force in the typical ith bar.

It is easy to prove that the physical meaning of $\delta_{pq} = \Sigma F_{pi} F_{qi} R_i$ is the relative displacement of the two sides of the section in the pth bar when $X_q = 1$ (the dummy load group in the section of the qth bar) is acting alone. Consequently, the meaning of, say, the second of Eqs. 1.8.11 is that the sum of the relative displacements of the two sides of the second fictitious section caused by the given external load, the unknown force X_1, the unknown force $X_2, \cdots$, and the unknown force X_n must be zero.

Moreover,

$$\delta_{pq} = \delta_{qp} \qquad 1.8.13$$

since the expression for δ_{qp} differs from that given in Eq. 1.8.12 for δ_{pq} only in the order of the factors in the products to be summed up. Equation 1.8.13 is a statement of Maxwell's reciprocal theorem for a framework; it is derived in a more general form in Art. 4.4.

1.9   Torsion of an Engine Mount for a Radial Airplane Engine

Statement of the Problem

Figure 1.9.1a is a pictorial representation of a welded steel-tube engine mount for a radial airplane engine.   The four rear points of attachment of the engine mount are bolted to fittings of the fuselage, nacelle, or wing

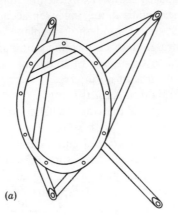

(a)

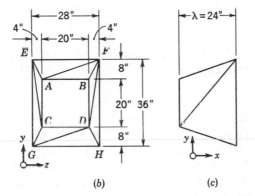

(b)   (c)

Fig. 1.9.1   Engine mount for radial airplane engine.

spar.   The air-cooled radial engine is fastened to the engine mount ring at the nine points of attachment in the front plane.   In many installations these nine attachments are provided with rubber shock absorbers in order to prevent the transmission of engine vibrations to the airplane.

The loading of the engine mount consists of the vertical shear force and the bending moment in a vertical plane caused by the weight of the engine in horizontal flight or at landing, the thrust of the propeller, the gyroscopic lateral force and the bending moment in a horizontal plane due to sudden maneuvers, and the reaction torque of the engine.   Only the last loading will be considered in this article, since the effect of the others can be calculated with little difficulty if the symmetry properties of structure and loading are duly taken into account.

For the purpose of determining the tensions in the bars, it is convenient to represent the actual space framework by the line diagrams of Figs. 1.9.1b and 1.9.1c.   In these the engine mount ring is replaced by four straight bars between points A, B, C, and D, and the rear plane of attachment of the engine mount by four straight bars connecting points E, F, G, and H.   The two planes are assumed to be braced by diagonals which are not shown because their exact location is of no interest.

The substitution of straight bars for the ring is, of course, not permissible when the necessary strength of the ring itself has to be calculated.   This problem will not be considered here.   In actual airplane design, the proportions of the ring are determined empirically, since the distribution of the loads among the rubber shock absorbers and the effect upon this distribution of the elastic properties of the engine casting are not easily accessible to analytical procedures.   It appears reasonable, however, to assume that the rubber mountings permit a small amount of warping of plane ABCD when the engine mount is twisted.   In this article it will also be assumed that plane EFGH is free to warp.   The effect of a rigid attachment in the rear plane will be investigated in detail in Art. 1.10.

Thus the complex problem of the torsion of the engine mount now appears stripped of its unessential features and reduced to the relatively simple problem of the torsion of the space framework of Figs. 1.9.1b and c. The framework consists of a number of bars in planes ABCD and EFGH, which are not all shown because they are of little interest, and of the eight bars connecting the two parallel planes.   The task set for this article is the determination of the tensions in the eight connecting bars when the loading of the structure is a counterclockwise torque of 10,000 in.-lb applied to plane ABCD, and an equal and opposite reaction torque applied to plane EFGH.   (The term counterclockwise is used here from the standpoint of the observer looking at Fig. 1.9.1b.)

Tension Coefficients

The unknown forces F in the bars must be calculated from the conditions of equilibrium at each joint.   At joint A, for instance, force F_{AF} in bar AF has to be considered in three equations enforcing equilibrium in the x, y, and z directions.   The component F_{AFx} of force F_{AF} in the x direction is

$$F_{AFx} = F_{AF} \cos (AF, x) = F_{AF}(24/L_{AF})   1.9.1$$

where cos (AF, x) is the cosine of the angle subtended by the x axis and the direction of bar AF, and L_{AF} is the true length of bar AF:

$$L_{AF} = (24^2 + 8^2 + 24^2)^{1/2} = 34.9 \text{ in.} \qquad 1.9.2$$

Similarly, the y and z components of the force F_{AF} are

$$F_{AFy} = F_{AF} \cos (AF, y) = F_{AF}(8/L_{AF})$$
$$\qquad 1.9.3$$
$$F_{AFz} = F_{AF} \cos (AF, z) = F_{AF}(24/L_{AF})$$

It is rather cumbersome to use direction cosines in the conditions of equilibrium. The work of writing can be reduced, and a considerably clearer insight into the equilibrium of the forces in a space framework can be gained, if the tension coefficient T is introduced in the calculations. The typical tension coefficient T_{NM} for bar NM is defined by the equation

$$T_{NM} = F_{NM}/L_{NM} \qquad 1.9.4$$

where F_{NM} is the force in bar NM, and L_{NM} is the true length of bar NM. Of course, the tension coefficient T_{AF} is unknown as long as the tension F_{AF} is unknown. But the conditions of equilibrium at the joints can be expressed more conveniently in terms of the tension coefficients T than in terms of the forces F. When the values of the tension coefficients become known from a solution of the equations of equilibrium, the forces F can be computed from Eq. 1.9.4.

The rule for the computation of the force components in the direction of the axes follows from Eqs. 1.9.1 and 1.9.3:

$$F_{AFx} = T_{AF} L_{AFx}$$
$$F_{AFy} = T_{AF} L_{AFy}$$
$$F_{AFz} = T_{AF} L_{AFz}$$

where L_{AFx}, L_{AFy}, and L_{AFz} are the components of the length of bar AF in the x, y, and z directions, respectively. In general,

The rectangular component of a force in a bar in any   **Theorem 7**
arbitrary direction is equal to the tension coefficient times
the component of the length of the bar in the same
direction.

With the aid of the tension coefficients it is now possible to proceed to the analysis of the space framework. It has already been mentioned that the applied load is a torque in plane $ABCD$ and an equal and opposite torque in plane $EFGH$. Because of the assumptions made earlier, there are no external forces (or reactions) acting perpendicularly to these planes.

The distribution of the loads which add up to the torques in the two planes is not specified because it is not needed in the analysis.

The condition of equilibrium at joint B in the x direction will be considered first. With T_{BF} the unknown tension coefficient in bar BF, the x component of the force in bar BF is $\lambda T_{BF} = 24 T_{BF}$ where λ is the distance between the front and rear planes. Bars AB and BD as well as any possible bracings of plane $ABCD$ lie entirely in plane $ABCD$; for this reason they cannot have components in the x direction. Because the external loads are also in the same plane, the condition of equilibrium of the forces in the x direction is

$$\lambda T_{BF} = 0$$

Since λ is not zero, the tension coefficient T_{BF} must be zero. In the same manner it can be shown that the condition of equilibrium in the x direction at joint H requires that

$$T_{DH} = 0$$

These conclusions can be generalized:

> In a space framework, consisting of two braced parallel   **Theorem 8**
> planes and a number of bars connecting these two planes,
> which is loaded exclusively by forces in the two parallel
> planes, the tension coefficient, and consequently the
> force, is zero in every bar that connects the two planes
> and is attached to a joint in one of the parallel planes to
> which no other bar is attached that connects the two
> planes.

Thus bars BF and DH can be disregarded entirely in the analysis since they do not carry loads. The remaining six bars connecting the two parallel planes constitute a continuous broken line running from joint A successively through joints F, D, G, C, and E back to A. Two bars of the continuous line meet at each joint. At joint A the condition of equilibrium of the forces in the x direction is

$$\lambda T_{AE} + \lambda T_{AF} = 0 \qquad\qquad 1.9.5$$

In equations between tension coefficients the correct sign can be determined in the following manner: Assume that the unknown force in the bar is a tensile force. If a tensile force in the bar pulls the joint in the direction of the positive x axis, the component force is positive in the equation of equilibrium of the x components of the forces. Similar rules hold, of course, for the y and z directions. If the actual force in the bar happens to be compressive, and hence the assumption is incorrect, solution of the equation yields a negative quantity for the tension coefficient.

In the case of joint A, tensile forces in bars AE and AF exert a pull on the joint in the positive x direction. For this reason, both terms in Eq. 1.9.5 are positive. It follows from Eq. 1.9.5 that

$$T_{AF} = -T_{AE}$$

In exactly the same manner, consideration of the equilibrium of the forces in the x direction at joint F gives

$$T_{DF} = -T_{AF}$$

since $T_{BF} = 0$. Following the continuous line of connecting bars, one finds that the tension coefficient has the same magnitude in each of the connecting bars and that the signs alternate.

Determination of the Forces in the Connecting Bars

With the ratios of the tension coefficients known, the only task that remains is the determination of the magnitude of a tension coefficient.

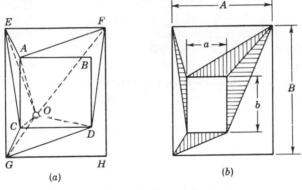

(a)   (b)

Fig. 1.9.2. Determination of the torque

For this purpose the equilibrium of the moments acting on one of the parallel planes must be considered. Figure 1.9.2a is again a front view of the space framework, but the connecting bars that do not carry loads are omitted. A reference point O is chosen arbitrarily in plane $ABCD$, and the moments of all the forces transmitted from the connecting bars to plane $ABCD$ are calculated with reference to point O.

Let it be assumed that the force in bar CE is tensile and the tension coefficient is $+T_{CE}$. The force in CE can be resolved at joint C into an x component of a magnitude λT_{CE}, and into a component lying in plane $ABCD$. According to the rule given for the calculation of the component forces from the tension coefficients, the magnitude of the planar component is T_{CE} times the length of the projection of bar CE in plane $ABCD$. This projection appears in the front view of the space framework in Figs. 1.9.1*b* and 1.9.2*a*. The moment of the x component λT_{CE} with respect to point O is zero, since λT_{CE} is perpendicular to the plane in which the moment is calculated. The moment of the component of the force in plane $ABCD$ is equal to this component times the perpendicular distance of line CE from O in Fig. 1.9.2*a*. Consequently the moment of the force in bar CE is equal to the length CE in Fig. 1.9.2*a* times the perpendicular distance of line CE from O times the tension coefficient T_{CE}. This is the same as T_{CE} times the double area of the triangle OCE. Since the force in bar CE is assumed to be tensile, it pulls joint C toward joint E and the moment about O is clockwise.

A similar calculation would show that the moment exerted by the force in bar AE at joint A with respect to O is equal to the double area of the triangle OAE in Fig. 1.9.2*a* times the tension coefficient in bar AE. As the sign of the forces alternates in the connecting bars, the force in bar AE must be compressive if that in bar CE is tensile. The tension coefficient itself is $T_{AE} = -T_{CE}$. A compressive force in bar AE tends to push joint A in the direction from E to A, and thus the moment of the force about O is again clockwise. Continuing the argument, one finds that the sum of the moments exerted about O by all the forces in the connecting bars is equal to the tension coefficient T_{CE} times the double area of the polygon $AFDGCEA$. Since the moment exerted by the connecting bars must be in equilibrium with the external moment applied to plane $ABCE$, the equation

$$M + 2A_p T_{CE} = 0 \qquad\qquad 1.9.6$$

must hold, where M is the applied torque, A_p the area of the polygon $AFDGCEA$ in Fig. 1.9.2*a*, and T_{CE} the tension coefficient in bar CE.

The reasoning given is, of course, valid for any continuous sequence of bars connecting two parallel planes when the loading consists of torques applied to the parallel planes. For space frameworks of simple shape it is easy to find analytical expressions for the area A_p. This will now be done for the case when the points of attachment of the connecting bars are arranged in rectangles with parallel sides but otherwise in arbitrary relative positions, as shown in Fig. 1.9.2*b*. The area of the two horizontally shaded triangles is altogether $\frac{1}{2}b(A - a)$; the areas of the two vertically shaded triangles add up to $\frac{1}{2}a(B - b)$; and the area of the small rectangle is ab. As the area A_p is defined as the sum of these, addition and algebraic manipulations yield

$$2A_p = aB + bA \qquad\qquad 1.9.7$$

Substitution of the numerical values gives

$$2A_p = 20 \times 36 + 20 \times 28 = 1280 \text{ sq in.}$$

Since M was assumed to be a counterclockwise (negative) moment of 10,000 in.-lb, solution of Eq. 1.9.6 leads to

$$T_{CE} = M/2A_p = 10,000/1280 = 7.8 \text{ lb per in.}$$

The computation of the forces in the connecting bars is carried out in Table 1.9.1.

Table 1.9.1.   Forces in Engine Mount Caused by Torque

Bar	AF	BF	DF	DH	DG	CG	CE	AE
T lb per in.	7.8	0	−7.8	0	7.8	−7.8	7.8	−7.8
L in.	34.9	25.6	37.1	25.6	34.9	25.6	37.1	25.6
F lb	272	0	289	0	272	−200	289	−200

It is interesting to note that the tension coefficients are independent of the relative positions of the two parallel planes.   Equation 1.9.7 gives the same value for $2A_p$ after any translation of plane $ABCD$ in the x, y, or z direction.   The forces in the connecting bars, however, do not remain the same since the true length of the bars changes.

1.10   The Effect of Restraint of Warping on the Forces in the Bars of the Engine Mount

Definition and Measure of Warping

At the beginning of the calculation of the forces in the engine mount, it was assumed that there were no forces perpendicular to plane $EFGH$ transmitted to the engine mount at the attachments E, F, G, and H. Under the action of a torque, however, the engine mount is likely to warp and, when it is elastically deformed, the points of attachment are not likely to lie in a plane.   If these points are bolted to a rigid structure, the restraint of warping will introduce in the engine mount forces perpendicular to the original plane $EFGH$.   In order to evaluate the magnitude of these restraining forces it is necessary to calculate the amount of warping.

The warping w is defined as the perpendicular distance of any one of the points E, F, G, and H from the plane passing through the remaining three points.   It is immaterial through which three of the four points the plane is laid, as can be proved easily when $EFGH$ is a rectangle.   Let the plane pass, say, through the displaced positions of points E, F, and G.   The volume of the pyramid having the triangle EFG as its base and point H as its apex is equal to one third of the product of the area of triangle EFG by the height of the pyramid, where the height is the perpendicular distance w of point H from triangle EFG.   On the other hand, if the plane is laid through points F, G, and H, and the perpendicular distance of point E from the base is denoted by w', the volume of the pyramid is one third of the product of w' by the area of the triangle FGH.   Since the area of triangle EFG is the same as that of triangle FGH (except for second-order small quantities), w' must be equal to w; otherwise two different values would be obtained for the volume of the same pyramid.

The warping can be calculated with the aid of the principle of virtual displacements if a suitable group of dummy loads is assumed.   Figure 1.10.1a shows the engine mount and four parallel unit forces applied

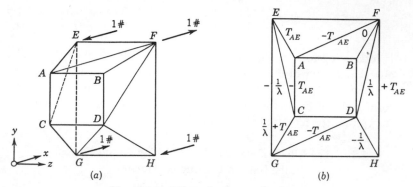

(a)   (b)

Fig. 1.10.1   Effect of unit warping group.

perpendicularly to plane $EFGH$ at points E, F, G, and H.   The four unit forces add up to zero resultant force and zero resultant moment; they are called a warping group.   The warping group does zero work during any translation or rotation of plane $EFGH$ and during any displacement of the four points in their own plane.   If plane $EFGH$ warps, however, the sum of the work done by the four unit forces is different from zero.

In the proof of this statement relative displacements of the points in plane $EFGH$ may be disregarded.   During such planar displacements obviously none of the unit forces does work, since displacement and force are perpendicular to each other.   It is permissible therefore to consider the most general displacement pattern of points E, F, G, and H with the exclusion of relative displacements of the points in plane $EFGH$. Obviously the final position of point E can be reached through suitable small translations of plane $EFGH$ in the x, y, and z directions.   During a positive x translation the positive work done by the forces at F and G is equal numerically to the negative work done by the forces at E and H. During y and z translations none of the forces does work.   The final position of point F can thereafter be reached through a rotation of plane $EFGH$ about the y direction through E, followed by a rotation about the x

direction through E.   During the first of these rotations points E and G are not displaced, and the work done by the forces at F and H is equal and opposite.   During the second rotation none of the forces does work. The final position of G can then be reached through rotating plane $EFGH$ about the final position of line EF.   The work done during this motion is again zero for reasons similar to those discussed before.

Finally the fourth point H can be brought into its final position through displacing it in the (positive or negative) direction of the unit force at H. The work done by this unit force is then numerically equal to the displacement which is the warping w.   The sign of the work is positive if the displacement is in the direction of the unit force at H, and negative if opposite to it.

Naturally it is possible to calculate the forces caused by the warping group in the bars of the space framework.   The forces in the bars and the warping group constitute an equilibrium system.   The work done by the system is therefore zero for any virtual displacement.   As the virtual displacements, the real displacements corresponding to the torque may be chosen, and the work W_i done by the internal forces corresponding to the warping group during the virtual displacements can be calculated. Because of Lemma 1 this internal work must be equal in magnitude and opposite in sign to the external work W_e done by the four unit forces of the warping group.   But W_e was shown to be equal to the warping w. Hence

$$w = -W_i \qquad\qquad 1.10.1$$

Forces Caused by the Unit Warping Group

The next task is the calculation of the forces in the engine mount caused by the warping group.   This dummy loading consists of the four unit forces at points E, F, G, and H which are perpendicular to plane $EFGH$.   There are no external loads applied to points A, B, C, and D. At point B there is just one bar, namely BF, that is not in plane $ABCD$. The condition of equilibrium of the forces in the x direction leads to

$$T_{BF} = 0$$

At joint A two bars connecting the parallel planes meet.   They are bars AE and AF.   Equilibrium of the x components requires that

$$\lambda T_{AE} + \lambda T_{AF} = 0$$

hence

$$T_{AF} = -T_{AE} \qquad\qquad (a)$$

At point F an external load of 1 lb is applied in the x direction.

Consequently

$$1 - \lambda T_{AF} - \lambda T_{DF} = 0$$

Substitution from Eq. a and solution for T_{DF} give

$$T_{DF} = (1/\lambda) + T_{AE} \tag{b}$$

At point H one has

$$-1 - \lambda T_{DH} = 0$$

or

$$T_{DH} = -1/\lambda \tag{c}$$

At point D the equation of equilibrium of the x components is

$$\lambda T_{DF} + \lambda T_{DH} + \lambda T_{DG} = 0$$

After substitutions and solution one gets

$$T_{DG} = -T_{AE} \tag{d}$$

At point G

$$1 - \lambda T_{CG} - \lambda T_{DG} = 0$$

Consequently

$$T_{CG} = 1/\lambda + T_{AE} \tag{e}$$

At point C

$$\lambda T_{CG} + \lambda T_{CE} = 0$$

Therefore

$$T_{CE} = -1/\lambda - T_{AE} \tag{f}$$

At point E

$$-1 - \lambda T_{CE} - \lambda T_{AE} = 0$$

Thus

$$T_{AE} = -(1/\lambda) - T_{CE} = -(1/\lambda) + (1/\lambda) + T_{AE}$$

The last equation is a check on the correctness of the calculation. It shows that the tension coefficients obtained satisfy all the conditions of equilibrium of the forces in the x direction. In the y and z directions the bracings of the two parallel planes can cope with any loads that constitute planar equilibrium systems. In plane $EFGH$, consideration of the three equations of equilibrium of the planar system of forces would only serve the purpose of determining the reactions at the rigid support, which are of no interest at present. In plane $ABCD$ the two conditions of force equilibrium are fulfilled, as can be easily ascertained. The equilibrium of moments, however, must be discussed in some detail.

In Fig. 1.10.1b the tension coefficients just calculated are written next to the bars. They can be divided into two groups; in the first are all the coefficients of magnitude T_{AE}, and in the second those of magnitude $1/\lambda$. It was shown in Art. 1.9 that pure torsion causes a set of tension coefficients corresponding to the first group.   The magnitude of the torque M of this group follows from Eqs. 1.9.6 and 1.9.7:

$$M = (aB + bA)T_{AE} \qquad\qquad 1.10.2$$

If T_{AE} is positive, the torque transmitted to plane $ABCD$ in Fig. 1.10.1b is counterclockwise.

The forces corresponding to the second group of tension coefficients also add up to a couple transmitted to plane $ABCD$.   At point D the x and z components of the forces in bars DF and DH cancel each other in accordance with the rule given in Art. 1.9 for the calculation of force components from tension coefficients.   The y components of the forces in the two bars add up to a vertical upward pull of $(1/\lambda)B$ transmitted to plane $ABCD$ at point D.   Similarly, the action of the connecting bars at point C corresponds to a vertical downward pull of a magnitude $(1/\lambda)B$. These two equal and opposite forces constitute a couple M' given by

$$M' = (1/\lambda)aB \qquad\qquad 1.10.3$$

This couple is also counterclockwise.   In the loading under investigation no external forces act on plane $ABCD$.   For this reason the forces transmitted to the plane from the connecting bars must add up to zero force and zero couple.   Because of the latter requirement $M = -M'$, and

$$T_{AE} = -(1/\lambda)[aB/(aB + bA)] \qquad\qquad 1.10.4$$

With the numerical values given in Fig. 1.9.1 one obtains

$$T_{AE} = -0.02345$$

$$(1/\lambda) = 0.0417$$

Warping Caused by Torque

In column 2 of Table 1.10.1 the numerical values of the tension coefficients are listed.   They are denoted by T_1 to indicate that they are caused by the unit warping group.   With these values the warping w can be calculated from Eq. 1.10.1.   In this equation the internal work

$$W_i = -\Sigma F_0 F_1 R \qquad\qquad 1.10.5$$

where F_0 is the force in a connecting bar caused by the torque of 10,000 in.-lb, F_1 that caused by the unit warping group, and $R = L/EA$. Equation 1.10.5 can be rewritten to contain tension coefficients rather than forces:

$$W_i = -\Sigma T_0 T_1 L^2 R = -\Sigma T_0 T_1 L^3/EA \qquad\qquad 1.10.6$$

where T_0 is the tension coefficient caused by the torque of 10,000 in.-lb, and T_1 that caused by the unit warping group.

Table 1.10.1. Computation of Warping

[1]	[2]	[3]	[4]	[5]	[6]	[7]	[8]
Bar	T_1, 1/in.	T_0, lb/in.	T_0T_1, psi	L^3/EA, in.^{3}/lb	[4]×[5] in.	$T_1{}^2$, 1/in.^{2}	[5] × [7] in./lb
AF	0.02345	7.8	0.183	0.01417	0.00259	0.00055	0.00000780
BF	0	0	0	0.00557	0	0	0
DF	0.01825	−7.8	−0.1425	0.01700	−0.00242	0.00033	0.00000565
DH	−0.0417	0	0	0.00557	0	0.00173	0.00000965
DG	0.02345	7.8	0.183	0.01417	0.00259	0.00055	0.00000780
CG	0.01825	−7.8	−0.1425	0.00557	−0.00079	0.00033	0.00000184
CE	−0.01825	7.8	−0.1425	0.01700	−0.00242	0.00033	0.00000565
AE	−0.02345	−7.8	0.183	0.00557	0.00102	0.00055	0.00000306
					0.00057		0.00004145

The computation of the warping is carried out in Table 1.10.1.   The values of T_0 are taken from Table 1.9.1. (In this table T was listed without the subscript.)   The cross-sectional area of each tube is assumed to be 0.1 sq in., and the elastic modulus of the material 30×10^6 psi. Column 4 contains the products T_0T_1, column 5 the values of L^3/EA, and column 6 the products of the corresponding values in columns 4 and 5. The sum of these products is the warping:

$$w = 0.00057 \text{ in.}$$

Forces Caused by Prevention of Warping

The value obtained for the warping is very small but not negligible in its effects.   Evidently plane $EFGH$ cannot warp if the structure to which the engine mount is attached is perfectly rigid.   The force needed to pull the fourth attachment back into the original plane of the four attachments, and the three reactions of this force at the other three attachments, constitute a warping group.   The magnitude X of the forces of the warping group can be determined from the requirement that the warping caused by the warping group must be equal to -0.00057 in.

If the external load is X times the unit warping group, T_0 must be replaced by XT_1.   The product T_0T_1 becomes $XT_1{}^2$.   In column 7 of Table 1.10.1 the values of $T_1{}^2$ are listed.   The products $T_1{}^2L^3/EA$ are entered in column 8.   They add up to a warping w^* equal to

$$w^* = 0.00004145 \text{ in. per lb}$$

Since the warping caused by the actual warping group must be -0.00057 in., the unknown warping force becomes

$$X = -0.00057/0.00004145 = -13.8 \text{ lb}$$

The computation of the actual force P in each bar is completed in Table 1.10.2.

Table 1.10.2. Forces in Engine Mount When Warping is Prevented

Bar		AF	BF	DF	DH	DG	CG	CE	AE
T_1	1/in.	0.02345	0	0.01825	-0.0417	0.02345	0.01825	-0.01825	-0.02345
XT_1	lb/in.	-0.323	0	-0.251	0.573	-0.323	-0.251	0.251	0.323
L	in.	34.9	25.6	37.1	25.6	34.9	25.6	37.1	25.6
F_w	lb	-11.3	0	-9.3	14.7	-11.3	-6.4	9.3	8.3
F_0	lb	272	0	-289.	0	272	-200	289	-200
P	lb	260.7	0	-289.3	14.7	260.7	-206.4	289.3	-191.7
F'_w	lb	-122	0	-101	160	-122	-70	101	90
P'	lb	150	0	-390	160	150	-270	390	-110

Critical Remarks

It might be said that correction of the values F_0 through the addition of the forces F_W caused by the prevention of warping means additional work but no significant increase in the accuracy of the calculation. This is true in the example just completed but is not so with many other engine mounts. In most engine mounts of the type shown in Fig. 1.9.1 the diagonals AF and DG are of a smaller cross section than the other members of the framework, since they carry very small loads in the important symmetric cases of loading which have not been discussed here. If their cross section is assumed to be, for example, one-third that of the other bars, the importance of warping increases materially. Under such conditions L^3/EA is 0.04251 for bars AF and DG, and the warping w increases from 0.00057 to 0.01093 in. At the same time the warping w' caused by the unit warping group also changes but not in the same ratio. The value of w' increases from 0.00004145 to 0.00007265 in. Consequently X becomes $-0.01093/0.00007265 = -150$ lb. The row marked F' in Table 1.10.2 contains the new values of the forces caused in the bars by the prevention of warping. The final forces in the framework with the lighter diagonals are listed in the row marked P'. They show significant deviations from the values obtained for free warping.

The internal work done in the bars of planes $ABCD$ and $EFGH$ was not taken into account in the calculations. This is the correct procedure if the ring in plane $ABCD$ and the structure between points E, F, G, and H are perfectly rigid in their planes and if at the same time the rubber shock absorbers are very easily deformed perpendicularly to the planes. The work in the bars of the two planes is then zero, and the internal work in the rubber is negligibly small. If the points of attachment of the engine mount give elastically and permit some warping of the plane $EFGH$, the forces in the bars must be somewhere between the values corresponding to free warping and those obtained when warping was prevented. The forces F_0 and P (or P') represent therefore the limiting values.

One more note is of some importance. The final values of the forces in the engine mount were calculated from the equations of equilibrium and from one more equation derived from a consideration of the deformations. These equations suffice for the complete solution of the problem if the structure is simply redundant. This can be shown to be true of the engine mount discussed. The ring was assumed to be rigid in its plane and very weak perpendicularly to it. Although a closed ring is threefold redundant in its plane, and three equations of deformations must be taken into account to calculate the stresses in it, the assumption of perfect rigidity eliminated the necessity of determining the details of the force transmission through the ring.   Because of the perfect rigidity every stress distribution is equivalent from the standpoint of deformations and every one corresponds to zero internal work.   The ring therefore merely fixes the distances of the end points of the connecting bars in plane $ABCD$, and this task can be performed just as well by the four straight bars AB, BD, CD, and AC in Fig. 1.9.1b and one diagonal in the same plane.   To these 5 bars must be added the 8 connecting bars.   There are therefore 13 bars available to attach the 4 points A, B, C, and D to the rigid structure of the wing, nacelle, or fuselage.   The attachment of each point requires 3 non-coplanar bars from which it follows that altogether 12 bars are needed.   Consequently there is 1 bar more in the structure than required, and 1 equation of deformation suffices therefore for the solution of the problem.

On the other hand, the structure would have 2 redundancies if the attachment to the engine did not permit relative displacements of points A, B, C, and D perpendicularly to plane $ABCD$.   When the engine and its attachments are very rigid, they compel points A, B, C, and D to remain in a plane and thus introduce an unknown warping group at the 4 points of attachment of the engine.   The calculation of the forces in the 8 connecting bars under the simultaneous action of the applied torque and the 2 warping groups can be carried out easily enough, but some additional work of computation is, of course, required.   The principles involved in the solution of problems with several unknowns, rather than a single one, are stated at the end of Art. 1.8.

1.11   Forces Caused by a Warping Group in a Highly Redundant Space Framework

The Effect of Symmetry and Antisymmetry

Figure 1.11.1 shows a space framework which may be thought of as representing a transmission tower or an airplane fuselage of the welded steel-tube type.   The framework consists of 4 cells, each of which is a cube.   Each square face of every cube is braced with 2 crossed diagonals.   The diagonals in the bottom and in the rear planes are not shown in the figure in order to avoid overcrowding.   Every bar except the diagonals is assumed to have the same cross-sectional area, and each diagonal is assumed to have one-half the cross-sectional area of the other bars.

The number j of joints in the framework is 20, while the number n of bars connecting the joints is 78.   According to Eq. 1.6.2 the necessary number of bars is

$$3j - 6 = 54$$

There are therefore $78 - 54 = 24$ redundant bars in the structure.

The loading of the structure consists of a warping group of 100 lb applied to bulkhead 0. Such loads can arise at the attachment of the lighter rear portion of the fuselage to the heavier middle portion when a torque is applied to the rear end of the fuselage; the situation then resembles that discussed in Art. 1.10. In a transmission tower, warping groups exist between cells when a cable breaks and the tower is subjected to torsion. In the present article only the forces in the members of the space framework caused by the given warping group are calculated; the problem of finding the forces caused by the torque is not discussed.

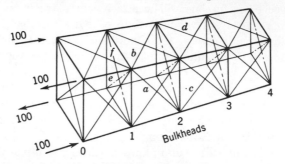

Fig. 1.11.1   Space framework with warping group.

The loading of the structure shown in Fig. 1.11.1 is symmetric with respect to each diagonal in bulkhead 0. Since the framework is also symmetric with respect to the two planes passing through the two groups of parallel bulkhead diagonals, the forces caused in the symmetrically situated bars of the framework must be equal in magnitude and sign. Consequently the forces must be the same in any two surface diagonals of the same cell if they intersect on an edge; such diagonals are, for example, bars a and b, or bars c and d.

The framework is also symmetric with respect to a horizontal plane which cuts the entire structure into an upper and a lower half. The loads, however, are antisymmetric with respect to this plane. Forces are called antisymmetric if in symmetric locations they are equal in magnitude and opposed in sign. For instance, the load applied to the top front longitudinal and the one applied to the bottom front longitudinal are 100 lb each. Their points of application are symmetrically situated with respect to the horizontal plane of symmetry. Since the upper force points to the left and the lower to the right, the two forces are antisymmetric with respect to the horizontal plane of symmetry of the framework. For similar reasons the forces applied to the rear longitudinals are also antisymmetric.

Just as one cannot conceive of a symmetric cause having a non-symmetric effect, so an antisymmetric cause cannot have a symmetric effect. For this reason the forces in the bars of the framework must be antisymmetric with respect to the horizontal plane of symmetry of the structure. Among other things this means that the forces in the two crossed diagonals of any bulkhead must be equal in magnitude and opposite in sign. If there is a force of $+10$ lb (tension) acting in bar e, there must be a force of -10 lb (compression) acting in bar f, since f is the image of e mirrored in the horizontal plane of symmetry.

Therefore, the conclusion can be drawn that it is necessary to calculate the forces in only one-half the number of the diagonal bars; in the other half they can be determined from considerations of symmetry. The number of longitudinals and uprights in the structure totals 36, while one-half the number of the diagonals is 21. Thus the total number of unknown forces in the bars is 57, only 3 more than the number of available equations of static equilibrium. Consideration of the symmetry properties of structure and loading reduced the number of effective redundancies from 24 to 3.

The Choice of the Redundant Quantities

The stress analyst has some freedom in selecting the unknown quantities. For instance, he may consider the forces in the cross-bracings of the intermediate bulkheads (bulkheads 1, 2, and 3) as the redundant elements. Another choice might be the cross-bracings in bulkheads 0, 2, and 4. A third possibility is discussed below in detail because it leads to calculations involving the least amount of numerical work.

In these calculations the framework is imagined to be broken up into 4 individual cells, as shown in Fig. 1.11.2. In the resulting system the bars of the bulkheads of the actual framework (Fig. 1.11.1) are drawn twice. This means that the diagonals e and f, for example, are considered part of cell 1 when the forces in this cell are determined, and they are taken as members of cell 2 when the equilibrium of the forces in cell 2 is calculated. In reality the diagonals e and f exist only once, not twice, in the framework, and thus the actual force in them is the sum of the forces needed for equilibrium in cells 1 and 2.

Each cell by itself is a space framework with 6 redundant bars, as may be verified by counting the joints and bars and substituting the numbers in Eq. 1.6.2. The calculation of the tensions in the bars of any one cell is nevertheless simple because symmetry considerations provide 6 equations in addition to the equations of static equilibrium. Cell 1 is under the action of the self-equilibrating warping group of 100 lb. Under these loads, cell 1 is certainly in equilibrium, and it is possible to calculate the forces caused in its bars by the warping group. The other three cells are not loaded. (The warping groups marked X, Y, and Z in the figure should be disregarded for the time being.) A state of static equilibrium exists therefore in which forces are acting in the bars of cell 1, and the bars of all the other cells are free of force. The question is whether such forces give rise to consistent deformations, or, in other words, whether the elastic deformations of the bars caused by this statically possible system of forces do not lead to gaps or discontinuities in the framework.

It was shown in Art. 1.10 that under the action of a warping group the end face of a cell warps. Consequently points A, B, C, and D of cell 1 do not remain in one plane. On the other hand, cell 2 is not loaded and thus the plane of its face $A'B'C'D'$ is not distorted. Obviously the warped

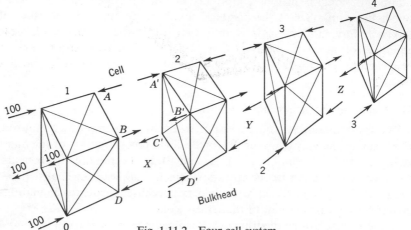

Fig. 1.11.2   Four-cell system.

surface $ABCD$ and the plane $A'B'C'D'$ cannot fit together. If the three pairs of points A and A', C and C', and D and D' are brought together, points B and B' will be some distance apart. They can be forced together by pulling simultaneously on B and B', that is, by exerting forces of equal magnitude and opposite sense at B and B'. Let X denote the magnitude of the unknown longitudinal force that is necessary to bring points B and B' into coincidence.

If a longitudinal force X is applied to cell 2 at point B', and the cell is attached to cell 1 at points A', C', and D', reactions are caused at the points of attachment. X and the three reactions are in equilibrium if they form the warping group shown in Fig. 1.11.2. The force X acting on cell 1 at point B and its reactions at points A, C, and D also constitute a warping group. Thus the effort to connect all the four corners of cells 1 and 2 results in alternating tensile and compressive forces of equal magnitude acting between the two cells at the four corners. The set of these forces may be designated a twin warping group, or an internal warping group. The twin warping group X causes forces in the bars of cell 1 which are superimposed upon the forces originally calculated. Thus the attachment of cell 2 to cell 1 changes the original forces in cell 1. Moreover, the twin warping group X gives rise to forces in the bars of cell 2 which originally were without forces. Under these forces cell 2 must warp, and consequently its end face cannot fit the corresponding face of cell 3. If the argument is continued it is found that twin warping groups Y are needed (see Fig. 1.11.2) to bring into coincidence the four corresponding corners of cells 2 and 3, and twin warping groups Z are necessary to achieve a similar result at the third bulkhead.

The following conclusions can therefore be drawn from these considerations: The conditions of static equilibrium can be satisfied if the forces equilibrating the applied 100-lb warping group in cell 1 alone are calculated and the forces in the bars of the other cells are set equal to zero. However, the deformations of the cells caused by these forces are not compatible. Since obviously the framework of Fig. 1.11.1 remains a continuous unit until the external loads cause failure, the force distribution in the bars just defined is impossible, in spite of the fact that it satisfies the requirements of static equilibrium. The actual forces in the bars can be calculated by adding to these statically correct forces the forces caused by the twin warping groups of as yet unknown magnitude designated as groups X, Y, and Z. The magnitude of these warping groups must be determined from the condition that the warping of the common faces of adjacent cells must be identical. The warping, on the other hand, can be calculated with the aid of the principle of virtual displacements as discussed in Art. 1.10.

In the derivations no effort has been made to determine whether the deformations of face $ABCD$ were identical with those of face $A'B'C'D'$ in the plane of the face when cells 1 and 2 were brought together. In the absence of the unknown warping groups X, Y and Z, face $A'B'C'D'$ does not deform, while face $ABCD$ is transformed from a square into a parallelogram under the action of the 100-lb external warping group. This must be so since the two bulkhead diagonals AD and BC are subjected to forces of equal magnitude and opposite sense, as was found earlier when the antisymmetry of the loading was discussed. At the same time no forces can act in the uprights AB, BC, CD, and DA as a consequence of the antisymmetry. Fortunately, the shearing of the bulkhead is proportional to its warping, and thus the deformations in the planes of the bulkheads are equalized when the warping out of the plane is equalized through the application of twin warping groups of suitable magnitude.

The Forces in the Bars of an Individual Cell

Before the warping of the cells can be calculated, it is necessary to determine the forces caused by a warping group in the bars of a cell.

This will now be done with the aid of Fig. 1.11.3.   Extensive use will be made of the properties of symmetry and antisymmetry.

If the tension coefficient notation is again introduced, the equation of equilibrium of the forces in the x direction at joint F can be written as

$$-aT_{BF} - aT_{AF} - aT_{DF} = 0 \qquad 1.11.1$$

But

$$T_{AF} = T_{DF} \qquad 1.11.2$$

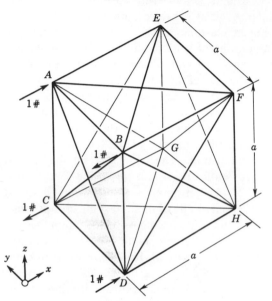

Fig. 1.11.3   Cell with warping group.

since the loading is symmetric with respect to plane $BCGF$.   Substitution of T_{DF} from Eq. 1.11.2 in Eq. 1.11.1 and solution for T_{AF} give

$$T_{AF} = -\tfrac{1}{2}T_{BF} \qquad 1.11.3$$

The equation of equilibrium of the x components at joint B can be written as

$$aT_{BF} + aT_{BE} + aT_{BH} - 1 = 0$$

But for reasons of symmetry

$$T_{BE} = T_{BH} \qquad 1.11.4$$

and thus

$$T_{BE} = \tfrac{1}{2}[(1/a) - T_{BF}] \qquad 1.11.5$$

On the other hand, the loading is antisymmetric with respect to the xz plane of symmetry of the structure.   The forces in the diagonals AF and BE must therefore be equal in magnitude and opposed in sign since the two diagonals are situated symmetrically with respect to the xz plane of symmetry. Equations 1.11.3 and 1.11.5 yield

$$-\tfrac{1}{2}T_{BF} = -\tfrac{1}{2}[(1/a) - T_{BF}]$$

or

$$T_{BF} = \tfrac{1}{2}(1/a) \qquad\qquad 1.11.6$$

Substitution of this value in the expressions found previously gives

$$T_{AF} = -\tfrac{1}{4}(1/a) \qquad\qquad 1.11.7$$

$$T_{BE} = \tfrac{1}{4}(1/a) \qquad\qquad 1.11.8$$

The condition of equilibrium of the x components of the forces at joint E is

$$-aT_{AE} - aT_{BE} - aT_{CE} = 0$$

Because of the symmetry of the structure and loading about plane $ADEH$

$$T_{CE} = T_{BE}$$

Consequently

$$T_{AE} = -2T_{BE}$$

and thus

$$T_{AE} = -\tfrac{1}{2}(1/a) \qquad\qquad 1.11.9$$

The tension coefficients in all the other bars connecting the two bulk-heads can be written without further calculation because of the symmetry of structure and loading about planes $ADEH$ and $BCFG$. Some of the values are shown in the right-hand part of Fig. 1.11.4 for the case when the length a of the edge of the cube is unity. The tension coefficients in the bars of the invisible sides of the cube are, in general, omitted to prevent the overcrowding of the figure. They have the same values as the tension coefficients in the bars that are located symetrically with respect to planes $ADEH$ and $BCFG$. The tension coefficients in the bars contained in the bulkheads can now be calculated with the aid of Fig. 1.11.5.

At joint B the only vertical force transmitted to the bulkhead is due to the tension in bar BH. This tension exerts a downward pull of a magnitude aT_{BH} on the joint. With $a = 1$ in. and $T_{BH} = 0.25$ lb per in., the force in question is 0.25 lb. Similarly the only horizontal force exerted by the connecting bars on the bulkhead at joint B has a magnitude of 0.25 lb and is directed to the left since it is caused by T_{BE}. If the equilibrium of the forces in the plane of the front bulkhead is investigated at each joint, the forces shown in Fig. 1.11.5a are found to act on the bulkhead.

It is readily seen that these forces constitute an equilibrium system. At joint A the two forces shown add up to a resultant of a magnitude of $0.25\sqrt{2}$ lb. This resultant is balanced by the resultant at joint D, and the

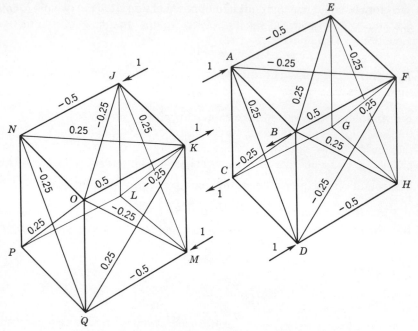

Fig. 1.11.4   Tension coefficients due to unit twin warping group
when edge length is unity.

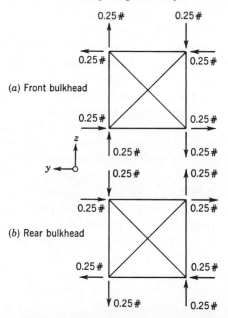

Fig. 1.11.5   Forces in the planes of the bulkheads.

two resultants give rise to a tensile force of equal magnitude in diagonal AD. Since the length of the diagonal is $\sqrt{2}$ in., the tension coefficient

$$T_{AD} = 0.25 \text{ lb per in.} \qquad\qquad 1.11.10$$

The value of the tension coefficient in the other diagonal can be found in a similar manner.   It is

$$T_{BC} = -0.25 \text{ lb per in.} \qquad\qquad 1.11.11$$

Figure 1.11.5 also contains the forces acting on the rear bulkhead.   Again the tension coefficients in the diagonals are found to have the magnitude 0.25 lb per in., while the edge members are not subjected to loads.   The tension coefficients so calculated are also shown in the right-hand part of Fig. 1.11.4.

The left-hand part of Fig. 1.11.4 contains the tension coefficients in a cube on which the reactions of the warping group just considered are acting.   If the effect of a twin unit warping group acting in the middle bulkhead of a double cell is sought, the two cells of Fig. 1.11.4 have to be brought together so that bulkheads $ABCD$ and $JKLM$ coincide.   The tension coefficients in this middle bulkhead have then the values $+0.50$ and -0.50 lb per in.

The total tension coefficients caused in the fuselage by the simultaneous action of the 100-lb warping group in bulkhead 0 and the unknown warping groups X, Y, and Z in bulkheads 1, 2, and 3 can now be obtained with the aid of the principle of superposition. The results of the superposition of the individual tension coefficients are presented in Fig. 1.11.6.   The magnitudes of the groups X, Y, and Z must be determined from the requirement that common faces of the individual cells must warp equally.

Enforcement of the Conditions of Compatible Deformations

It was shown in Art. 1.10 that the work done by the four forces of a unit warping group during any virtual displacement of the joints of the bulkhead on which the forces are acting is numerically equal to the distance traveled by any one of the four corners of the bulkhead perpendicularly to the plane of the remaining three corners.   For this reason, in Fig. 1.11.4 the work of the unit warping group acting on bulkhead $ABCD$ can be interpreted as the displacement of joint D to the right out of plane ABC. At the same time the virtual work done by the warping group acting on bulkhead $JKLM$ can be considered as the displacement to the left of point M out of the plane JKL.   The sum of the work done by the two groups is therefore the gap between points D and M when points A, B, and C coincide with points J, K, and L, respectively.   If the bulkheads $ABCD$ and $JKLM$ represent the same bulkhead of the framework of Fig. 1.11.1

drawn twice, as shown in Fig. 1.11.2, then, of course, the gap cannot exist
as long as the tower or fuselage does not break.   Hence, the work done by
any of the twin warping groups X, Y, and Z of Fig. 1.11.2 must be equal
to zero for any virtual displacement pattern.

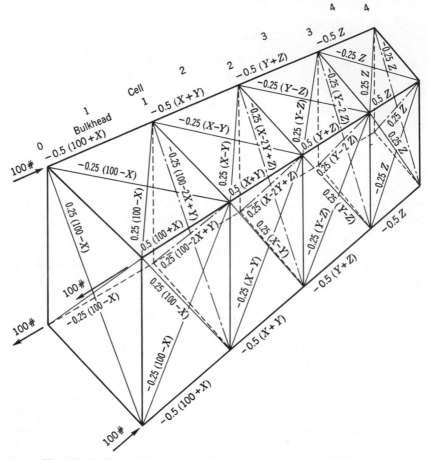

Fig. 1.11.6   Total tension coefficients in pounds per inch when $a = 1$ in.

The work done by any of the twin warping groups is numerically equal
to the internal work done by the forces caused by the respective twin
warping group in the members of the framework during the elongations
of the members.   The internal work was given in Eq. 1.10.6 as

$$W_i = -\Sigma T_0 T_1 L^3 / EA$$

In the present problem T_1 is the tension coefficient caused by any of the twin warping groups X, Y, Z. Its value can be taken from Fig. 1.11.4. Since the equality of warping is to be established for the actual displacements caused by the simultaneous action of the 100-lb warping group as well as the twin warping groups X, Y, and Z, T_0 has the values shown in Fig. 1.11.6. The summation must be extended over all the members of the space framework.

As far as the factor L^3/EA is concerned, it should be noted that all members but the diagonals have the same length a and the same cross-sectional area A^*. The length of any diagonal is $\sqrt{2}a$, while its cross-sectional area is $A^*/2$. Hence

$$(L^3/EA)_{\text{diag}} = 2\sqrt{2}a^3/(EA^*/2)$$

and the multiplying factor in Eq. 1.10.6 of the product of the tension coefficients is $4\sqrt{2}$ times greater for diagonals than for the other members of the framework.

The calculation of the internal work done by the twin warping groups $X = 1$ can now be undertaken. In the computations the factors -1 and a^3/EA^* will be disregarded for the time being, since they are common to all the terms in the summation indicated in Eq. 1.10.6. Moreover, the summation has to be extended only over cells 1 and 2 including, of course, bulkheads 0 and 2, since the group X does not cause any forces in the rest of the space framework. The values of T_1 can be taken from Fig. 1.11.4 if plane $ABCD$—$JKLM$ is identified with bulkhead 1 of Fig. 1.11.6. The values of T_0 are shown in Fig. 1.11.6.

In the front top longitudinal of the first cell $T_0 = 0.5(100 + X)$ and $T_1 = 0.5$. Because of the symmetry the same values also prevail in the rear bottom longitudinal. In the front bottom and rear top longitudinals the values are equal numerically, but they are negative. Hence the product

$$T_0T_1 = 0.25(100 + X)$$

for each longitudinal of the first cell. Since there are four longitudinals in the first cell

$$\Sigma T_0T_1 = 100 + X$$

In the second cell one obtains

$$\Sigma T_0T_1 = 4 \times 0.5 \times 0.5(X + Y) = X + Y$$

Consequently the contribution of all the longitudinals is

$$\sum_{\text{long.}} T_0T_1 = 100 + 2X + Y \qquad (a)$$

In the surface diagonals of the first cell the signs of T_0 and T_1 are opposed, and in those of the second cell they are equal.   Consequently

$$\sum_{\text{s.diag}} T_0 T_1 = -8 \times 0.25 \times 0.25(100 - X) + 8 \times 0.25 \times 0.25(X - Y)$$

$$= -50 + X - 0.5Y \tag{b}$$

In the transverse diagonals one gets

$$\sum_{\text{t.diag}} T_0 T_1 = -2 \times 0.25 \times 0.25(100 - X) - 2 \times 0.5 \times 0.25(100 - 2X + Y)$$

$$+ 2 \times 0.25 \times 0.25(X - 2Y + Z)$$

$$= -37.5 + 0.75X - 0.5Y + 0.125Z \tag{c}$$

The total internal work is therefore

$$W_i = -(a^3/EA^*)[100 + 2X + Y + 4\sqrt{2}(-50 + X - 0.5Y)$$

$$+ 4\sqrt{2}(-37.5 + 0.75X - 0.5Y + 0.125Z)]$$

$$= -(a^3/EA^*)(-395 + 11.9X - 4.66Y + 0.707Z) \tag{d}$$

As was shown before, the displacements of adjacent cells are compatible only if $W_i = 0$.   The conditions of equal warping at bulkhead 1 can be written therefore as

$$11.9X - 4.66Y + 0.707Z = 395 \tag{1.11.12}$$

The conditions at bulkheads 2 and 3 can be found in a similar manner. T_0 is always given by Fig. 1.11.6, and Fig. 1.11.4 can be used in determining the value of T_1.   The plane $ABCD-JKLM$ must in turn be identified with bulkheads 2 and 3 in order to obtain the conditions corresponding to $Y = 1$ and $Z = 1$.   The continuity conditions at bulkheads 2 and 3 become

$$-4.66X + 11.9Y - 4.66Z = -70.7$$

$$0.707X - 4.66Y + 11.9Z = 0 \tag{1.11.13}$$

Equations 1.11.12 and 1.11.13 can be solved for the unknown quantities. The solution is

$$X = 36.4, \qquad Y = 8.85, \qquad Z = 1.31 \tag{1.11.14}$$

If these values are substituted in the expressions listed in Fig. 1.11.6, and the tension coefficients so obtained are multiplied by the length of the members to which they refer, the forces shown in Fig. 1.11.7 result.

These forces represent the solution of the problem stated at the beginning of this article.

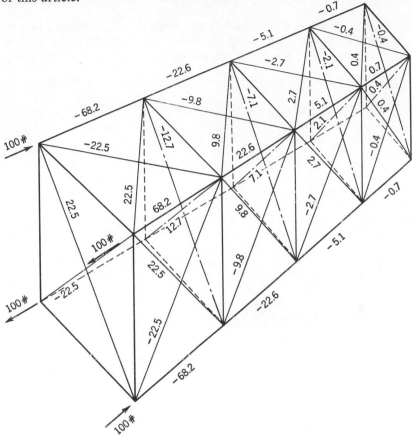

Fig. 1.11.7   Forces in members of framework fuselage, in pounds.
From author's paper in *J. Aero. Sciences.*

Space Framework without Transverse Bracing

The tension coefficients listed in Fig. 1.11.6 can also be used to solve a modified problem in which the internal transverse bracings are missing from the tower or fuselage. In such a structure, of course, there are no forces in the non-existent internal transverse bracings. For this reason the expressions shown next to the diagonals of bulkheads 1, 2, and 3 in Fig. 1.11.6 must vanish. The resulting equations follow:

$$2X - Y = 100$$
$$X - 2Y + Z = 0 \qquad\qquad 1.11.15$$
$$Y - 2Z = 0$$

The solution is

$$X = 75, \qquad Y = 50, \qquad Z = 25 \qquad\qquad 1.11.16$$

Substitution of these values in the expressions listed in Fig. 1.11.6 and multiplication by the length of the members yield the forces of the fuselage without internal bracings.   The values obtained are shown in Fig. 1.11.8.

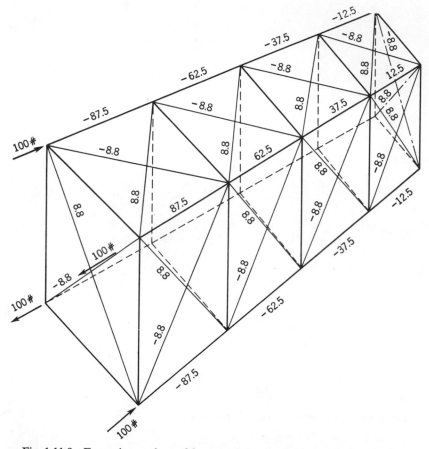

Fig. 1.11.8   Forces in members of framework fuselage without internal bracing, in pounds.   From author's paper in *J. Aero. Sciences.*

Comparison of Figs. 1.11.7 and 1.11.8 shows the beneficial effect of the internal bracings.   The maximum force in the first and fourth bays is 87.5 and 12.5 lb, respectively, when the internal bracing elements are absent from bulkheads 1, 2, and 3, and 68.2 and 0.7 lb when they are present.

1.12   Deflections of Beams

It is known from experiment that the relative rotation $d\phi$ of two sections of a beam a small distance dx apart can be expressed by the formula

$$d\phi = kM\,dx$$

if the material of the beam follows Hooke's law.   In the equation, M is the bending moment acting in the sections of the beam and k a factor of

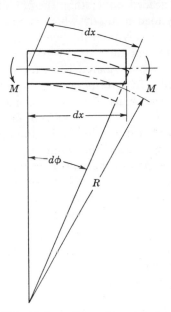

Fig. 1.12.1   Deformations of a segment of beam.

proportionality.   The equation is an expression of Hooke's law for bending; it states that the rotation is proportional to the applied moment. The value of the proportionality factor can be calculated from the Bernoulli-Navier assumption that plane sections before bending remain plane after bending.   It is proved in strength of materials that the factor is $1/EI$, where E is Young's modulus of the material and I the moment of inertia of the cross section:

$$d\phi = (M/EI)\,dx \qquad\qquad 1.12.1$$

Figure 1.12.1 shows the deformations of a segment of a beam of infinitesimal length.

When the variation of the bending moment and of the bending rigidity along the beam is known, Eq. 1.12.1, or some other equation if the beam is not perfectly elastic, defines the deformations of all the beam elements. The calculation of the deflection of any section of the beam relative to the supports, or relative to any other section, is then a problem in geometry; this is true whether the beam is simply supported or statically indeterminate. An ingenious and convenient graphic procedure is available for the construction of the deformed shape of the beam; it is Mohr's method of the string polygon in which the deflections are drawn to a scale much larger than the scale of the longitudinal dimensions. When a single deflection quantity is needed rather than the complete deflected shape of the beam, the dummy load method† which is based on mechanics yields an answer much more rapidly than the geometric construction.

In the dummy load method the virtual work done by the internal stresses due to the dummy loads is calculated when the beam is deformed by the real loads of the system. In agreement with the notation employed in the calculation of the displacements of joints in frameworks, the subscript 0 will be used to designate quantities related to the real loading, and subscript 1 will be introduced to refer to the dummy load system. In the calculation of the work quantities, the infinitesimal element extending from section x to section $x + dx$ of a beam is considered isolated. When the actual moments in the end sections of the infinitesimal beam are M_0 and $M_0 + (dM_0/dx)\, dx$, the actual relative rotation of the two sections is approximately

$$d\phi_0 = (M_0/EI)\, dx$$

The expression is approximate because the small increment dM_0 in the moment has been disregarded.

If a dummy load system causes moments M_1 and $M_1 + (dM_1/dx)\, dx$ in the same end sections of the infinitesimal beam, the work done by the dummy moments during the actual deformations is approximately

$$dW = M_1\, d\phi_0 = (M_1 M_0/EI)\, dx$$

The expression obtained is again approximate because the small changes dM_0 and dM_1 in the real and dummy moments have been disregarded. From the standpoint of the infinitesimal beam element, the moments are external, and thus the work calculated is the external work. Naturally the internal stresses in the element also do work during the displacements; since the sum of the work done by the internal and the external forces must vanish in accordance with Lemma 1, the internal work must be

$$dW_i = -(M_0 M_1/EI)\, dx$$

When a beam of a finite length L is under the action of real moments M_0 and dummy moments M_1, the virtual work W_i done by the internal

† The dummy load method is described in Art. 1.7.

stresses due to the dummy moments during the real displacements corresponding to the moments M_0 is obtained by integration:

$$W_i = - \int_0^L \frac{M_0 M_1}{EI} \, dx \qquad 1.12.2$$

This expression is accurate because in the limiting process of the integral calculus the changes dM_0 and dM_1 vanish.

With the rule for the calculation of the internal work established it is possible now to proceed to the analysis of the deflections of beams subjected

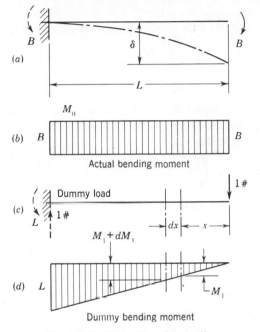

Fig. 1.12.2   Deflection of a cantilever.

to arbitrary loads. As a first example, the deflection of the end point of a cantilever beam of uniform section will be determined when the loading of the beam is a concentrated bending moment B applied at the free end of the cantilever. The beam is shown in Fig. 1.12.2a, and the constant bending moment $M_0 = B$ is represented in part b of the figure. In the calculation of the deflection it is advisable to find a dummy load whose work during the deformations of the beam is numerically equal to the displacement sought. For this reason a vertical dummy force of unit magnitude is chosen, as shown in Fig. 1.12.2c. The bending moment M_1 caused by the unit dummy load can be obtained from the basic laws of statics. The moment diagram is drawn in part d of the figure.

The dummy load does work only when the free end of the cantilever beam is displaced in the vertical direction. If the vertical downward displacement of the end point is denoted by δ, the external work W_e is

$$W_e = 1 \times \delta$$

The reaction force and the reaction moment do no work because the left end of the beam shown in Fig. 1.12.2 is rigidly fixed and consequently cannot move. It follows from the principle of virtual displacements

$$W_e + W_i = 0$$

that the end deflection of the cantilever can be expressed as

$$\delta = \int_0^L \frac{M_0 M_1}{EI} \, dx \qquad\qquad 1.12.3$$

The actual deflection can now be calculated without difficulty. The bending moment M_0 caused by the real loading has the constant value B. The bending moment M_1 caused by the dummy load is $M_1 = x$. Substitution and integration yield

$$\delta = \int_0^L \frac{Bx}{EI} \, dx = \frac{B}{EI} \int_0^L x \, dx = \frac{BL^2}{2EI}$$

Consequently the deflection of a cantilever beam loaded by an end moment B is

$$\delta = BL^2/2EI \qquad\qquad 1.12.4$$

It is of some interest to check the value of this deflection by a calculation based on geometry. It follows from Eq. 1.12.1 and Fig. 1.12.1 that the radius of curvature R of the beam is $R = dx/d\phi = EI/M$. Consequently the radius of curvature of the cantilever beam of Fig. 1.12.2a is constant when the moment M_0 acting on it is constant. The cantilever is bent to an arc of a circle of a radius

$$R = EI/B$$

It can be seen from Fig. 1.12.3 that the end deflection δ can be expressed from the geometry of the figure as

$$\delta = R - R \cos \phi$$

As shown in any text book on the calculus, the Taylor series expansion of the cosine function is

$$\cos \phi = 1 - \phi^2/2 + \phi^4/24 - \cdots$$

Consequently

$$\delta = R(1 - 1 + \phi^2/2 - \phi^4/24 + \cdots)$$

Because the elastic deformations of structural materials are small, ϕ must be a small quantity. Its fourth power can be neglected therefore in comparison to its second power. Hence

$$\delta = \tfrac{1}{2}R\phi^2$$

Since according to Fig. 1.12.3 the angle ϕ is

$$\phi = L/R$$

the deflection becomes

$$\delta = \tfrac{1}{2}(L^2/R) = BL^2/2EI$$

This agrees with the result found earlier and given in Eq. 1.12.4. The agreement could be expected because the deformations of the beam are uniquely defined by the bending-moment distribution together with Hooke's law for bending (Eq. 1.12.1). The use of the principle of virtual displacements in the calculation of displacement quantities is just a convenient routine procedure through which geometric considerations can be avoided.

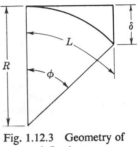

Fig. 1.12.3   Geometry of deflections.

It may be seen therefore that the basic principle involved in the calculation of the deformations of a bent beam is exactly the same as that used in the determination of the displacements of the joints of frameworks. First a suitable dummy loading is found. The choice is made in such a manner that the external work done by the dummy loads is equal numerically to the deflection quantity sought. Clear distinction must be made between two entirely different systems of forces and moments: One corresponds to the dummy loading and represents a suitable equilibrium system, and the other consists of the actual loads and the internal moments corresponding to them. The deformations of the elements of the beam are calculated from the real load system, and the work done by the fictitious dummy load system during the real deformations is determined. From the condition that the sum of the external work and the internal work must vanish, the integral of Eq. 1.12.3 is obtained for the deflection quantity sought.

The procedure developed will now be applied to simply supported and cantilever beams under various loading conditions. It follows from Eq. 1.12.3 that the deflection is obtained by means of an integration of the product M_0M_1 if the beam has a constant cross section. In the preceding example both bending moments M_0 and M_1 were linear functions of x. Since this is true in many other problems also, it is worth while to develop a general expression for the integral for cases when both the M_0 and M_1 moment diagrams are bounded by straight lines. In Fig. 1.12.4 two such

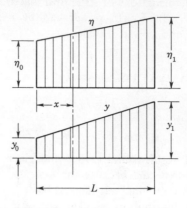

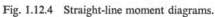

Fig. 1.12.4   Straight-line moment diagrams.

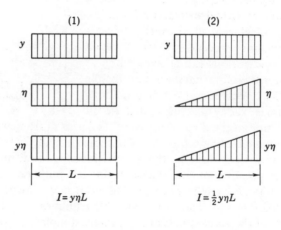

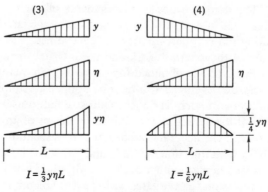

Fig. 1.12.5   Particular moment diagrams.

straight-line diagrams are shown.   The integral I to be evaluated can be written as

$$I = \int_0^L \eta y \, dx$$

Since

$$\eta = \eta_0 + (\eta_1 - \eta_0)(x/L)$$
$$y = y_0 + (y_1 - y_0)(x/L)$$

integration, substitution of the limits, and algebraic simplifications yield

$$I = (L/6)[y_0(2\eta_0 + \eta_1) + y_1(2\eta_1 + \eta_0)] \qquad 1.12.5$$

Often recurring particular cases of the two straight-line diagram, together with the value of the integral, are given in Fig. 1.12.5.   With the

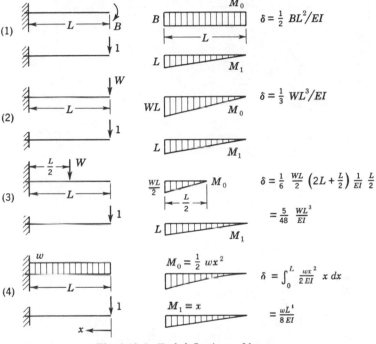

Fig. 1.12.6   End deflections of beams.

aid of these expressions many deflection problems of beams can be solved with very little work.   The manner in which such calculations can be carried out and the final expressions for the deflections are presented in Figs. 1.12.6, 1.12.7, and 1.12.8.   The first one of these figures presents end deflections, the second middle deflections, and the third end rotations. The diagrams are self-explanatory, but a few additional remarks may be of value.   Whenever a deflection is sought, the dummy load chosen is a unit force.   When the deflection quantity required is a rotation, the dummy load is a unit moment applied in the section whose angle of rotation is required.   The expressions given in Eq. 1.12.5 and in Fig. 1.12.5 can be used to advantage in most of the problems presented.   It is important, however, to realize that they are valid only for those segments of the beam in which the moment diagrams can be represented by a single straight line.

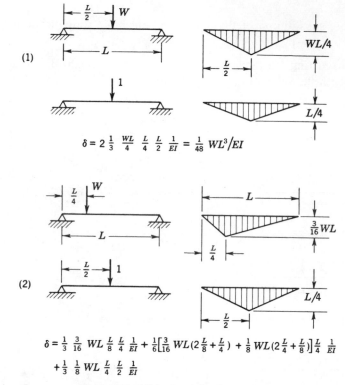

Fig. 1.12.7   Midpoint deflections of beams.

Consequently, in problem 3 of Fig. 1.12.6, in problems 1 and 2 of Fig. 1.12.7, and in problem 2 of Fig. 1.12.8, the beam must be broken up into segments and the formulas of Fig. 1.12.5 have to be applied separately to each segment.   The deflection is then computed from the sum of the integrals obtained.   If the deflection is calculated by carrying out analytically the integration indicated in Eq. 1.12.3, the beam must be broken up into segments in exactly the same manner since the integrand is represented by different expressions in the various segments of the beam.

The last problem of Fig. 1.12.6 and the last two problems shown in Fig. 1.12.8 must be solved by integration since the moment diagrams

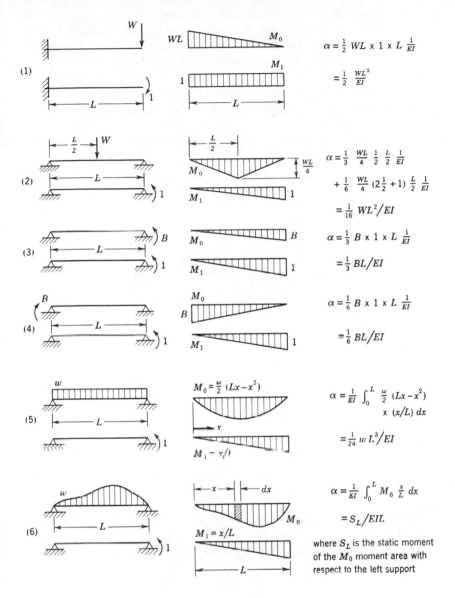

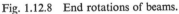

Fig. 1.12.8   End rotations of beams.

corresponding to the actual loads are not linear. The result obtained in problem 6 of Fig. 1.12.8 is of particular interest. It shows that the end rotation of a simply supported beam at the right-hand support is proportional to the static moment of the moment area of the actual loading with respect to the left-hand support. This result is valid for any distributed loading and for any number of concentrated loads. A corresponding expression for the end rotation at the left-hand support can be obtained by interchanging left and right and vice versa in the sentence before the last.

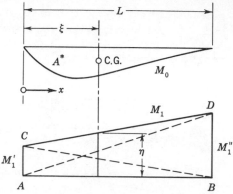

Fig. 1.12.9 Semigraphic method of displacement calculation.

The rule can be generalized. If the M_0 diagram is curved and the M_1 diagram is a straight line, as in Fig. 1.12.9, it is advantageous to think of diagram $ACDB$ as the sum of diagrams ADB and ACB. One can write

$$\frac{1}{EI}\int_0^L M_0 M_1\,dx = \frac{1}{EI}\int_0^L M_0 M'_1\frac{L-x}{L}\,dx + \frac{1}{EI}\int_0^L M_0 M''_1\frac{x}{L}\,dx$$

$$= \frac{1}{EI}(M'_1 S_R + M''_1 S_L) \qquad\qquad 1.12.6$$

where S_R is the static moment of the M_0 area with respect to the right support and S_L that with respect to the left support. If the x coordinate of the centroid of the M_0 area is designated by ξ, naturally

$$S_L = A^*\xi, \qquad S_R = A^*(L-\xi)$$

where A^* is the area under the M_0 curve. Consequently

$$\frac{1}{EI}\int_0^L M_0 M_1\,dx = \frac{A^*}{EI}[M'_1(L-\xi) + M''_1\xi] \qquad\qquad 1.12.7$$

But from Fig. 1.12.9

$$M'_1(L - \xi) + M''_1\xi = \eta$$

which is the ordinate of the M_1 diagram corresponding to ξ. This means that

$$\frac{1}{EI}\int_0^L M_0 M_1 \, dx = \frac{A^*\eta}{EI} \qquad 1.12.8$$

When the end rotation of a simply supported beam or the end deflection of a cantilever beam of constant moment of inertia is sought under a complex system of distributed or concentrated loads, the first task is to find the centroid of the M_0 area. Next the ordinate η of the proper M_1 diagram has to be measured at the centroid. Substitution of this value and of the area A^* under the M_0 curve into Eq. 1.12.8 yields the displacement quantity sought.

When the moment of inertia, or more generally the bending rigidity, of the beam is not constant, the only difference in the calculations is that the product EI cannot be taken before the integral sign in Eq. 1.12.3, while such a procedure was permissible and was indeed followed in the preceding examples. When the analytic expression representing the variation of the bending rigidity is known, the integration can be carried out occasionally in closed form. When such an integration is not possible, or is too difficult, or when the variation of EI is given graphically or numerically for a number of sections of the beam, the displacement quantity can always be obtained by numerical or graphical integration.

The procedure that led to Eq. 1.12.8 can be modified to take care of variations in the moment of inertia I, the elastic modulus E, or both. Instead of starting out with the M_0 diagram, one has to draw the M_0/EI curve. Equation 1.12.8 is then replaced by the equation

$$\int_0^L \frac{M_0}{EI} M_1 \, dx = A^{**}\eta' \qquad 1.12.9$$

where A^{**} is the area under the M_0/EI curve and η' is the ordinate of the M_1 curve corresponding to the centroid of the area under the M_0/EI curve.

A final remark on the calculation of the deflections of beams concerns the effect of shear forces. It follows from simple considerations of the equilibrium of a beam element that the bending moment can change along the beam only if shear forces are also acting in the sections of the beam. Nevertheless the effect of the shear forces on the deformations has been entirely disregarded in this article. This is general practice in structural engineering because in most problems the shearing deformations are small compared to the bending deformations. The problem is examined in more detail in Art. 2.1.

1.13   Statically Indeterminate Problems Involving Bending

Figure 1.13.1 represents a beam of uniform cross section on three rigid supports.   As is usual in structural engineering, it is assumed that one of the supports holds the beam fixed in the horizontal as well as in the vertical direction while the other two supports permit horizontal displacements of any magnitude.   The structure clearly represents a singly redundant

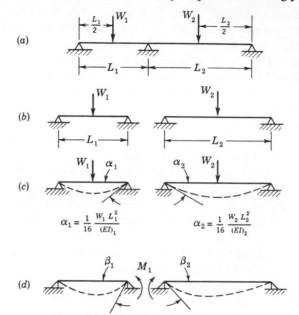

Fig. 1.13.1   Statically indeterminate beam.

problem.   If any one of the three supports is removed, the shear forces and bending moments in the beam can be determined by using the principles of statics alone.   The magnitude of the vertical reaction at the third support however, can be calculated only if the elastic deformations of the beam are taken into account.

It is possible therefore to transform the beam into a statically determinate one by removing, say, the middle support.   The reaction at the middle support can be determined then from the requirement that it must cause a vertical upward deflection equal in magnitude to the vertical downward deflection of the beam calculated for the beam with its middle support removed.   This requirement permits the determination of the unknown reaction at the middle support.   This is not the only manner, however, in which the statically indeterminate problem of the beam on three supports can be solved. Another procedure, which is usually more convenient, consists of imagining the beam cut at the middle support. In such a manner the original beam is transformed into two independent beams each of which rests on two simple supports and is statically determinate (Fig. 1.13.1b). Obviously, the structure obtained differs greatly from the actual structure, but the difference between the two can be eliminated if a bending moment M_1 of suitable magnitude is introduced in the section at the middle support.

Of course, the magnitude and the sign of M_1 are not known. They can be determined from the requirement that the rotations of the two artificially created independent beams should be the same at the middle support. This can also be expressed by stating that the actual beam must remain continuous over the middle support after it has been cut and the moment M_1 applied in the section.

The requirement of continuity can be formulated mathematically with the aid of the formulas developed in Art. 1.12. According to problem 2 of Fig. 1.12.8, the rotation of the first beam at the middle support is

$$\tfrac{1}{16}W_1L_1{}^2/E_1I_1$$

Similarly the rotation of the second beam at the middle support is

$$\tfrac{1}{16}W_2L_2{}^2/E_2I_2$$

The unknown moment M_1 at the middle support causes a rotation equal to

$$\tfrac{1}{3}M_1L_1/E_1I_1$$

in the first beam and a rotation equal to

$$\tfrac{1}{3}M_1L_2/E_2I_2$$

in the second beam. It should be noted, however, that a positive rotation of the first beam at the middle support is counterclockwise while a positive rotation of the second beam at the middle support is clockwise. In order to prevent a gap from opening up over the middle support, it is necessary that the positive rotation of the first beam be equal to the negative rotation of the second beam. This can be expressed in the following equation:

$$\tfrac{1}{16}(W_1L_1{}^2/E_1I_1) + \tfrac{1}{3}(M_1L_1/E_1I_1) = -\tfrac{1}{16}(W_2L_2{}^2/E_2I_2) - \tfrac{1}{3}(M_1L_2/E_2I_2)$$

$$1.13.1$$

This equation can be solved for the unknown quantity M_1. When the cross section and the material of the beam are constant, the solution is

$$M_1 = -\tfrac{3}{16}(W_1L_1{}^2 + W_2L_2{}^2)/(L_1 + L_2) \qquad 1.13.2$$

The three moment equations generally used in the calculation of the bending moments in beams on several supports can be developed easily with the aid of considerations similar to the foregoing ones. Figure 1.13.2 shows three spans of a beam which is resting on a number of supports. Only the $(n-2)$nd, $(n-1)$st, nth, and $(n+1)$st supports are shown. It is assumed that the loading of each span is distributed uniformly and that the distributed load of the nth span is denoted by the symbol w_n. The continuous beam is transformed again into a statically determinate structure by imagining sections through it at the location of each support. The rotation of the end section of each simply supported beam so obtained

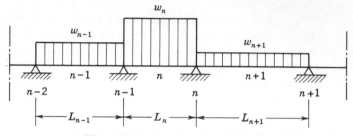

Fig. 1.13.2 Beam on several supports.

can be calculated as in the preceding example. The angle of rotation caused by the uniformly distributed load is given as item 5 of Fig. 1.12.8. Consequently the total angle of rotation $\alpha_{n,n}$ of the nth beam at the nth support, caused by the distributed load w_n of the nth span, the unknown bending moment M_{n-1} at the $(n-1)$st support and the unknown bending moment M_n at the nth support can be written as

$$\alpha_{n,n} = \tfrac{1}{24}(wL^3/EI)_n + \tfrac{1}{3}(ML/EI)_n + \tfrac{1}{6}(M_{n-1}L_n/E_nI_n)$$

In a similar manner the total angle of rotation $\alpha_{n,n+1}$ of the $(n+1)$st beam at the nth support caused by the simultaneous action of the distributed load w_{n+1} of the $(n+1)$st span, the unknown bending moment M_n at the nth support, and the unknown bending moment M_{n+1} at the $(n+1)$st support is equal to

$$\alpha_{n,n+1} = \tfrac{1}{24}(wL^3/EI)_{n+1} + \tfrac{1}{3}(M_nL_{n+1}/E_{n+1}I_{n+1}) + \tfrac{1}{6}(ML/EI)_{n+1}$$

As in the preceding example the requirement of consistent deformations or, in other words, of the preservation of the continuity of the beam over the nth support, can be expressed by the equation

$$\alpha_{n,n} = -\alpha_{n,n+1}$$

Substitutions yield

$$\frac{M_{n-1}L_n}{6(EI)_n} + \frac{M_nL_n}{3(EI)_n} + \frac{M_nL_{n+1}}{3(EI)_{n+1}} + \frac{M_{n+1}L_{n+1}}{6(EI)_{n+1}}$$

$$= -\frac{1}{24}\frac{w_nL_n^3}{(EI)_n} - \frac{1}{24}\frac{w_{n+1}L^3_{n+1}}{(EI)_{n+1}} \qquad 1.13.3$$

If the section of the beam is constant over all the spans, the equation can be simplified by multiplying through by $6EI$ and rearranging the terms:

$$M_{n-1}L_n + 2M_n(L_n + L_{n+1}) + M_{n+1}L_{n+1}$$
$$= -\tfrac{1}{4}(w_nL^3_n + w_{n+1}L^3_{n+1}) \qquad 1.13.4$$

This is the three-moment equation in its usual form. It is applied to the continuous beam of Fig. 1.13.3 to illustrate its use in the solution of statically indeterminate beam problems. Fictitious sections at supports

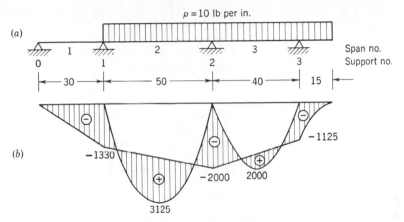

Fig. 1.13.3   Beam on four rigid supports.
Distances in inches, moments in inch-pounds.

1 and 2 transform the continuous beam into an assembly of three shorter beams, each of which is simply supported. The bending moments caused by the distributed loads in the individual spans can therefore be calculated with the aid of the laws of statics. The first span is not loaded. In the second the bending-moment curve is a parabola with a maximum

$$M_{\text{max}} = \tfrac{1}{8}pL^2 = 3125 \text{ in.-lb}$$

In the third span the parabola of the bending moment has a maximum value

$$M_{\text{max}} = 2000 \text{ in.-lb}$$

The moment distribution over the cantilever is also parabolic, and the maximum is reached over the third support. It is

$$M_3 = -\tfrac{1}{2}pL^2 = -1125 \text{ in.-lb}$$

The sign is negative since with downward loads the cantilever, fixed at support 3, assumes a bent shape that is convex when viewed from above, while under the downward loads the simply supported independent beams over spans 2 and 3 become concave as seen from above.

Since the bending moment M_0 at support 0 is obviously zero as long as the beam rests on simple supports, the moments M_1 and M_2 at supports 1 and 2 are the only unknown quantities. They can be determined by writing the three-moment equation twice and solving the two simultaneous equations for the unknown moments.

The three-moment equation for spans 1 and 2 is obtained from Eq. 1.13.4 by replacing the subscripts $n-1$, n, and $n+1$ by the subscripts 0, 1, and 2, respectively. Substitution of the values already calculated yields

$$0 \times 30 + 2M_1(30 + 50) + M_2 \times 50 = -\tfrac{1}{4}(0 \times 30^3 + 10 \times 50^3)$$

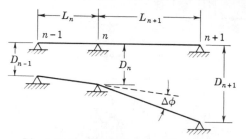

Fig. 1.13.4   Displacements of supports.

When the computations are carried put, this becomes

$$160M_1 + 50M_2 = -312{,}500 \tag{a}$$

For spans 2 and 3 the three-moment equation reads

$$M_1 \times 50 + 2M_2(50 + 40) - 1125 \times 40 = -\tfrac{1}{4}(10 \times 50^3 + 10 \times 40^3)$$

that is

$$50M_1 + 180M_2 - 45{,}000 = -472{,}500 \tag{b}$$

Solution of the two simultaneous equations a and b gives

$$M_1 = -1330 \text{ in.-lb}$$
$$M_2 = -2000 \text{ in.-lb}$$

A convenient graphic representation of the moments along the continuous beam is obtained if the positive moments in the spans and the negative moments at the supports are laid out in the same direction; this is shown in Fig. 1.13.3b. The differences between the ordinates of the parabolas and those of the straight lines connecting the points that represent the moments at the supports are the resultant bending moments. They are shown in the figure by shading.

The principle of virtual displacements made the derivation of the three-moment equation a simple task. It is just as easy to generalize the equation to apply to beams whose supports are displaced in consequence of yielding or settling. Figure 1.13.4 shows two spans of a beam which rests on several supports.   The displacements D_{n-1}, D_n, and D_{n+1} indicated in the figure are assumed to be known.   If the displaced positions of the supports are connected by straight lines, the angle $\Delta\phi$ subtended by the two straight lines representing the new positions of the two beams is

$$\Delta\phi = (D_{n+1} - D_n)/L_{n+1} - (D_n - D_{n-1})/L_n$$

The requirement of continuity over the nth support can be written as

$$\alpha_{n,n} = -(\alpha_{n,n+1} + \Delta\phi)$$

After substitution and rearrangement, this equation becomes

$$\frac{M_{n-1}L_n}{6(EI)_n} + \frac{M_nL_n}{3(EI)_n} + \frac{M_nL_{n+1}}{3(EI)_{n+1}} + \frac{M_{n+1}L_{n+1}}{6(EI)_{n+1}}$$

$$= -\frac{1}{24}\frac{w_nL^3_n}{(EI)_n} - \frac{1}{24}\frac{w_{n+1}L^3_{n+1}}{(EI)_{n+1}} - \Delta\phi \qquad 1.13.5$$

When the bending rigidity of the beam is constant over all the spans, the equation reduces to

$$M_{n-1}L_n + 2M_n(L_n + L_{n+1}) + M_{n+1}L_{n+1}$$

$$= -\tfrac{1}{4}(w_nL^3_n + w_{n+1}L^3_{n+1}) - 6EI\Delta\phi \qquad 1.13.6$$

Equation 1.13.6 is used in Art. 1.14 in the calculation of the moments in a beam on elastic supports.   At the end of Art. 1.17 it is shown how the equation can be applied when an end section of a beam is rigidly clamped.

Another statically indeterminate problem of interest is the one represented by Fig. 1.13.5.   The uniformly loaded beam would be statically determinate were its two ends free to rotate.   The rigid fixation deprives the beam of two degrees of freedom of motion and transforms the statically determinate beam into a doubly redundant one.   Without the rigid fixation, the ends of the beam would undergo rotations through an angle

$$\alpha = \tfrac{1}{24}(wL^3/EI)$$

as is calculated in item 5 of Fig. 1.12.8.   The end fixation eliminates these rotations by providing such reaction moments at supports 1 and 2 as cause rotations of equal magnitude and opposite sense.   According to item 3 of Fig. 1.12.8, the rotation β_2 at support 2 caused by M_2 (see Fig. 1.13.5c) is

$$\beta_2 = \tfrac{1}{3}M_2L/EI$$

Similarly, in agreement with problem 4 of Fig. 1.12.8, the rotation β_1 at support 2 caused by M_1 is

$$\beta_1 = \tfrac{1}{6}M_1L/EI$$

Because of the symmetry of both structure and loading the two end moments must be equal in magnitude:

$$M_1 = M_2 = M_{end}$$

Hence the total rotation due to the end moments is

$$\beta = \beta_1 + \beta_2 = (\tfrac{1}{3} + \tfrac{1}{6})M_{end}L/EI = \tfrac{1}{2}M_{end}L/EI$$

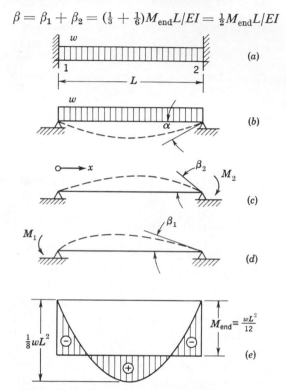

Fig. 1.13.5   Clamped beam with distributed load.

The clockwise rotation $\beta = \beta_1 + \beta_2$ due to M_1 and M_2 must be equal to the counterclockwise rotation α due to w.   Therefore

$$\tfrac{1}{24}wL^3/EI = \tfrac{1}{2}M_{end}L/EI$$

From this requirement the magnitude of the end moment can be calculated:

$$M_{end} = \tfrac{1}{12}wL^2 \qquad\qquad 1.13.7$$

The moment diagram of the redundant beam can be obtained by super-position of the individual moment diagrams of the simply supported beam corresponding to the distributed load w and the end moments M_{end}.   The diagram is shown in Fig. 1.13.5e.

If the loading of the beam is concentrated at the middle rather than distributed uniformly over the span, the end moments can be obtained in a similar manner. With a concentrated load W at midspan

$$\alpha = \tfrac{1}{16} WL^2/EI$$

as is calculated in problem 2 of Fig. 1.12.8. The requirement of a vanishing end rotation can be given therefore in the form

$$\tfrac{1}{16} WL^2/EI = \tfrac{1}{2} M_{end} L/EI$$

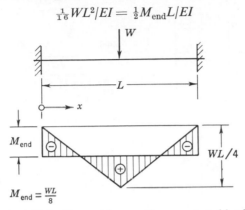

Fig. 1.13.6   Clamped beam with concentrated load.

Hence the end moment is

$$M_{end} = \tfrac{1}{8} WL \qquad\qquad 1.13.8$$

The moment diagram is shown in Fig. 1.13.6.

1.14   Aileron on Yielding Supports

Ailerons are movable surfaces extending along the trailing edge of a wing of an airplane near the wing tips. Their controls are so arranged that the left aileron is lowered when the right is lifted. Ailerons are used to control the rotation of the airplane about its longitudinal axis. As a rule, each aileron is attached to the rear spar of the wing by three or more hinges (see Fig. 1.14.1). The bending-moment distribution in the aileron spar constitutes therefore a statically indeterminate problem which is usually solved by means of the three-moment equation. It is general practice to neglect the displacement of the supports in the calculation. This practice may lead, however, to erroneous results when the wing is heavily loaded and the aileron is relatively rigid.

To illustrate the principle rather than to present detailed calculations related to an actual airplane, the following simplified problem will now be discussed: The aileron is replaced by its main girder, or spar, and the wing by its rear spar. The two are connected by the three hinges indicated in

Fig. 1.14.1*a.* The aileron spar is a steel tube of constant section (Fig. 1.14.1*c*) while the rear wing spar has two aluminum-alloy flanges of constant section and a reinforced aluminum-alloy shear web as shown in

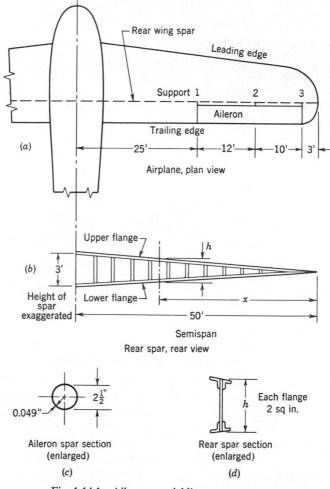

Fig. 1.14.1   Aileron on yielding supports.

Fig. 1.14.1*d.* It is assumed that the uniform loading of the wing spar is 200 lb per ft and that of the aileron spar 25 lb per ft.

Because the total load on the wing is 10,000 lb while the total load on the aileron is only 550 lb., in a first approximation the displacements of the hinge points can be assumed to be independent of the aileron and its loading. They can be calculated therefore from the actual bending moments M_0 of the wing spar caused by the uniformly distributed load w with the aid of the dummy load method.   The work of calculation can be reduced materially if it is observed that the bending-moment distribution in the aileron spar does not depend on the absolute values of the support deflections but only on the relative displacement of the middle support with

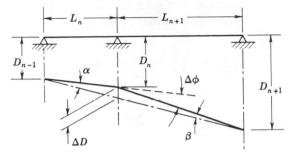

Fig. 1.14.2   Support deflections.

respect to the straight line connecting the two end supports.   This relative displacement ΔD, shown in Fig. 1.14.2, determines the curvature of the aileron spar caused by the deflections of the supports; it also suffices for the calculation of the angle $\Delta\phi$ needed in the three-moment equation (Eq. 1.13.6).

It follows from the laws of plane geometry and the sketch of Fig. 1.14.2 that

$$\Delta\phi = \alpha + \beta$$

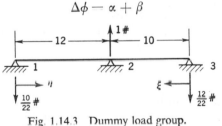

Fig. 1.14.3   Dummy load group.

On the other hand, as long as the deflections D are small,

$$\alpha = \Delta D / L_n \quad \text{and} \quad \beta = \Delta D / L_{n+1}$$

Consequently

$$\Delta\phi = (\Delta D / L_n) + (\Delta D / L_{n+1}) \tag{a}$$

But the calculation of ΔD involves less work than that of the displacements D_{n-1}, D_n, and D_{n+1} if a suitable dummy load is chosen.

The dummy load group is shown in Fig. 1.14.3.   It is easy to prove that the work done by this particular dummy load group is zero for any rigid-body displacement and for any elastic deformation of the wing spar that does not include a relative displacement ΔD of the middle support with respect to the straight line connecting the end supports. When, however, ΔD is not zero, the external work W_e done by the dummy load group is $1 \times \Delta D$ ft-lb (if ΔD is measured in feet). The next task is therefore the calculation of the internal work W_i done by the moments M_D caused by the dummy load group during the deformations of the wing spar corresponding to the actual bending moments M_0. Then, by the principle of virtual displacements,

$$W_e + W_i = 0$$

and thus

$$\Delta D = W_e = -W_i \qquad (b)$$

From Fig. 1.14.3 the bending moment M_D caused by the dummy load group is

$$M_D = \tfrac{12}{22}\xi \quad \text{in the 10-ft span}$$

and

$$M_D = \tfrac{10}{22}\eta \quad \text{in the 12-ft span}$$

If x is measured in feet from the end of the wing spar as indicated in Fig. 1.14.1b,

$$M_D = \tfrac{12}{22}(x - 3) \quad \text{when} \quad 3 \leq x \leq 13$$

$$M_D = \tfrac{10}{22}(25 - x) \quad \text{when} \quad 13 \leq x \leq 25$$

and the bending moment M_D is given in foot-pounds.

The bending moment M_0 caused in the wing spar by the actual uniformly distributed load $w = 200$ lb per ft is

$$M_0 = wx^2/2 = 100x^2$$

The internal work W_i can be calculated from Eq. 1.12.2:

$$W_i = - \int_0^L \frac{M_0 M_D}{EI} \, dx \qquad (c)$$

where the moment of inertia I of the wing spar is variable (see Fig. 1.14.1d):

$$I = 2 \times 2(h/2)^2 = h^2 = (\tfrac{3}{50})^2 x^2 \text{ in.}^2\text{-ft}^2$$

If the moment of inertia is expressed in ft⁴, the above equation becomes

$$I = \tfrac{1}{144}(\tfrac{3}{50})^2 x^2 = 2.5 \times 10^{-5} x^2 \text{ ft}^4$$

The internal work vanishes in the regions where $0 \leq x \leq 3$ and $25 \leq x \leq 50$

since there M_D vanishes.   The definite integral in Eq. *c* reduces therefore to

$$W_i = -\int_3^{13} 100x^2\,\frac{12}{22}\,\frac{x-3}{2.5 \times 10^{-5}x^2 E}\,dx$$

$$-\int_{13}^{25} 100x^2\,\frac{10}{22}\,\frac{25-x}{2.5 \times 10^{-5}x^2 E}\,dx$$

$$= -2.4 \times \frac{10^8}{E}$$

This value is in foot-pounds if the dimension of Young's modulus E is pounds per square foot.   With $E = 144 \times 10.5 \times 10^6$ psf, substitution into Eq. *b* yields

$$\Delta D = 0.159 \text{ ft} = 1.9 \text{ in.}$$

From Eq. *a* the angular difference $\Delta\phi$ is then

$$\Delta\phi = 0.159/12 + 0.159/10 = 0.0291 \text{ radian}$$

The bending moments in the aileron spar are now determined by imagining the spar cut at the middle support.   The bending moments caused in each span by the uniformly distributed load can then be calculated from the laws of statics alone.   The moment curves are parabolas, and the maximum moment M_{12} in the 12-ft span is

$$M_{12} = qL^2_{12}/8 = 25 \times 12^2/8 = 450 \text{ ft-lb} = 5400 \text{ in.-lb}$$

Similarly the maximum moment M_{10} in the 10-ft span is

$$M_{10} = qL^2_{10}/8 = 312.5 \text{ ft-lb} = 3750 \text{ in.-lb}$$

In the calculation of the moment M_2 acting in reality in the section at the middle support Eq. 1.13.6 is used, and the subscripts $n-1$, n, $n+1$ are replaced by 1, 2, and 3, respectively; for the load w the value 25 lb per ft is substituted.   The moments M_1 and M_3 are zero since no external moments are applied at the simply supported ends of the aileron spar. Consequently

$$2M_2(12+10) = -\tfrac{1}{4}\,24(12^3+10^3) - 6EI \times 0.0291$$

With $E = 144 \times 29 \times 10^6$ psf and $I = 0.2834$ in.$^4 = 0.2834/12^4$ ft^4 the equation becomes

$$44M_2 = -17{,}050 - 10{,}000 = -27{,}050$$

from which the bending moment is

$$M_2 = -615 \text{ ft-lb} = -7380 \text{ in.-lb}$$

The bending-moment diagram of the aileron is shown in Fig. 1.14.4. Without support displacements, that is, with $\Delta\phi = 0$, the moment at support 2 would be

$$M_2 = -4660 \text{ in.-lb}$$

The difference is considerable, and the analysis disregarding support displacements leads to unconservative design.

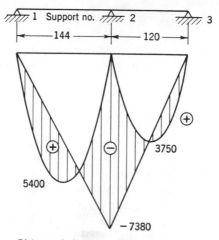

Distances in inches, moments in in.-lb

Fig. 1.14.4   Bending-moment diagram of aileron spar.

The solution of the aileron problem was obtained on the assumption that the reactions of the aileron did not influence the deflections of the wing spar. When such an assumption is not justified, the three-moment equation can still be used if the connection between support reaction and support displacement is known. In many cases of practical interest, the deflection D_n of a support is proportional to the reaction R_n prevailing there:

$$D_n = kR_n \qquad\qquad 1.14.1$$

where k is a factor of proportionality. With the aileron just discussed the deflection of any support depends on the reactions prevailing at all the three supports:

$$D_1 = k_{11}R_1 + k_{12}R_2 + k_{13}R_3$$
$$D_2 = k_{21}R_1 + k_{22}R_2 + k_{23}R_3 \qquad\qquad 1.14.2$$
$$D_3 = k_{31}R_1 + k_{32}R_2 + k_{33}R_3$$

where the proportionality factors k_{pq} are known as *influence coefficients*.

As simple equilibrium considerations result in the formula for the reaction R_n

$$R_n = \tfrac{1}{2}(w_n L_n + w_{n+1} L_{n+1}) + (1/L_n)(M_{n-1} - M_n) + (1/L_{n+1})(M_{n+1} - M_n)$$

$$1.14.3$$

Equation 1.13.6, together with Eqs. 1.14.1 to 1.14.3, suffices for the determination of the bending moments in a continuous beam on several supports when the supports undergo elastic displacements because of the reactions.

1.15   Deformations of Statically Indeterminate Beams

Figure 1.15.1 shows a beam whose left end is rigidly fixed while the right end is simply supported. The beam is statically indeterminate to the first degree since it becomes a statically determinate structure, namely, a simply supported beam, when the rigid fixation is replaced by a simple support. Alternatively the rigid fixation may be maintained and the simple support at the right end removed. The resulting structure is a cantilever beam which is also statically determinate. The beam is loaded with a uniformly distributed load of intensity w. The problem is to determine the end rotation of the beam at the simple support.

The following procedure is adopted for the solution of the problem: First the reaction at the right support is calculated by transforming the beam into a cantilever through removal of the right support, and by equating the downward displacement of the cantilever due to the distributed load to its upward displacement due to the unknown reaction. An expression is established for the bending moment along the cantilever beam as caused by the simultaneous action of the distributed load and the reaction force at the right support. This is the bending moment M_0 in the actual statically indeterminate structure under the loads specified. Next a suitable dummy load is chosen, the moment M_1 caused by it is calculated, and the end rotation is found by evaluating the integral in the right-hand member of Eq. 1.12.3.

The downward deflection of the right end of the cantilever under the action of the distributed load w is, according to problem 4 of Fig. 1.12.6,

$$\delta_w = \tfrac{1}{8} w L^4 / EI$$

The upward deflection because of the unknown reaction R is, according to problem 2 of the same figure,

$$\delta_R = -\tfrac{1}{3} R L^3 / EI$$

Since in reality the beam is not a cantilever but is fixed at its right end in the vertical direction by the support, the displacement must vanish; that is,

$$\delta_w + \delta_R = 0$$

Substitution and solution for R give

$$R = \tfrac{3}{8}wL$$

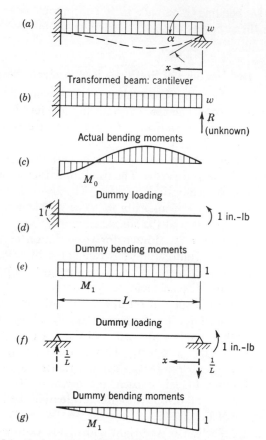

Fig. 1.15.1   End rotation of indeterminate beam under distributed load.

The actual bending moment M_0 is therefore

$$M_0 = -\tfrac{1}{2}wx^2 + \tfrac{3}{8}wLx$$
$$= (w/8)(3Lx - 4x^2) \qquad\qquad 1.15.1$$

The dummy load can be conveniently chosen to be an external moment of 1 in.-lb magnitude acting at the right support. Its work will be numerically equal to the rotation sought. However, it is onerous and unnecessary to calculate the bending moments M_1 caused by this concentrated moment in the statically indeterminate beam. Instead the dummy load may be imagined as acting on the statically determinate cantilever shown in Fig. 1.15.1d. Then

$$M_1 = 1 \qquad\qquad (a)$$

Such an arbitrary change in the end conditions is perfectly permissible since the only purpose of the dummy loading is the establishment of a system of external and internal forces that is in equilibrium and whose external work is numerically equal to the displacement quantity sought. These requirements are obviously satisfied. The external moment of 1 in.-lb magnitude, the internal bending moments represented in Fig. 1.15.1e, and the reaction moment at the rigid fixation (shown dashed in the figure) are in equilibrium, the work done by the applied moment is $1 \times \alpha$, and the reaction moment does no work because the rigid fixation at the left end does not permit any rotation. Consequently, the rotation α is numerically equal to the internal work done by the dummy moments M_1 during the deformations of the beam corresponding to the actual moments M_0:

$$\alpha = \frac{1}{EI} \int_0^L M_0 M_1 \, dx$$

$$= \frac{1}{EI} \frac{w}{8} \int_0^L (3Lx - 4x^2) \, dx = \frac{wL^3}{48EI} \qquad\qquad 1.15.2$$

The choice of the end conditions for the dummy load just made is, of course, not the only possible one. In another, although slightly less convenient, approach, the beam is imagined to be simply supported as shown in Fig. 1.15.1f. Again the applied moment, the reaction forces shown dotted, and the internal bending moments represented in Fig. 1.15.1g are in equilibrium. Moreover, the work done by the external moment is $1 \times \alpha$ while the reactions do no work. Hence, with

$$M_1 = 1 - x/L \qquad\qquad (b)$$

the rotation becomes

$$\alpha = \frac{1}{EI} \int_0^L M_0 M_1 \, dx$$

$$= \frac{1}{EI} \frac{w}{8} \int_0^L (3Lx - 4x^2) \left(1 - \frac{x}{L}\right) dx$$

Evaluation of the definite integral again yields

$$\alpha = wL^3/48EI$$

It will be proved now that the same result is obtained when the dummy load is applied to the actual indeterminate structure. The indeterminate beam is shown in Fig. 1.15.2a while part b of the figure presents the cantilever obtained from it by omitting the right support. It follows from the formulas given in Fig. 1.12.6 that the deflections of the right end are

$$\delta_B = -\tfrac{1}{2}BL^2/EI$$
$$\delta_R = \tfrac{1}{3}RL^3/EI$$

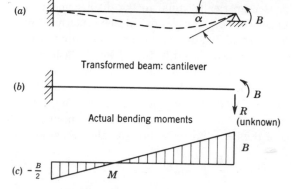

Fig. 1.15.2   End rotation of indeterminate beam under end moment.

Solution of the equation

$$\delta_B + \delta_R = 0$$

yields

$$R = \tfrac{3}{2}(B/L)$$

Consequently the moment distribution corresponding to the beam and loading shown in Fig. 1.15.2a is

$$M = B[1 - \tfrac{3}{2}(x/L)] \qquad\qquad 1.15.3$$

If a dummy load of 1 in.-lb magnitude is applied in place of B, Eq. 1.15.3 becomes

$$M_1 = 1 - \tfrac{3}{2}(x/L) \qquad\qquad (c)$$

Integration of the right-hand member in the equation

$$\alpha = \frac{1}{EI} \int_0^L M_0 M_1 \, dx$$

$$= \frac{1}{EI} \frac{w}{8} \int_0^L (3Lx - 4x^2) \left(1 - \tfrac{3}{2}\frac{x}{L}\right) dx$$

yields

$$\alpha = wL^3/48EI$$

which is again the result first given in Eq. 1.15.2.

In Art. 1.17 the end rotation of a beam rigidly fixed at one end and simply supported at the other will be needed when the load is a concentrated moment applied at the simply supported end. The bending moment corresponding to these end conditions and loading has just been calculated and is given in Eq. 1.15.3. The dummy load for the calculation of the end rotation is suitably chosen to be a unit moment at the simply supported end; then the bending moment caused by the dummy load is expressed by Eq. *c*. Hence the rotation is

$$\alpha = \frac{1}{EI} \int_0^L B\left(1 - \frac{3}{2}\frac{x}{L}\right)\left(1 - \frac{3}{2}\frac{x}{L}\right) dx$$

$$= \frac{1}{4}\frac{BL}{EI} \tag{1.15.4}$$

It is easy to prove that in this case again the same value is obtained for the rotation if the moments caused by the dummy load are chosen as shown in Fig. 1.15.1e or 1.15.1g.

Finally the midpoint deflection will be calculated for the clamped beam of Figs. 1.13.5 and 1.13.6. The dummy load of 1 lb is applied at the midspan, and the moments caused by it are calculated for the simply supported beam. The dummy moment is then

$$M_1 = \tfrac{1}{2}x, \qquad 0 \le x \le L/2 \tag{1.15.5}$$

provided x is measured along the beam from one of the supports as the origin. When the load is uniformly distributed, the actual bending moment is (see Fig. 1.13.5e)

$$M_0 = -\tfrac{1}{12}wL^2 + \tfrac{1}{2}wLx - \tfrac{1}{2}wx^2 \tag{1.15.6}$$

As Eq. 1.15.5 is valid only in one half of the beam and the moment distribution is symmetric with respect to the midpoint of the beam, the integration can be carried out from $x = 0$ to $x = L/2$, and the result obtained can be doubled. Since Eq. 1.12.3 gives the deflection

$$\delta = \int_0^L \frac{M_0 M_1}{EI} dx$$

substitutions and integration yield

$$\delta = \frac{2}{EI} \int_0^{L/2} \left(-\frac{1}{12}wL^2 + \frac{1}{2}wLx - \frac{1}{2}wx^2\right)\frac{1}{2}x \, dx$$

$$= \frac{1}{384}\frac{wL^4}{EI} \tag{1.15.7}$$

When the actual load is concentrated in the middle of the beam, the bending moment is, in agreement with Fig. 1.13.6,

$$M_0 = -\tfrac{1}{8}WL + \tfrac{1}{2}Wx \qquad\qquad 1.15.8$$

provided

$$0 \leq x \leq L/2$$

The midpoint deflection is then

$$\delta = 2\,\frac{1}{EI}\int_0^{L/2}\left(-\frac{1}{8}\,WL + \frac{1}{2}\,Wx\right)\frac{1}{2}\,x\,dx$$

$$= \frac{1}{192}\,\frac{WL^3}{EI} \qquad\qquad 1.15.9$$

1.16   The Moment-Area Method

The principle of virtual displacements can be used to obtain a well-known relationship between the moment diagram of a beam and the relative rotation of any two of its cross sections. If M_0 is the bending moment, as shown in Fig. 1.16.1a, and the moment of inertia of the beam varies, the first step in the calculation is to draw a curve representing M_0/EI (see Fig. 1.16.1b). Next the unit dummy moments indicated in Fig. 1.16.1c are assumed in the cross sections whose relative rotation is sought. Obviously the dummy moments do no work during any rigid-body displacement of the portion of the beam on which they are acting, nor do they work when section x_2 is displaced relative to section x_1 without rotation. When, however, the two sections are rotated relative to each other through an angle $\Delta\alpha$, the work done by the two dummy moments is

$$W_e = 1 \times \Delta\alpha \qquad\qquad 1.16.1$$

The internal work done by the stresses caused by the dummy moments during the actual rotations of the cross sections due to the moments M_0 is given by the integral

$$W_i = -\int_{x_1}^{x_2}\frac{M_0}{EI}\,M_D\,dx = -\int_{x_1}^{x_2}\frac{M_0}{EI}\,dx = -A \qquad 1.16.2$$

where A is the area enclosed by the M_0/EI curve, the two verticals corresponding to $x = x_1$ and $x = x_2$, and the base line of the M_0/EI diagram.

As the total work must vanish, one has

$$W_e + W_i = 0 = \Delta\alpha - A$$

or

$$\Delta\alpha = A \qquad\qquad 1.16.3$$

It can be stated therefore that

> The relative rotation of two sections x_2 and x_1 of a bent   **Theorem 9**
> beam, or the difference in the slope of the beam at x_2 and
> x_1, is equal to the area bounded by the M_0/EI curve,
> the two verticals corresponding to x_2 and x_1, and the
> base line of the M_0/EI curve.

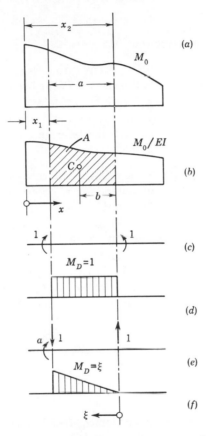

Fig. 1.16.1   Proof of the moment-area method.

When the displacement of section x_2 relative to that of section x_1, measured perpendicularly to the tangent of the deflected beam at x_1, is sought, the dummy loading chosen is that shown in Fig. 1.16.1e. It is easy to prove that work is done by this loading only when section x_2 is displaced perpendicularly to the tangent of the beam at x_1, and the numerical value of the work is δ, if δ designates the relative displacement. The internal work is

$$W_i = - \int_0^a \frac{M_0}{EI} M_D \, d\xi = - \int_0^a \frac{M_0}{EI} \xi \, d\xi = -Ab \qquad 1.16.4$$

where $a = x_2 - x_1$, ξ is measured as indicated in Fig. 1.16.1f, and b is the horizontal distance of the centroid of area A from section x_2. As $W_e + W_i = 0$, solution for δ yields

$$\delta = Ab \qquad 1.16.5$$

It has been proved therefore that

> The displacement of section x_2 relative to section x_1,   **Theorem 10**
> measured perpendicularly to the tangent of the beam at
> x_1, is equal to the first moment with respect to $x = x_2$ of
> the area comprised between the M_0/EI curve, the two
> verticals $x = x_1$ and $x = x_2$, and the base line of the
> M_0/EI diagram.

These two theorems are often used in the calculation of slopes and deflections, and the method of calculation is known as the moment-area method. Two simple examples should suffice to show how the method is applied to particular problems. Problem 1 in Fig. 1.12.8 deals with a cantilever beam upon whose free end a transverse load W is acting. The moment diagram is a triangle. If the rotation of the free end relative to the horizontal tangent of the fixed end is sought, the area A enclosed by the M_0/EI diagram, the verticals $x = 0$ and $x = L$, and the horizontal base line must be determined. With $EI = $ constant, the area is $WL^2/2EI$, and this is the rotation of the free end of the beam.

The distance of the centroid of area A from the free end is $\frac{2}{3}L$, and thus the static moment with respect to a vertical through the free end is $(WL^2/2EI)\frac{2}{3}L = WL^3/3EI$. This is the end deflection, as quoted in problem 2 of Fig. 1.12.6.

1.17 The Hardy Cross Moment-Distribution Method

The Hardy Cross moment-distribution method is a procedure of step-by-step approximations, which can be used in place of the three-moment equations for the calculation of the moments in continuous beams on several supports. It is particularly useful in the solution of rigid-frame problems when the structure is so highly redundant that its analysis by classical methods is too cumbersome or even impossible for practical purposes. The moment-distribution method is not an approximate method; it is a method of successive approximations, the accuracy of which can be increased at will by increasing the number of steps under-taken.   In most problems an accuracy sufficient for engineering purposes can be reached by carrying out a comparatively small number of operations.

The method will be explained by means of the example of a beam on four supports loaded over one span, as shown in Fig. 1.17.1.   First, the two ends of the middle span are imagined to be fixed rigidly against rota-tion at supports 1 and 2.   Then the middle span is in equilibrium under the simultaneous action of the distributed load w, the reaction forces at supports 1 and 2, and the reaction moments exerted on the span by the fictitious constraints that provide the end fixation.   The magnitude of these end moments was calculated in Art. 1.13:

$$M_{\text{fe}} = \tfrac{1}{12}wL^2 \qquad\qquad 1.17.1$$

M_{fe} is known as the fixed-end moment in the Hardy Cross procedure.

Since action is equal to reaction, the middle span of the beam exerts moments of magnitude M_{fe} upon the fictitious constraints.   These moments tend to rotate the constraint at support 1 clockwise and that at support 2 counterclockwise (see Fig. 1.17.1b).   Because the constraints are fictitious, the system in reality is unbalanced.   The unbalanced moments are $+M_{\text{fe}}$ at support 1 and $-M_{\text{fe}}$ at support 2 if clockwise moments acting on the constraints are considered as positive.

The constraints can be imagined as rigid vertical bars attached to the beam (Fig. 1.17.1c) and clamped to some perfectly rigid wall behind the beam.   Through the artifice of this clamping device the equilibrium of the beam is maintained in the deflected position shown.   However, the purpose of this investigation is the determination of the moment equi-librium and of the deflections of the actual beam which is not provided with artificial constraints.   Hence, the constraints must be released, and the beam must be permitted to rotate at supports 1 and 2.

As a first step in this direction the clamps are removed at support 1 while those at support 2 are maintained.   The gradual removal of the constraint at support 1, loaded by $+M_{\text{fe}}$, is equivalent to a gradual application of a moment $+M_{\text{fe}}$ to the beam at support 1.   Obviously a clockwise rotation of the beam at support 1 will result from the transfer of moment M_{fe} from the constraint to the beam.   Part of M_{fe} will be transferred to the right end of span 0–1 and the rest to the left end of span 1–2.   The determination of the ratio of these two moments is the next step in the calculation.

The ratio is determined from the requirement of continuity at support 1. As long as the beam does not fracture, the rotation of the right end of span 0–1 must be the same as that of the left end of span 1–2.   It is known from the calculations carried out in Art. 1.15 that, under the action

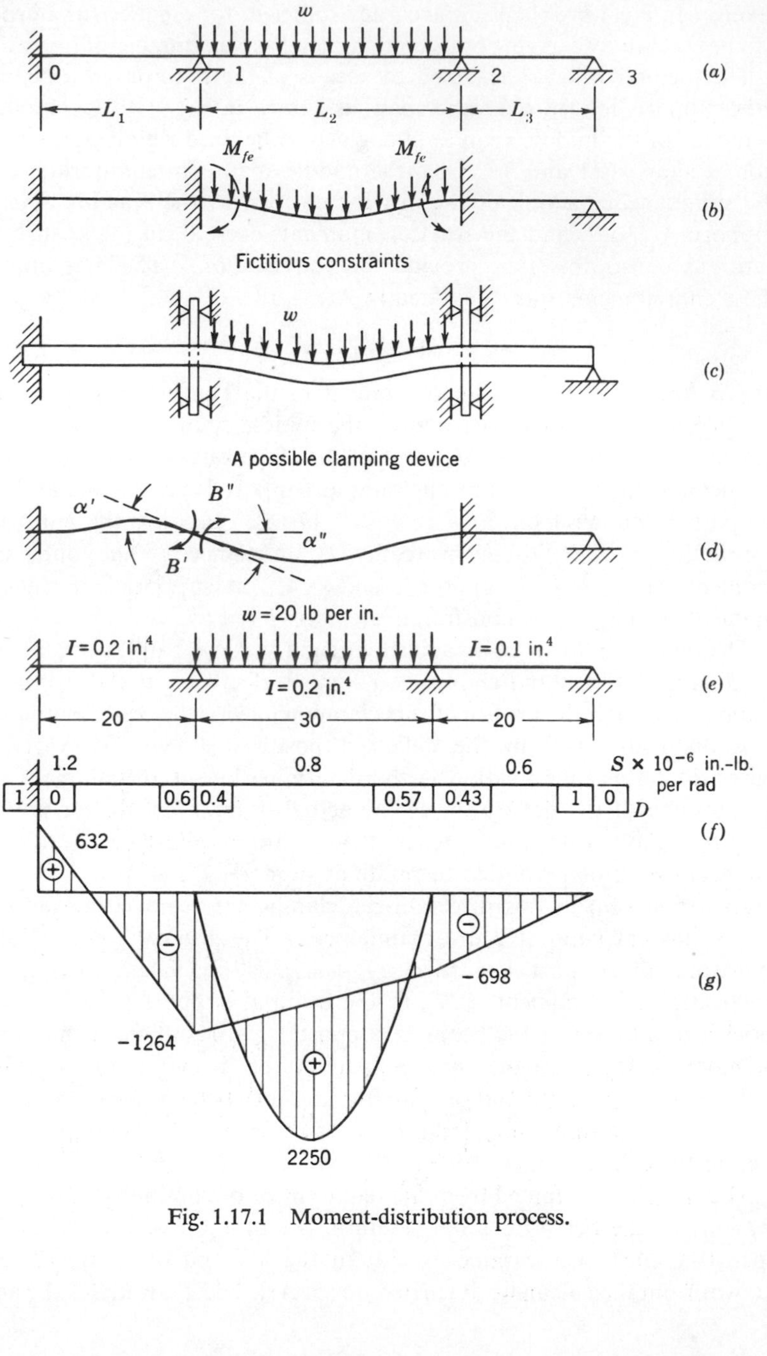

Fig. 1.17.1   Moment-distribution process.

of a bending moment B' applied to it at support 1, the rotation α' of the right end of beam 0–1 is

$$\alpha' = \tfrac{1}{4}B'L_1/(EI)_1 \qquad\qquad (a)$$

since the left end of beam 0–1 is rigidly fixed and its right end simply supported (see Eq. 1.15.4).

Similarly, the rotation α'' of the left end of beam 1–2 is, under the action of a bending moment B'' applied to it at support 1,

$$\alpha'' = \tfrac{1}{4}B''L_2/(EI)_2 \qquad\qquad (b)$$

when the right end of beam 1–2 is rigidly fixed against rotation by the fictitious clamping device at support 2.

The sum of the two moments B' and B'' must be equal to the unbalanced moment M_{fe}:

$$B' + B'' = M_{\text{fe}} \qquad\qquad (c)$$

while the requirement of consistent deformations can be stated in the form

$$\alpha' = \alpha'' = \alpha \qquad\qquad (d)$$

Solution of Eqs. a to d yields

$$B' = (EI/L)_1 M_{\text{fe}}/[(EI/L)_1 + (EI/L)_2]$$

$$B'' = (EI/L)_2 M_{\text{fe}}/[(EI/L)_1 + (EI/L)_2] \qquad\qquad (e)$$

$$\alpha = \tfrac{1}{4}M_{\text{fe}}/[(EI/L)_1 + (EI/L)_2]$$

These equations can be put into a more convenient form if use is made of the concept of the stiffness coefficient of a beam. The stiffness coefficient S is defined as the moment necessary at the simply supported end of the beam to cause a unit rotation there while the other end is rigidly fixed against rotation. The rotation α of a beam having one end fixed and the other simply supported is given in Eq. 1.15.4:

$$\alpha = \tfrac{1}{4}BL/EI$$

Solution of this equation for the moment B gives

$$B = 4\alpha EI/L$$

When α is unity, B becomes

$$B = 4EI/L = S \qquad\qquad 1.17.2$$

As a rule, the stiffness coefficient S is such a large moment that the beam would fail under it, but this is of no importance because it is not applied to the beam in reality. In its definition the angle α could have been taken as one thousandth of a radian. This would have introduced, to no purpose, a factor of $1/1000$ in all the equations.

If the stiffness coefficient of beam 0–1 is denoted by S_1 and that of beam 1–2 by S_2, Eqs. e can be rewritten in the form

$$B' = [S_1/(S_1 + S_2)]M_{fe}$$
$$B'' = [S_2/(S_1 + S_2)]M_{fe}$$

1.17.3

$$\alpha = M_{fe}/(S_1 + S_2)$$

1.17.4

The multiplying factors of M_{fe} in Eqs. 1.17.3 are called distribution factors. They are designated by the letter D according to the equations

$$D_1 = S_1/(S_1 + S_2)$$
$$D_2 = S_2/(S_1 + S_2)$$

1.17.5

Any unbalanced moment at support 1 must be distributed in proportion to the distribution factors.

Summarizing, one may say that the release of the fictitious constraint at support 1, and the consequent transfer of the unbalanced moment M_{fe} to the beam, results in a rotation α of the beam at support 1. The angle of rotation is given in the last of Eqs. e and in Eq. 1.17.4. The unbalanced moment is distributed between beams 0–1 and 1–2 in accordance with the first two of Eqs. e and with Eqs. 1.17.3. Since the left end of beam 0–1 is rigidly fixed, and the right end of beam 1–2 is imagined to be so, the application of the moments B' and B'' at support 1 must obviously cause moments to appear at the far ends of the two beams, that is, at supports 0 and 2. The magnitude of these far end moments can be determined if reference is made to Art. 1.15. The last problem discussed there is the calculation of the moment diagram and the end rotation of a beam, one end of which is rigidly fixed while the other is simply supported, when the external load consists of a moment B applied to the simply supported end. The moment diagram of Fig. 1.15.2 shows that the moment at the far end is $-\frac{1}{2}B$.

The sign convention used in drawing the moment diagram of Fig. 1.15.2 is known as the beam convention; according to it, a moment is positive if it causes a horizontal beam to assume a bent shape that is concave when seen from above. In the moment-distribution method it is more convenient to use the rigid-frame convention according to which a clockwise moment is positive. The two conventions differ materially, since a clockwise moment applied to the right end of a horizontal beam causes a convex curvature when viewed from above, while a clockwise moment applied to the left end of the same beam causes a concave curvature. Consequently a clockwise moment at the right end of a beam is positive according to the frame convention but negative according to the beam convention, while a clockwise moment at the left end is positive according to either convention.

In Fig. 1.15.2 the moment B applied to the right end of the beam is positive according to the beam convention but negative according to the frame convention. The far end moment is negative according to either convention. Hence it may be stated that application of a moment at the simply supported end gives rise to a far end moment at the rigidly fixed end which is one-half that of the applied moment and is of equal sign according to the frame convention. In other words, a clockwise moment B applied to a beam at its simply supported end necessitates a clockwise moment $\frac{1}{2}B$ at the far end if that end is not to rotate. In the parlance of the Hardy Cross moment-distribution method this is expressed by saying that one half of the applied moment is carried over to the far end, or that the carry-over factor C is 1/2. Of course, the calculations in connection with Fig. 1.15.2 were carried out for a constant section beam. For a beam of variable section C need not be 1/2.

The moment $\frac{1}{2}B'$ required to act on beam 0–1 at support 0 in order to keep the beam fixed in its horizontal position at support 0 is available in reality because of the rigid end fixation. On the other hand, at support 2 the reaction moments are furnished only by a purely fictitious constraint. On this constraint the reaction of the carry-over moment $\frac{1}{2}B''$ is now acting, as well as the original fixed-end moment $-M_{\text{fe}}$. The actual beam is therefore unbalanced at support 2.

In the next step of the moment-distribution method the fictitious constraint at support 2 is gradually released, and the beam is allowed to rotate. At the same time the fictitious clamps are tightened at supports 1 and 3, and thus the far ends of beams 1–2 and 2–3 are rigidly fixed against rotation. The situation at support 2 is now very much the same as that which prevailed at support 1 when it was allowed to rotate under the action of the moment $B' + B''$. Consequently the stiffness and distribution factors can be determined according to the definitions and formulas developed earlier. The angle of rotation at support 2 becomes

$$\alpha = (\tfrac{1}{2}B'' + M_{\text{fe}})/(S_2 + S_3)$$

where S_3 denotes the stiffness factor of beam 2–3. The moment distributed to beam 1–2 is

$$-(\tfrac{1}{2}B'' + M_{\text{fe}})S_2/(S_2 + S_3)$$

and that distributed to beam 2–3

$$-(\tfrac{1}{2}B'' + M_{\text{fe}})S_3/(S_2 + S_3)$$

Equilibrium is thus established at support 2, but the carry-over moments at the far ends of beams 1–2 and 2–3 have to be resisted by the fictitious constraints at supports 1 and 3. In the real structure therefore these supports are now unbalanced, although support 3 was in equilibrium at the beginning of the procedure while support 1 was balanced when it was allowed to rotate.

It is possible to proceed from one unbalanced joint to another, release each in turn, and permit it to rotate until the unbalanced moments at all supports become sufficiently small to be negligible in engineering calculations. Moreover, the number of steps required is reasonably small since any unbalanced moment is distributed at each support and only one half of the distributed moment is carried over to the far end of the beam.

Since in engineering the magnitude of the bending moment is usually of primary interest rather than that of the rotation, in the Hardy Cross procedure, account is kept only of the moments. Details of the bookkeeping are shown in connection with a numerical analysis of the example given in Fig. 1.17.1.

Figure 1.17.1e contains all the numerical data needed in the calculations. First, the stiffness factors are computed according to Eq. 1.17.2, assuming $E = 30 \times 10^6$ psi as Young's modulus for steel. They are denoted by S_1, S_2, and S_3 for beams 0–1, 1–2, and 2–3, respectively:

$$S_1 = 4 \times 30 \times 10^6 \times 0.2/20 = 1.2 \times 10^6 \text{ in.-lb per radian}$$

$$S_2 = 0.8 \times 10^6 \text{ in.-lb per radian}$$

$$S_3 = 0.6 \times 10^6 \text{ in.-lb per radian}$$

The values of the distribution factors at support 1 follow from Eq. 1.17.5:

$$D_1 = 1.2 \times 10^6/(1.2 \times 10^6 + 0.8 \times 10^6) = 0.6$$

$$D_2 = 0.8 \times 10^6/(1.2 \times 10^6 + 0.8 \times 10^6 = 0.4$$

In the same manner, the distribution factors are found to be 0.57 and 0.43 at support 2. The left side of support 0 is a rigid wall; it can take all the unbalanced moments transferred to support 0. For this reason the distribution factor 1 is entered to the left of support 0 in Fig. 1.17.1f, while to the right of the support the distribution factor 0 is listed. The situation is just about reversed at support 3, where to the right of the support there is nothing to take over moments, and thus all unbalanced moments must be transferred to the beam at the left.

The actual distribution of the moments is carried out in Table 1.17.1. In the first row the distribution factors are listed, and in the second the fixed-end moments as they act on the middle beam, that is, with signs opposite to those shown in Fig. 1.17.1b. (In the figure the moments act on the constraints.) The fixed-end moments are computed from Eq. 1.17.1. Support 1 is balanced first by the application of a moment +1500 in.-lb, which is distributed in the ratio 0.6 : 0.4 between left and right. The distributed moments are listed in row 3, and the figures are underlined to signify that support 1 is now in equilibrium. Indeed, the fixed-end moment and the distributed moments add up to zero.

Table 1.17.1.   Moment Distribution

Support No.	0		1		2		3	
1	1	0	0.6	0.4	0.57	0.43	1	0
2				−1,500	1,500			
3		450	900	600	300			
4				−512	−1,025	−775	−386	
5						193	386	
6				−55	−110	−83	−42	
7		170	341	226	113			
8						21	42	
9				−38	−76	−58	−29	
10		12	23	15	7			
11		632	1,264	1,264		15	29	
					709	−687	0	

Row 3 also contains the moments carried over to supports 0 and 2. As mentioned earlier, the moment at support 0 is balanced automatically by the reaction of the rigid wall. On the other hand, at support 2 the original fixed-end moment and the carried-over moment add up to an unbalanced moment of 1800 in.-lb. This is now balanced by the application of a moment of −1800 in.-lb distributed in the ratio 0.57 : 0.43. The distributed moments are again underlined since support 2 is now in equilibrium. The distributed moments as well as the moments carried over to supports 1 and 3 are listed in row 4.

The unbalanced moment of −386 in.-lb at support 3 is balanced next. The balancing moment, +386 in.-lb, has to be applied in full to the right end of beam 2–3 since the distribution factor to the right of support 3 is zero. One half of the distributed moment is carried over to support 2. It can be seen from Table 1.17.1 that the supports are then balanced in the following order: 2, 1, 3, 2, 1, 3. When the nine balancing operations are completed, all the supports are in equilibrium except the second, and support 2 is unbalanced to the amount of 22 in.-lb. As this figure is less than 1.5 per cent of the original unbalanced fixed-end moment of 1500 in.-lb, the accuracy of the moment distribution is considered satisfactory for engineering purposes, and the balancing procedure is discontinued. The moments listed in each column are added up, and the sums are written under the double underlining of the last figure in the column. It may be seen that the moments on the two sides of supports 1 and 2 add up to zero, at support 1 accurately, and at support 2 approximately only because of the unbalanced moment left there. The vanishing sums signify equilibrium at each of these supports. Similarly, support 3 is in equilibrium because of the absence of any moment, and support 0 under the moment of 632 in.-lb is balanced by the rigid wall.

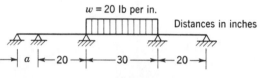

Fig. 1.17.2.   Beam with imaginary span.

In Fig. 1.17.1g the moment diagram is shown. Before it was drawn, the moments listed in Table 1.17.1 had to be changed to conform with the beam convention. Hence, the moment at the left end of each beam was left unchanged while that at the right end was multiplied by -1. The following moments were thus obtained: $+632$, -1264, -698, and 0 in.-lb at supports 0, 1, 2, and 3, respectively. At support 2 the average of the two values calculated was taken as the estimated best approximation to the correct answer. The maximum moment at the middle of the uniformly loaded beam was $wL^2/8 = 2250$ in.-lb when the beam was considered as simply supported. The diagram was drawn as in Fig. 1.13.3 by plotting downward negative moments at the supports and positive moments in the field. The shaded area represents the resultant moments along the beam.

In this case the moment distribution can be checked easily by means of the three-moment equations. The rigid-end fixation at support 0 can be accounted for by the artifice of an additional span of length a to the left of support 0 (see Fig. 1.17.2). When a approaches zero, support 0 and the fictitious one to its left fix two points of the beam that are an infinitesimal distance apart. This means that the tangent to the beam at support 0 is maintained horizontal as required by the rigid-end fixation.

The fictitious beam in Fig. 1.17.2 rests on five supports. If the variations in the moment of inertia of the beam are duly considered, the three independent three-moment equations can be written in the following simplified form:

$$2(20 + a)M_0 + 20M_1 = 0$$

$$20M_0 + 100M_1 + 30M_2 = -135,000$$

$$30M_1 + 140M_2 = -135,000$$

With $a = 0$ the first of these equations yields

$$M_0 = -\tfrac{1}{2}M_1$$

Substitution of this expression for M_0 in the second equation and solution of the last two simultaneous equations results in the following values of the bending moments:

$$M_0 = 635 \text{ in.-lb}, \qquad M_1 = -1270 \text{ in.-lb}, \qquad M_2 = -692 \text{ in.-lb}$$

These values are in good agreement with the ones obtained by the Hardy Cross method.

1.18 Secondary Stresses in an Engine Mount for an In-Line Engine

It is common practice to calculate the forces in the members of frameworks on the assumption that the members are attached to one another by means of ideal frictionless pin joints. The uniformly distributed stresses caused by the tensile or compressive forces so obtained are called primary stresses. In reality, however, the members are either welded or riveted; when they are riveted, gusset plates of considerable rigidity are used. It is therefore a closer approximation to reality to consider the attachment of the members as perfectly rigid rather than to assume the members to be free to rotate relative to one another. On the other hand, the analysis of the forces and moments in a rigid-jointed framework is extremely laborious and certainly not worth the trouble because experience has proved that the primary stresses in the members determined on the assumption of pin connections are in excellent agreement with test data. In addition to these primary stresses corresponding to pure tension or compression, stresses can be observed in the members of a rigid-jointed framework that are due to bending. The latter are known as secondary stresses. They can be calculated by means of the moment distribution method, as will be shown by the example of an engine mount for an in-line airplane engine (Fig. 1.18.1a).

The following procedure is adopted in the calculations: First the forces are determined in the members of the framework on the assumption of pin connections. Next the displacements of the joints of the fictitious pin-connected framework are calculated, using the principle of virtual displacements. It is assumed then that the members are rigidly connected with one another at the joints in their displaced positions. When the rigid connection is established, the joints are not rotated, and members a and b, for example, remain horizontal at joint 1 (see Fig. 1.18.1c). Obviously each bar is bent in the state of deformation thus established,

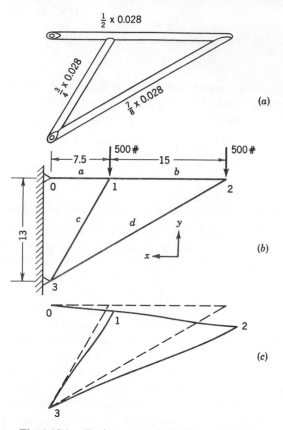

Fig. 1.18.1.   Engine mount for in-line engine.

and the curvature of the bars can be maintained only by end moments. These end moments are calculated next, and it is found that they do not add up to zero at each joint. The non-vanishing resultant moments are considered as an external loading of the rigid-jointed framework, and they are balanced by the moment distribution procedure. Finally check calculations are carried out to prove that the unbalanced *forces* caused by the balancing of the *moments* at all the joints are small enough to be neglected in engineering calculations.

Figure 1.18.1b shows the line diagram of one side of an engine mount for an in-line engine together with the forces acting on it.   On the assumption of pin connections, the forces in the four members of the framework can be determined from the following four tension coefficient equations (see Art. 1.9):

$$500 + 13T_d = 0$$

$$15T_b + 22.5T_d = 0$$

$$500 + 13T_c = 0$$

$$7.5T_a + 7.5T_c - 15T_b = 0$$

The solution of the four simultaneous equations is

$$T_a = 153.6, \qquad T_b = 57.6, \qquad T_c = -38.4, \qquad T_d = -38.4$$

From the figure the lengths of the four members are 7.5, 15, 15, and 26 in. Multiplication of the tension coefficient by the corresponding length gives the following forces in the bars:

$$F_a = 1153 \text{ lb}, \qquad F_b = 865 \text{ lb}, \qquad F_c = -577 \text{ lb}, \qquad F_d = -1000 \text{ lb}$$

The joint displacements are determined by the dummy load method (see Art. 1.7).   A vertical dummy load of 1-lb magnitude acting downward at joint 1 causes the following forces:

$$F_a = 0.577 \text{ lb}, \qquad F_b = 0, \qquad F_c = -1.154 \text{ lb}, \qquad F_d = 0$$

A vertical dummy load of 1-lb magnitude acting downward at joint 2 causes the following forces:

$$F_a = 1.73 \text{ lb}, \qquad F_b = 1.73 \text{ lb}, \qquad F_c = 0, \qquad F_d = -2 \text{ lb}$$

The engine mount is welded of steel tubes which have the sectional properties given in Table 1.18.1.   If Young's modulus for steel is taken

Table 1.18.1. Sectional Properties of Tubes in Engine Mount

Member	Size, in.	Moment of inertia, in.^{4}	Area, sq in.
a	$\frac{1}{2} \times 0.028$	0.0012	0.0415
b	$\frac{1}{2} \times 0.028$	0.0012	0.0415
c	$\frac{3}{4} \times 0.028$	0.0041	0.0635
d	$\frac{7}{8} \times 0.028$	0.0067	0.0745

as $E = 29 \times 10^6$ psi, R is used to denote L/EA, and the symbols F_0, F_1, and F_2 refer to the forces in the members caused by the external loads, by the dummy load at joint 1, and by the dummy load at joint 2, respectively, the determination of the vertical displacements of joints 1 and 2 can be summarized in Table 1.18.2.

Table 1.18.2. Computation of Displacements

Bar	$10^6 R$, in. per lb	F_0, lb	F_1, lb	F_2, lb	$10^3 RF_0 F_1$	$10^3 RF_0 F_2$
a	6.23	1153	0.577	1.73	4.14	12.44
b	12.46	865	0	1.73	0	18.62
c	8.15	−577	−1.154	0	5.42	0
d	12.00	−1000	0	−2.00	0	24.00
				Summation	9.56	55.06

Consequently the vertical downward displacement of joint 1 is 0.00956 in., and that of joint 2 is 0.05506 in. The horizontal displacement of joint 1 is equal to the elongation of bar a which is 0.00719 in., while that of joint 2 is equal to the sum of the elongations of bars a and b which is 0.01796 in. The deformed shape of the framework is indicated in Fig. 1.18.1c on the assumption of no joint rotations.

The bending moments in a bar depend on the relative displacements of the end points perpendicular to the bar. This relative displacement is obviously 0.00956 in. for bar a, and 0.0455 in. for bar b, since the horizontal displacements of the joints do not have components perpendicular to these bars. For bar c the relative displacement D_c is

$$D_c = 0.00956(7.5/15) + 0.00719(13/15) = 0.01101 \text{ in.}$$

The relative displacement D_d of the end points of bar d is

$$D_d = 0.05506(22.5/26) + 0.01796(13/26) = 0.05668 \text{ in.}$$

The fixed-end moments necessary to keep the two end tangents of a bar unchanged during a relative displacement D of the two end points perpendicular to the bar can be calculated with the aid of Fig. 1.18.2. The deflected shape is antisymmetric with respect to the vertical AA passing through the midpoint of the bar. The left half of the bar is convex, and the right half concave when viewed from above. Hence, at the midpoint there must be an inflection point. Since the inflection point corresponds to zero moment in a bent bar, the state of equilibrium is not changed if a frictionless pin is inserted in the bar at the middle, as shown in Fig. 1.18.2b. The pin can transmit only a shear force V. Each half of the bar can therefore be considered a cantilever loaded with a concentrated force applied to its free end. A formula for the calculation of the deflection of a cantilever under such a load is given in problem 2 of Fig. 1.12.6. With the problem on hand the deflection is $D/2$, the force V, and the length $L/2$, and thus

$$D/2 = \tfrac{1}{3}V(L/2)^3/EI$$

or

$$V = 12EID/L^3$$

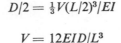

Fig. 1.18.2.   Fixed-end moments due to deflection.

It follows from the equilibrium conditions of either half of the beam in Fig. 1.18.2b that the fixed-end moment is

$$M_{fe} = VL/2$$

Consequently

$$M_{fe} = 6EI(D/L^2) \qquad\qquad 1.18.1$$

Equation 1.18.1 gives the fixed-end moment when the perpendicular relative displacement of the two end points of a bar is D and the two ends are prevented from rotating.   It will be recalled that in Art. 1.17 the fixed-end moments corresponding to a uniformly distributed load were calculated on the assumption that the end tangents of the beam were rigidly fixed although in reality only an elastic end fixation was available. In the same manner, in the present problem rigid-end fixation as regards rotation will be assumed at joints 1, 2, and 3, the fixed-end moments will be calculated from Eq. 1.18.1, and the unbalanced moments acting on the fictitious rigid constraints at the joints will be eliminated step by step by means of the Hardy Cross moment distribution method.

To accelerate the convergence of the moment-distribution method, the actual pin joint at joint 0 is not transformed into a fictitious rigid fixation. With its end at 0 free to rotate and the tangent at its end at 1 fixed in the original horizontal position, bar a is clearly a cantilever.   The connection between deflection D and shear force V is given therefore by the equation

$$D = \tfrac{1}{3}VL^3/EI$$

and thus the fixed-end moment $M_{\text{fe}} = VL$ at joint 1 is

$$M_{\text{fe}} = 3EI(D/L^2) \qquad\qquad 1.18.2$$

From Eqs. 1.18.1 and 1.18.2 the numerical values of the fixed end moments can be calculated; they are compiled in Table 1.18.3.

Table 1.18.3.   Fixed-End Moments in Bars of Framework

Bar	EI/L^2, lb	D, in.	EID/L^2	M_{fe}, in.-lb
a	619	0.00956	5.91	17.73
b	155	0.0455	7.05	42.25
c	529	0.01101	5.82	34.9
d	287	0.05668	16.25	97.5

As far as the sign of the fixed-end moments is concerned, inspection of Fig. 1.18.1c reveals that the deflected shape can be maintained only if the end moments act counterclockwise on the ends of the bars.   This is, of course, not true with every framework under every loading; it just happens

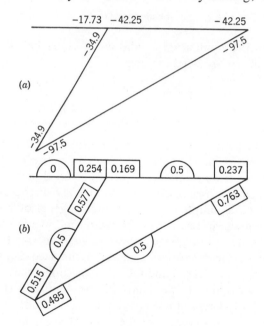

Fig. 1.18.3.   Data needed for moment distribution.

to be a peculiarity of the pattern of deflections in the present example. The fixed-end moments are listed in Fig. 1.18.3a.   It can be seen that at each joint they add up to considerable unbalances.

It can be concluded therefore that the determination of the forces and the joint displacements on the basis of a pin-joint theory leads to large unbalanced moments when the joints are prevented from rotation. In the next part of the calculations each joint is permitted to rotate in turn until equilibrium of moments is reached at each joint. As a consequence of the distribution of moments, shear forces have to be transmitted by the bars. They destroy the equilibrium of the forces established at each joint at the beginning of the calculation. If the unbalanced forces turn out to be large a second calculation of the forces in the ideal pin jointed framework may be necessary. Fortunately, this is seldom the case. In all frameworks commonly encountered in engineering, the effect of the shear forces in the bars on the force equilibrium is negligibly small.

Before the unbalanced moments can be disposed of by means of the Hardy Cross procedure, the stiffness factors and distribution factors of the framework must be calculated. The expressions developed in Art. 1.17 can be easily extended to include joints with more than two bars. Generalization of Eq. 1.17.4 gives for the angle α caused by a moment M

$$\alpha = M/\Sigma S \qquad 1.18.3$$

where ΣS denotes the sum of the stiffness factors of all the bars attached to the joint. Similarly, the distribution factor D_i pertaining to the ith bar is

$$D_i = S_i/\Sigma S \qquad 1.18.4$$

where S_i is the stiffness factor of the ith bar and the meaning of ΣS is the same as before. The stiffness factor S can be calculated from Eq. 1.17.2 for all the bars except bar a. For the latter a modified stiffness factor S' is defined as the moment required at one end of a simply supported beam in order to cause a unit rotation there. It follows from Fig. 1.12.8, part 3, that the angle of rotation α at the end of a simply supported beam at which a moment B is applied is

$$\alpha = \tfrac{1}{3}BL/EI$$

Hence, the moment B required to cause a rotation $\alpha = 1$ is

$$B = 3EI/L = S'$$

This moment is the stiffness factor S' corresponding to a simply supported rather than a rigidly fixed far end. Because of the simple support at the far end, no moment can appear there during the rotation of the beam; consequently the carry-over factor C is zero. In Table 1.18.4 the stiffness factors are collected for the numerical example.

The distribution factors are shown in Fig. 1.18.3b. They are enclosed in polygons. The carry-over factors are listed in the middle of each bar and are enclosed in semicircles. The moment distribution is carried out

Table 1.18.4. Stiffness Factors of Bars

Bar	$10^{-3}EI/L$ in.-lb	Stiffness Factor, in.-lb per radian
a	4.64	13,920
b	2.32	9,280
c	7.93	31,720
d	7.47	29,880

in Fig. 1.18.4 in a manner similar to that described in detail in connection with Table 1.17.1. Fixed-end, distributed, and carried-over moments are listed in columns beginning near the joints of a line diagram of the framework since a table would not be clear because of the two-dimensional

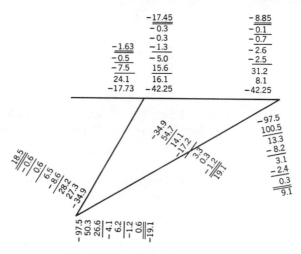

Fig. 1.18.4. Moment distribution.

arrangement of the bars. The order of balancing is: joints 1, 2, 3, 1, 2, 3, 2, 3, and 1. No moments are carried over from joint 1 to joint 0 because $C = 0$. After the completion of the nine balancing operations, the columns are added up to obtain the final bending moments. The moments are listed according to the rigid-frame convention and should add up to zero at each joint. The greatest deviation from a vanishing resultant occurs at joint 3 where the moment in bar c is 18.5 in.-lb and that in bar d is -19.1 in.-lb. An unbalance of 0.6 in.-lb is considered permissible, and for this reason the balancing is not continued.

The bending moments are now transformed to make them correspond to the beam convention; they are left unchanged at the left end of a bar and are multiplied by -1 at the right end. With inclined bars, of course, like bars c and d in the example, the designation of any end as the left end is somewhat arbitrary. The bending-moment distribution is shown in Fig. 1.18.5a.   In part b of the same figure the final deformed shape of the framework is indicated.

The force equilibrium can be checked with the aid of Fig. 1.18.5a.   The shear force transferred by a beam is equal to the difference of the end moments divided by the length of the beam.   In the present example the maximum shear force occurs in member c.   It is

$$V = (18.5 + 17.45)/15 = 2.5 \text{ lb}$$

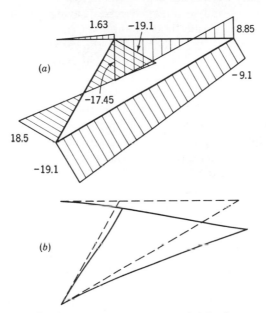

Fig. 1.18.5.   Final moments and deflections.

Obviously unbalanced forces of the order of magnitude of 2.5 lb can be safely neglected when each external load is 500 lb.

The maximum secondary stress occurs in bar b. With $M = -17.45$ in.-lb and $Z = 0.0048$ in.^{3} the stress is

$$\sigma_{\text{sec}} = 17.45/0.0048 = 3640 \text{ psi}$$

At the same time the primary stress in bar b is

$$\sigma_{\text{pr}} = F/A = 865/0.0415 = 20{,}800 \text{ psi}$$

The maximum secondary stress is therefore 17.5 per cent of the primary stress in the same bar.

As a rule, the secondary stresses are reasonably small when the framework contains slender bars which subtend large angles with one another at the joints. The stresses may become very large when many stocky bars are connected at any one joint. Since, however, even the 17.5 per cent additional stress of the present example is a sizable quantity, there must be some physical reason why secondary stresses are not often calculated in framework analyses.

The reason is that the primary stresses are necessary for equilibrium while the secondary stresses are not. An increase in the loads by a factor n results in primary and secondary stresses having n times the original values as long as the elastic limit of the material is not exceeded. But beyond the elastic limit the primary and the secondary stresses do not increase in the same ratio. An example will clarify the situation: Let it be assumed that, under an unexpected overload in a particular bar of a framework, the primary stress and the secondary stress each amount to 50 per cent of the yield stress. If the loads are increased a further 10 per cent, the primary stress increases to 55 per cent of the yield stress since otherwise the force equilibrium at the joints could not be maintained. The increased joint deflections will increase the curvature of the members, but the plastic flow of the material will continue without the total stress (primary plus secondary) exceeding the yield stress if the material has a pronounced yield point like mild steel. Consequently the secondary stress can be only 45 per cent of the yield stress under the increased loads. It follows from this argument that increased loads do not necessarily mean increased secondary stresses although they do involve increased deformations. In contrast the primary stresses increase proportionately with the loading.

Under static loads therefore only the magnitude of the primary stress need be considered when the load-carrying capacity of the framework is investigated except when large deformations, coupled with an unusually small ductility of the material, exhaust the capacity of the structure to deform. On the other hand, fatigue failure can be caused by secondary stresses. Engine mounts which are subjected to severe vibrations are known to have failed in fatigue because of secondary stresses.

If primary and secondary stresses are understood to mean necessary and additional stresses, bending stresses caused in members of a framework by transverse loads applied between the joints must be classified as primary stresses. Obviously such loads cannot be carried without the occurrence of bending stresses in the member. As in the definition of an ideal framework in Art. 1.6 load application at the joints was stipulated, frameworks in which some loads are applied between the joints resemble rigid frames more than ideal frameworks in this respect.

1.19 The Slope-Deflection Method and the Analysis of Rigid Frames

The Slope-Deflection Equations

Structural assemblies of bars are known as rigid frames if they rely for their strength and rigidity on the bending moments transmitted by their joints. If their rigid joints were replaced by pin connections, the assembly of bars would collapse. Although the material cannot be so well utilized in the elements of a rigid frame, which are subjected to variable bending moments, as in a pin-jointed framework, in whose members uniform tensile or compressive stresses are acting, space limitations and ease of

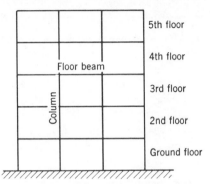

Fig. 1.19.1.   Sketch of welded steel frame of apartment house.

manufacture, as well as various other non-structural reasons, make the rigid frame one of the most common types of construction,   For instance, the welded steel frame of a modern apartment house is without diagonal bracing (see Fig. 1.19.1) and must be analyzed as a rigid frame.

One of the best-known methods of analysis for rigid frames is the slope-deflection method. It considers the rotations and displacements of the joints as the unknown quantities and expresses the bending moments and shear forces acting on the end sections of the beams and columns in terms of the unknown rotations and displacements. Next the equilibrium conditions are written; they furnish a sufficient number of linear algebraic equations to be solved for the unknown quantities.

The first step in the analysis is therefore the establishment of the relationship among end moments, displacements, and rotations. This can be done with the aid of Fig. 1.19.2 in which the positive directions of the bending moments are indicated in agreement with the rigid-frame sign convention. According to this convention, moments as well as rotations are counted positive if they are clockwise; the differences between the rigid frame and the beam conventions are discussed in some detail in Art. 1.17.

In rigid-frame analysis the rigid-frame convention is preferred to the beam convention because it permits an easier formulation of the equilibrium conditions at joints where several bars of different direction meet.

Bar AB is first assumed to be displaced as a rigid body into position $A'B'$ in consequence of the displacements of the joints. (The change in the length of the bar is disregarded; it is small of the second order if the displacements are small of the first order.) Under the action of bending moments M_A and M_B, the bar then assumes the curved shape indicated. It is shown in Fig. 1.12.8 that a moment M acting at one end of a simply supported beam of constant rigidity causes a rotation $ML/3EI$ to take

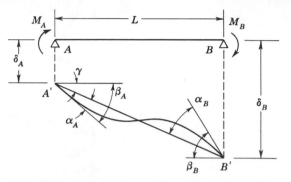

Fig. 1.19.2.   Displacements and rotations of bar AB.

place there, while the rotation at the other end is $ML/6EI$. These angles are given in accordance with the beam convention. If the rigid-frame convention is used, the two rotations must have opposite signs. According to this latter convention therefore

$$\alpha_A = (M_A L/3EI) - (M_B L/6EI) \qquad\qquad 1.19.1$$

$$\alpha_B = (M_B L/3EI) - (M_A L/6EI) \qquad\qquad 1.19.2$$

When the displacements are small, the angle γ of rigid-body rotation is equal to the tangent of the angle:

$$\gamma = (\delta_B - \delta_A)/L$$

Consequently the total angles β can be given as

$$\beta_A = (L/6EI)(2M_A - M_B) + (1/L)(\delta_B - \delta_A) \qquad 1.19.3$$

$$\beta_B = (L/6EI)(2M_B - M_A) + (1/L)(\delta_B - \delta_A) \qquad 1.19.4$$

Multiplication of Eq. 1.19.3 by 2 and addition to Eq. 1.19.4 lead to the following expression for the moment M_A:

$$M_A = (2EI/L)[2\beta_A + \beta_B - (3/L)(\delta_B - \delta_A)] \qquad 1.19.5$$

Similarly one obtains

$$M_B = (2EI/L)[2\beta_B + \beta_A - (3/L)(\delta_B - \delta_A)] \qquad 1.19.6$$

These moments act on the ends of the bar when the ends are displaced and rotated; as far as the physical facts are concerned, it is perhaps more accurate to say that the reactions of these moments are exerted by the bar on the joints in consequence of the displacements δ and rotations β. In the presence of external loads, bars that are members of rigid frames may be subjected to end moments even if their ends are prevented by some fictitious constraints from displacing and rotating. For instance, Fig. 1.13.5 shows the end moments arising from the clamping of the ends of a beam when the load is distributed uniformly over the span. Such clamping moments are designated as fixed-end moments.

The total moments acting on the ends of bar AB, when it is part of a rigid frame, are therefore

$$M_{A\cdot\text{tot}} = (2EI/L)[2\beta_A + \beta_B - (3/L)(\delta_B - \delta_A)] + M_{\text{fe}\cdot A} \qquad 1.19.7$$

$$M_{B\cdot\text{tot}} = (2EI/L)[2\beta_B + \beta_A - (3/L)(\delta_B - \delta_A) + M_{\text{fe}\cdot B} \qquad 1.19.8$$

where $M_{\text{fe}\cdot A}$ and $M_{\text{fe}\cdot B}$ are the fixed-end moments exerted on the bar by the fictitious constraints.† These two equations are designated as the slope-deflection equations. They were derived without consideration of the effects of axial loads on the moments and deflections. This is not objectionable because in structural engineering practice the axial forces caused in the members of rigid frames by the design loads are almost always so small that their influence on the deformations can be neglected. The effect of the shear forces on the deformations was also disregarded; this is in agreement with structural engineering practice as mentioned at the end of Art. 1.12.

Rigid Frame Subjected to Symmetric Loading

With the tools thus established the solution of a rigid-frame problem can now be undertaken. Figure 1.19.3a represents a simple rectangular bent subjected to a uniformly distributed load w. Without rigid connections at the joints the frame would sway to one side and collapse. With the rigid connections, however, the whole sequence of bars $ABCD$ is a single rigidly connected body and as such is statically indeterminate. If column CD were sawed through near point D, the remaining structure could still carry the load. Hence the unknown tension, the unknown shear, and the unknown bending moment in section D are all redundant quantities. Their magnitude could be established from the requirement that, under the simultaneous action of the external load w and the unknown

† Fictitious constraints are discussed at greater length in Art. 1.17.

redundant forces and moments at D, the relative displacements and rotation of the two sides of the section at D should vanish. This approach will be discussed in Art. 4.3, but here the slope-deflection method will be employed in the solution of the problem.

First it should be mentioned that use will be made of the symmetry

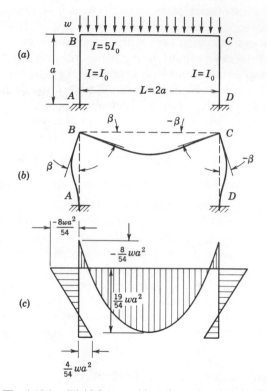

Fig. 1.19.3.   Rigid frame subjected to symmetric loading.

properties of the problem to simplify the analysis. The symmetrically loaded symmetric frame must naturally deform in a symmetric manner. The deflected shape is indicated in Fig. 1.19.3b. The initial right angles between columns and beam are assumed to remain unchanged during the deformations. It may be noted that in the figure the positions of joints B and C are the same after deformation as before. This means that the shortening of the columns under the compressive load wa is neglected. This is reasonable because the shortening is very small compared to the lateral deflections of the bent elements. Similarly the variation of the distance between the end points of the beam caused by the curvature is disregarded. These changes in length can be included in the calculations, but their effect upon the bending-moment distribution is negligibly small under ordinary conditions of rigid-frame design.

The fixed-end moments at B and C can be calculated with the aid of Fig. 1.13.5 if L is replaced by $2a$:

$$M_{\text{fe}\cdot B} = -wa^2/3$$

$$M_{\text{fe}\cdot C} = +wa^2/3$$

The signs are in accordance with the rigid-frame convention, and the moments are those exerted by the fictitious constraints on the beam. In Eq. 1.19.5 the subscripts A and B must be replaced by B and C when the beam of Fig. 1.19.3 is analyzed. The moment M_{BC} acting on beam BC from joint B is then

$$M_{BC} = (5EI_0/a)(2\beta - \beta) = (5EI_0/a)\beta \qquad \text{1.19.9}$$

because $\beta_B = -\beta_C = \beta$ and the displacements δ_B and δ_C of the joints are negligibly small. The total moment acting on the beam at B is from Eq. 1.19.7.

$$M_{BC\cdot\text{tot}} = (5EI_0/a)\beta - (wa^2/3) \qquad \text{1.19.10}$$

The moment acting on column AB from joint B can be obtained from Eq. 1.19.8:

$$M_{BA\cdot\text{tot}} = (2EI_0/a)(2\beta + 0) = (4EI_0/a)\beta \qquad \text{1.19.11}$$

Naturally the fixed-end moment is zero because no external loads are applied to the columns. The sum of $M_{BC\cdot\text{tot}}$ and $M_{AB\cdot\text{tot}}$ must vanish when joint B is in equilibrium:

$$(5EI_0/a)\beta - (wa^2/3) + (4EI_0/a)\beta = 0 \qquad \text{1.19.12}$$

Solution for β yields

$$\beta = \tfrac{1}{27}(wa^3/EI_0) \qquad \text{1.19.13}$$

Substitution of the value obtained for β in Eq. 1.19.10 leads to

$$M_{BC\cdot\text{tot}} = -\tfrac{4}{27}wa^2 \qquad \text{1.19.14}$$

according to the rigid-frame convention.

The end moments acting on the column can be calculated in a similar manner. From Eqs. 1.19.11 and 1.19.13 the moment at B is

$$M_{BA\cdot\text{tot}} = \tfrac{4}{27}wa^2 \qquad \text{1.19.15}$$

At A the moment is

$$M_{AB\cdot\text{tot}} = (2EI_0/a)(0 + \beta) = \tfrac{2}{27}wa^2 \qquad \text{1.19.16}$$

The signs in Eqs. 1.19.14 to 1.19.16 are given in accordance with the frame convention. They have to be changed to agree with the beam convention before moment diagrams can be drawn. If B is considered as the left end of beam BC, the sign there remains unchanged. If A is the left end of column AB, the sign of the moment there is unchanged, but at B the moment, that is, $M_{BA \cdot tot}$, must be multiplied by -1. The moment diagram is shown in part c of Fig. 1.19.3.

It can be said therefore that the slope-deflection method permits an easy solution of this rigid-frame problem. There is a single unknown to be determined while the more classical approach of consistent deformations would require the determination of three unknowns. A suitable choice of the redundant quantities can eliminate one unknown in the classical analysis, but the minimum number remaining is still two. Of course, in the slope-deflection method the total number of unknowns is in reality three because the vertical and horizontal displacements of joint B are not known in advance. Experience has shown, however, that these displacements do not influence noticeably the moment distribution in many practical frames under various conditions of loading. Whenever this is true, the only unknown quantities in the slope-deflection method are the rotations.

Rigid Frame Subjected to Antisymmetric Loads

The situation is a little more complex when the frame is subjected to side loads. In Fig. 1.19.4 the load P causes a side sway δ which is of the same order of magnitude as the lateral displacements of the beam and the columns due to flexure; it is not negligibly small like joint displacements under symmetric loads. Because of the antisymmetry of the system, there are only two unknown quantities in the slope-deflection equations, namely β and δ. When the moment acting on the beam at B is sought, Eq. 1.19.7 yields, upon replacement of A and B by B and C,

$$M_{BC \cdot tot} = (5EI_0/a)(2\beta + \beta) = 15(EI_0/a)\beta \qquad 1.19.17$$

The deflections δ_A and δ_B in the equation have to be measured perpendicularly to the center line of the beam, and thus they are zero for beam BC.

From Eq. 1.19.8 the end moment acting on the column at B is

$$M_{BA \cdot tot} = (2EI_0/a)[2\beta + 0 - (3/a)(\delta - 0)]$$
$$= (2EI_0/a)[2\beta - 3(\delta/a)] \qquad 1.19.18$$

For equilibrium, the sum of the moments $M_{BC \cdot tot}$ and $M_{BA \cdot tot}$ must vanish. Consequently

$$19\beta - 6(\delta/a) = 0 \qquad 1.19.19$$

This single equation does not suffice for the calculation of the two unknowns β and δ, and the condition of the equilibrium of moments at C leads to an equation identical with Eq. 1.19.19. The additional condition necessary for the solution of the problem is a statement of the equilibrium of the horizontal forces in the system.

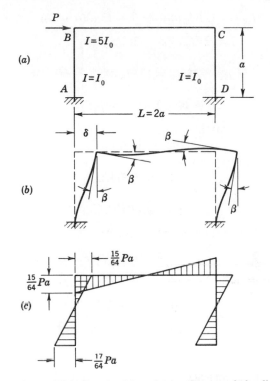

Fig. 1.19.4.   Rigid frame subjected to antisymmetric loading.

As long as the shortening of beam BC under the compressive load is neglected, the lateral displacement of the two columns is equal in magnitude. Hence each column must transmit a shear force $V = -P/2$. If a clockwise moment M_{BA} and the shear force V are acting on column AB at A, the clockwise moment acting on the foundation at A is $M_{BA} - Va$. The reaction exerted by the foundation on the column is of the same magnitude but counterclockwise. It must be considered negative therefore according to the frame convention. If it is designed by $-M_{AB}$, one has the connection

$$-M_{AB} = M_{BA} - Va$$

It follows that
$$V = (1/a)(M_{AB} + M_{BA}) \qquad 1.19.20$$

and with the present shear forces

$$(1/a)(M_{AB} + M_{BA}) = -P/2 \qquad 1.19.21$$

$M_{BA} = M_{BA \cdot \text{tot}}$ was calculated before. From Eq. 1.19.7 the moment $M_{AB} = M_{AB \cdot \text{tot}}$ becomes

$$M_{AB} = (2EI_0/a)[0 + \beta - (3/a)(\delta - 0)]$$
$$= (2EI_0/a)[\beta - 3(\delta/a)] \qquad 1.19.22$$

Addition gives

$$M_{AB} + M_{BA} = (2EI_0/a)[3\beta - 6(\delta/a)]$$

and thus

$$3\beta - 6(\delta/a) = -Pa^2/4EI_0 \qquad 1.19.23$$

If Eq. 1.19.19 is rewritten

$$19\beta - 6(\delta/a) = 0$$

subtraction is seen to yield

$$16\beta = Pa^2/4EI_0$$

that is

$$\beta = \tfrac{1}{64}(Pa^2/EI_0) \qquad 1.19.24$$

Equation 1.19.19 can then be solved for δ:

$$\delta = \tfrac{19}{6}\beta a = \tfrac{19}{384}(Pa^3/EI_0) \qquad 1.19.25$$

Finally substitution in the moment expressions leads to

$$M_{AB} = -\tfrac{17}{64}Pa$$
$$M_{BA} = -\tfrac{15}{64}Pa$$
$$M_{BC} = +\tfrac{15}{64}Pa$$

These are the moments acting on the bars, and their signs are given in agreement with the frame convention. Adoption of the beam convention and drawing of the moment diagrams result in Fig. 1.19.4c.

Rigid frames can also be analyzed with the aid of Castigliano's second theorem and by means of the Hardy Cross moment-distribution method. These topics are discussed in Art. 4.3.

PART TWO

The Minimum of the
Total Potential

2.1 Strain Energy

Normal Stresses

When the principle of virtual displacements is applied to the solution of problems that are more involved than those discussed in Part 1, it is necessary to make use of the concept of "strain energy." This concept can be explained with the aid of Fig. 2.1.1.

The bar shown in Fig. 2.1.1a is loaded with a force P_1 which is uniformly distributed over the end section of the bar. Under this load a uniform

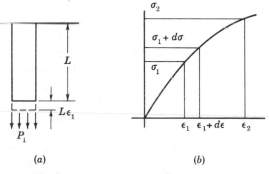

Fig. 2.1.1. Strain energy of tension.

stress $\sigma_1 = P_1/A$, where A is the cross-sectional area of the bar, prevails throughout the bar; similarly the strain ϵ_1 in the axial direction is the same throughout the bar if local disturbances near the attachment to the rigid wall are disregarded. Figure 2.1.1b represents the stress-strain diagram of the material, which is assumed to be perfectly elastic although it does not necessarily follow Hooke's law. In the figure the stress is not proportional to the strain, but the curve is assumed to show the relationship

121

between stress and strain both when the load is increased and when it is decreased. Nowhere has the stress-strain curve a vertical or a horizontal tangent.

Let it be assumed now that the force is increased from P_1 to $P_1 + dP$, entailing an increase in the stress from σ_1 to $\sigma_1 + d\sigma$. The corresponding increase $d\epsilon$ in the strain can be determined from Fig. 2.1.1b. During this change in the state of stress and strain the average value of the stress is greater than σ_1. It can be represented by the expression $\sigma_1(1 + k)$ if k is a properly chosen number. The average force is therefore $A\sigma_1(1 + k)$, and the work dW_e done by it during the elongation $L\,d\epsilon$ of the bar is

$$dW_e = A\sigma_1(1 + k)L\,d\epsilon = V\sigma_1(1 + k)\,d\epsilon \qquad 2.1.1$$

where $V = AL$ denotes the volume of the bar. The total work done by the applied force while the strain increases from zero to a final value ϵ_2 can be obtained by dividing the abscissa into a number of segments of length $d\epsilon$, calculating the work done during each increment of strain from Eq. 2.1.1, and summing up these work quantities:

$$W_e = V \sum \sigma(1 + k)\,d\epsilon$$

In this summation not only σ but also k varies from term to term. This is self-evident since, in order to obtain the correct average, k must be small for segments over which the stress-strain curve is almost horizontal, and it must be large for segments over which the stress-strain curve is steep.

The difficulties involved in the evaluation of k can be obviated if the number of segments between $\epsilon = 0$ and $\epsilon = \epsilon_2$ is greatly increased in such a manner that the length of each segment $d\epsilon$ becomes very small. When $d\epsilon$ approaches zero, the difference between the ordinates corresponding to ϵ and $\epsilon + d\epsilon$ diminishes and in the limit k also approaches zero. In the limiting case an indefinitely large number of work quantities must be added up, and the summation sign is replaced by the integral sign:

$$W_e = \lim_{d\epsilon \to 0} V \sum \sigma(1 + k)\,d\epsilon = V \int_0^{\epsilon_2} \sigma\,d\epsilon \qquad (a)$$

This is the work done by the external load. Because of the assumption that the material is perfectly elastic, the total work done can be regained if the load is gradually decreased to zero. The work is, therefore, stored in the elastically distorted body in the form of energy, and this energy is known as the strain energy, or elastic energy. In everyday life the capacity of steel to store elastic energy is utilized in a number of ways. For instance, the work done in winding a watch is stored in a spring. The energy is released slowly as the watch runs, and is used to overcome the friction in the mechanism.

The preceding derivations are valid only if the load is increased and decreased at such a low rate that static equilibrium is maintained during the whole process and vibrations are not excited.   Then at every moment during both loading and unloading the stresses in the interior of the body are in equilibrium with the external load.   The work done by the internal forces acting between the particles of the material plus the work done by the external load is equal to zero for a body in equilibrium according to the principle of virtual displacements.   Hence, during each infinitesimal increment $d\epsilon$ of the strain, the internal work is equal and opposite to the external work done by the applied load.   Consequently the total internal work is also equal and opposite to the total external work.   It can be readily understood why the internal work must be negative: The deformations are not caused by the internal forces but are imposed upon the body against them.   The internal forces resist the deformations and maintain equilibrium with the external load at every moment as long as the external load is increased at a low rate.

If the symbols U and W_i are used to denote the strain energy and the internal work, respectively, Eqs. 2.1.1 and a can be written as

$$dU = -dW_i = dW_e = V\sigma_1(1 + k)\,d\epsilon \qquad 2.1.1a$$

$$U = -W_i = W_e = V\int_0^{\epsilon_2}\sigma\,d\epsilon \qquad (b)$$

It is often convenient to have an expression for the strain energy per unit volume:

$$\frac{U}{V} = \frac{W_e}{V} = -\frac{W_i}{V} = \int_0^{\epsilon_2}\sigma\,d\epsilon \qquad 2.1.2$$

In the particular case when Hooke's law is valid,

$$\sigma = E\epsilon \qquad 2.1.3$$

where E is Young's modulus, substitution of the expression for σ yields

$$\frac{U}{V} = \int_0^{\epsilon_2}E\epsilon\,d\epsilon = \frac{1}{2}E\epsilon_2{}^2 \qquad 2.1.4a$$

The strain energy can also be given in terms of stress rather than strain. Substitution of σ_2/E for ϵ_2 in Eq. 2.1.4a yields

$$U/V = \tfrac{1}{2}\sigma_2{}^2/E \qquad 2.1.4b$$

In Eqs. 2.1.4 the subscript 2 refers to the end state of stress and strain for which the strain energy is sought.   If the stress and the strain in this state are designated by σ and ϵ without a subscript, the subscript 2 must be omitted from the equations.

Shearing Stresses

Work is also stored in the form of strain energy when an elastic body deforms under the action of shearing stresses. When the torque T_1 is applied by means of uniformly distributed shearing stresses to the end sections of the thin-walled circular cylinder shown in Fig. 2.1.2, the intensity of the shearing stress τ_1 is

$$\tau_1 = T_1/(r \times 2r\pi t) = T/2\pi r^2 t$$

where r is the radius of the median line of the cylinder wall. If the relationship between shear stress and shear strain is represented by the curve

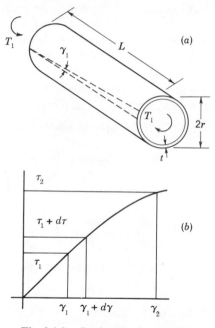

Fig. 2.1.2.   Strain energy of shear.

shown in Fig. 2.1.2b, the work done by the applied torque during the rotation caused by an increase of the torque from T_1 to $T_1 + dT$ can be calculated as was the work in connection with Fig. 2.1.1. The average value of the torque is

$$T_{\text{avg}} = 2\pi r^2 t \tau_1 (1 + k)$$

The change in shear strain is $d\gamma$ which causes a relative tangential displacement $L \tan (d\gamma)$ between the two end sections of the cylinder. When $d\gamma$ is small, its tangent can be replaced by the angle itself. Hence the relative angle of rotation between the two end sections is $(L\,d\gamma)/r$. The work dW_e done by the torque is consequently

$$dW_e = 2\pi r^2 t \tau_1 (1 + k) L\,d\gamma/r$$
$$= V\tau_1(1 + k)\,d\gamma$$

where $V = 2\pi rtL$ is the volume of the material in the cylinder. Summation of the work quantities corresponding to all the increments $d\gamma$ from zero to γ_2 and a limiting process in which $d\gamma$ and k approach zero and the summation sign is replaced by the integral sign yield

$$W_e = V \int_0^{\gamma_2} \tau\,d\gamma$$

If the material is perfectly elastic, the total amount of work is stored in the form of strain energy. The specific strain energy is

$$\frac{U}{V} = \frac{W_e}{V} = -\frac{W_i}{V} = \int_0^{\gamma_2} \tau\,d\gamma \qquad 2.1.5$$

In the particular case when the material follows Hooke's law for shear,

$$\tau = G\gamma \qquad 2.1.6$$

where G is the shear modulus of elasticity, integration yields the following two equations·

$$U/V = \tfrac{1}{2}G\gamma_2{}^2 \qquad 2.1.7a$$

$$U/V = \tfrac{1}{2}(\tau_2{}^2/G) \qquad 2.1.7b$$

Again the subscript 2 may be omitted if it is desired that the stress and the strain in the end state be designated by τ and γ without a subscript.

Strain Energy in Bars and Beams

When the normal strain, or the shear strain, varies in a body, it is convenient to write the strain-energy expression for an infinitesimal element of volume dV. It follows from Eqs. 2.1.4 and 2.1.7 that

$$dU = \tfrac{1}{2}E\epsilon^2\,dV = \tfrac{1}{2}(\sigma^2/E)\,dV \qquad 2.1.8$$

$$dU = \tfrac{1}{2}G\gamma^2\,dV = \tfrac{1}{2}(\tau^2/G)\,dV \qquad 2.1.9$$

where the stress and the strain in the state for which the strain energy is calculated are denoted by symbols without a subscript.

Equations 2.1.8 and 2.1.9 suffice for the calculation of the strain energy in a body whose material satisfies Hooke's law and in which a simple state of stress exists.   It is convenient, however, to have formulas expressing the strain energy directly in terms of the loads.   In a tension bar, as shown in Fig. 2.1.1a, or in a compressed column, integration gives

$$U = \tfrac{1}{2}(\sigma^2/E)V = \tfrac{1}{2}P^2L/EA \qquad\qquad 2.1.10$$

since the stress is constant.   Similarly in a twisted thin-walled circular tube, as shown in Fig. 2.1.2a,

$$U = \tfrac{1}{2}(\tau^2/G)V = \tfrac{1}{2}T^2L/GI_p \qquad\qquad 2.1.11$$

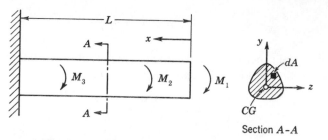

Fig. 2.1.3.   Beam subjected to pure bending.

where T is the applied torque and I_p the polar moment of inertia of the cross section of the tube.

In a bent beam the normal stress is not distributed uniformly.   If Hooke's law is valid and the Bernoulli-Navier assumptions are made, the normal stress prevailing in the section of the beam shown in Fig. 2.1.3 can be expressed as

$$\sigma = (M/I)y$$

where M is the bending moment, I the moment of inertia of the section about the z axis, and the coordinates y and z are measured from the centroid of the section in the directions of the principal axes of inertia of the section. The moments M_1, M_2, and M_3 shown in the figure are concentrated moments applied to the beam as loads.   The strain energy stored in an infinitesimal element of volume $dV = dA\,dx$ is, according to Eq. 2.1.8,

$$dU = \tfrac{1}{2}(\sigma^2/E)\,dV = \tfrac{1}{2}(M^2/I^2E)y^2\,dA\,dx$$

The strain energy in the entire beam can be obtained by integrating first with respect to A and then with respect to x:

$$U = \frac{1}{2}\int_0^L \int_A \frac{M^2}{I^2E}\,y^2\,dA\,dx$$

During the first integration M, I, and E are constants.   Moreover

$$\int_A y^2 \, dA = I$$

as is known from strength of materials.   Hence

$$U = \frac{1}{2} \int_0^L \frac{M^2}{IE} \, dx \qquad\qquad 2.1.12$$

The integration indicated in Eq. 2.1.12 can be carried out when M is given as a function of x.   The strain energy in a bent beam can also be expressed in a different form if use is made of the known relation between bending moment and curvature $(1/\rho)$, where ρ is the radius of curvature:

$$1/\rho = M/EI \qquad\qquad 2.1.13$$

Equation 2.1.13 is a statement of Hooke's law for a bent beam.   When the deflections η of the beam are known for all values of the axial coordinate, the curvature can be calculated from the formula

$$\frac{1}{\rho} = \frac{d^2\eta/dx^2}{[1 + (d\eta/dx)^2]^{3/2}}$$

In most applications the deflections η and the slopes $d\eta/dx$ are so small that the square of the latter quantity can be neglected when it is added to unity. Consequently

$$1/\rho = d^2\eta/dx^2 \qquad\qquad 2.1.14$$

and

$$EI(d^2\eta/dx^2)$$

can be substituted for M in Eq. 2.1.12 with the result that

$$U = \frac{1}{2} \int_0^L EI \left(\frac{d^2\eta}{dx^2}\right)^2 dx \qquad\qquad 2.1.15$$

Ratio of Shear Strain Energy to Normal Strain Energy in Beams

In Eqs. 2.1.12 and 2.1.15 the strain energy of bending was calculated as if the beam had been subjected to normal stresses alone.   In reality, whenever the bending moment varies, the bending is accompanied by shear, and shear stresses act upon each section of the beam perpendicular to the axis.   These shear stresses give rise to an additional shear strain-energy quantity.   This, however, is very small compared to the bending strain energy calculated earlier, except when the beam is unusually weak in shear.   Moreover, there is no interaction between normal and shearing stresses.   The normal stresses cause a rotation of the sections, and thus the shear force applied to the section does no work during the deformations corresponding to the normal stresses. The shear stresses only give rise to shearing translations and, since these are perpendicular to the normal stresses due to bending, no work can be done by the normal stresses during the deformations caused by the shear stresses. Hence the bending strain energy corresponding to the normal stresses and the shear strain energy can be calculated independently.

A numerical example will show the unimportance of the shear strain energy in a conventional beam. If the cross section is square with a side

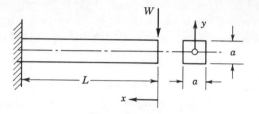

Fig. 2.1.4.   Cantilever beam.

length of a, the moment of inertia is $a^4/12$. Assuming a cantilever arrangement with an end load W and a beam length L (see Fig. 2.1.4), one obtains from the bending strain-energy formula Eq. 2.1.12

$$U_b = \frac{6}{Ea^4} \int_0^L W^2 x^2 \, dx = \frac{2W^2 L^3}{Ea^4}$$

It is shown in the strength of materials that in a beam of solid square section the shear stress is distributed according to the parabolic law

$$\tau = \tfrac{3}{2}(W/a^2)[1 - 4(y/a)^2]$$

Substitution in Eq. 2.1.9 and integration yield

$$U_s = \frac{1}{2G} \int \tau^2 \, dV = \frac{a}{2G} \int_0^L \int_{-a/2}^{a/2} \tau^2 \, dy \, dx$$

$$= \frac{9W^2}{8Ga^3} \int_0^L \int_{-a/2}^{a/2} \left[1 - 4 \left(\frac{y}{a}\right)^2\right]^2 dy \, dx$$

$$= \frac{9W^2 L}{8Ga^3} \int_{-a/2}^{a/2} \left[1 - 8 \left(\frac{y}{a}\right)^2 + 16 \left(\frac{y}{a}\right)^4\right] dy$$

$$= \frac{9}{15} \frac{W^2 L}{Ga^2}$$

The ratio of the bending strain energy due to normal stress to the shear strain energy is

$$U_b/U_s = \tfrac{3.0}{9}(G/E)(L/a)^2$$

Since for structural metals $G/E = 1/2.6$, there results

$$U_b/U_s = 1.28(L/a)^2$$

Consequently the ratio is 32, 128, and 800 when the length of the beam divided by its depth is 5, 10, and 25, respectively. This proves that the shear strain energy is negligibly small in beams of conventional design; in a built-up beam with a very small cross-sectional area of the web it can be significant if the beam is very short.

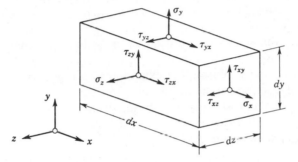

Fig. 2.1.5. Stresses in parallelepiped.

Three-Dimensional State of Stress

Next the strain energy stored in the infinitesimal parallelepiped shown in Fig. 2.1.5 will be determined in order to obtain formulas valid when the most general state of stress exists. The parallelepiped is subjected to the normal stresses σ_x, σ_y, and σ_z and to the shear stresses τ_{yz}, τ_{zx}, and τ_{xy}, all of which act simultaneously. The products of these stresses by the areas upon which they are acting can be considered the external loads of the parallelepiped. The strain energy is calculated by determining the work done by all the applied loads during the deformations.

If the normal stress σ_x in the x direction is applied first, the work done by it during the elongation of the edge dx is easily determined from Eq. 2.1.8. Next the normal stress σ_y might be applied and the work corresponding to an elongation in the y direction obtained. However, simultaneously with the elongation in the y direction the parallelepiped contracts in the perpendicular x and z directions (Poisson's effect), and, since stresses are present in the x direction, they also do work during this contraction.

Similarly, when the normal stress is finally applied in the z direction, the work quantities corresponding to the elongation in the z direction and to the contractions in the x and y directions must be evaluated. On the other hand, the work quantities due to the three shear stresses τ_{xy}, τ_{yz}, and τ_{zx} are independent of one another and of the normal stresses when the body is isotropic. For example, the shear stresses τ_{xy} cause only shearing deformations in the xy plane, but no shearing deformations in the yz and zx planes or normal strains in any of the three coordinate directions.

If any other loading schedule is followed, the individual work quantities will differ, but their sum total will be the same as in the case just described, as long as the material of the parallelepiped is perfectly elastic. This statement must be true by virtue of the first fundamental principle of thermodynamics, verified by centuries of practical experience, according to which it is not possible to construct an engine that operates in a cycle and continuously produces work from nothing.

The proof of this statement can be given in the following manner: If A stands for the work done by the stress σ_x when it is applied first, and B for that done by σ_x and σ_y during the extension in the y direction and the corresponding contraction in the x direction caused by a subsequent application of σ_y, the total work done during the loading process is $A + B$. This work is stored in the elastic body in the form of strain energy, and it can be recovered completely if σ_y is removed first and σ_x afterward. Similarly $C + D$ may represent the strain energy stored in the parallelepiped if σ_y is applied first and σ_x afterward. This work quantity can be regained by removing σ_x first and σ_y second. Let us now prescribe the following cycle of operations: The normal stress σ_x is applied to the body, and this is followed by an application of σ_y; then σ_x is removed, and finally σ_y is taken off. The work done during the loading operation is $A + B$ while the work recovered is $C + D$. If $C + D$ is greater than $A + B$ and the cycle is repeated indefinitely, an engine is available that creates work out of nothing. If $C + D$ is smaller than $A + B$, a change in the order of operations results in a similar perpetual-motion engine. Since perpetual motion is impossible, $A + B$ must be equal to $C + D$. In other words, the strain energy cannot depend on the order of application of the loads. This can also be expressed by saying that the strain energy must be a single-valued function of the loads.

Consequently the strain energy stored in the parallelepiped may be calculated for any arbitrary process of loading. It is convenient to choose that in which all the six stress components are increased simultaneously from zero to their final values σ_x, σ_y, σ_z, τ_{yz}, τ_{zx}, and τ_{xy}. During this loading the six strain components will also increase from zero to their final values ϵ_x, ϵ_y, ϵ_z, γ_{yz}, γ_{zx}, and γ_{xy}. The total work done can now be calculated if it is observed that the force corresponding to the stress σ_x is $\sigma_x\, dy\, dz$, and the elongation corresponding to the strain ϵ_x is $\epsilon_x\, dx$, and that similar relations exist for the other stress and strain components. Then

$$dU = \int_0^{\epsilon_x} \sigma_x\, dy\, dz\, d\epsilon_x\, dx + \int_0^{\epsilon_y} \sigma_y\, dz\, dx\, d\epsilon_y\, dy + \int_0^{\epsilon_z} \sigma_z\, dx\, dy\, d\epsilon_z\, dz$$

$$+ \int_0^{\gamma_{yz}} \tau_{yz}\, dz\, dx\, d\gamma_{yz}\, dy + \int_0^{\gamma_{zx}} \tau_{zx}\, dx\, dy\, d\gamma_{zx}\, dz$$

$$+ \int_0^{\gamma_{xy}} \tau_{xy}\, dy\, dz\, d\gamma_{xy}\, dx$$

$$= \left(\int_0^{\epsilon_x} \sigma_x\, d\epsilon_x + \int_0^{\epsilon_y} \sigma_y\, d\epsilon_y + \int_0^{\epsilon_z} \sigma_z\, d\epsilon_z \right.$$

$$\left. + \int_0^{\gamma_{yz}} \tau_{yz}\, d\gamma_{yz} + \int_0^{\gamma_{zx}} \tau_{zx}\, d\gamma_{zx} + \int_0^{\gamma_{xy}} \tau_{xy}\, d\gamma_{xy} \right) dV$$

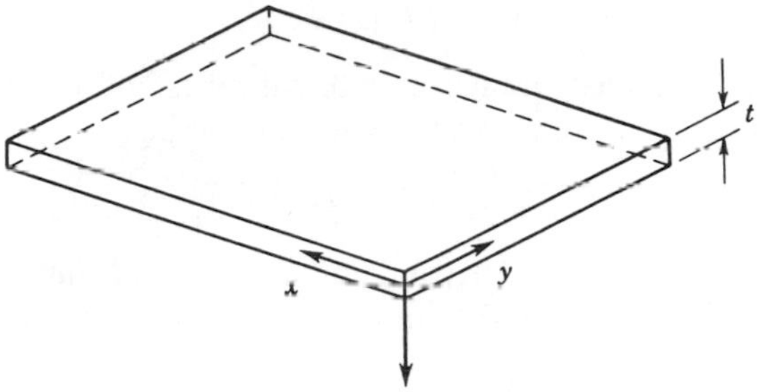

Fig. 2.1.6.   Thin plate.

When Hooke's law is valid and thus the stresses are proportional to the strains, we get

$$dU = \tfrac{1}{2}(\sigma_x\epsilon_x + \sigma_y\epsilon_y + \sigma_z\epsilon_z + \tau_{yz}\gamma_{yz} + \tau_{zx}\gamma_{zx} + \tau_{xy}\gamma_{xy})\, dV$$

Hence the total strain energy stored in a finite body is

$$U = \frac{1}{2} \int_V (\sigma_x\epsilon_x + \sigma_y\epsilon_y + \sigma_z\epsilon_z + \tau_{yz}\gamma_{yz} + \tau_{zx}\gamma_{zx} + \tau_{xy}\gamma_{xy})\, dV \qquad 2.1.16$$

where the integration has to be extended over the entire volume of the body. It should be remembered that in this equation ϵ_x depends on σ_y and σ_z as well as on σ_x and that similar statements can be made in regard to ϵ_y and ϵ_z.

For particular bodies and cases of loading Eq. 2.1.16 can be replaced by expressions more convenient in calculations by substituting displacement quantities for stress and strain. For instance, in a thin plate as shown in Fig. 2.1.6, the strain energy is

$$U = \frac{1}{24} \frac{Et^3}{1-v^2} \int\int \left\{ \left(\frac{\partial^2 w}{\partial x^2} + \frac{\partial^2 w}{\partial y^2} \right)^2 - 2(1-v) \left[\frac{\partial^2 w}{\partial x^2} \frac{\partial^2 w}{\partial y^2} - \left(\frac{\partial^2 w}{\partial x \, \partial y} \right)^2 \right] \right\} dx \, dy$$

2.1.17

where w is the deflection of any point on the median surface of the thin plate perpendicularly to the plate, and v is Poissons' ratio. Equation 2.1.17 is developed in textbooks on the theory of elasticity and is given here without proof, by way of example. It is analogous to Eq. 2.1.15 which expresses the strain energy in a bent beam in terms of the deflections.

Torsion of Bars of Arbitrary Section

The strain energy stored in a thin-walled circular tube subjected to a constant torque was given in Eq. 2.1.11. It is easy to prove that the formula

$$U = \tfrac{1}{2} T^2 L / G I_p$$

2.1.11a

also holds for thick-walled circular cylindrical tubes and for bars of solid circular section. For all other sections we may write

$$U = \tfrac{1}{2}(T^2 L / GC)$$

2.1.18

where the torsional rigidity GC depends on the shape of the cross section. For a solid rectangle b wide and h high

$$C = kb^3 h, \qquad b \leq h$$

2.1.19

and the value of the coefficient k depends on the ratio h/b. When this ratio is unity ($b = h$), k is 0.1406. As h/b increases, k decreases, and, for ratios exceeding 10, the value of k is 1/3 in good approximation. The same equations can be used for the different portions of a thin-walled built-up section, and the total torsional rigidity of the section is equal to the sum of the torsional rigidities of its elements.

As the angle of twist $d\phi$ between two sections dx apart is

$$d\phi = (T/GC) \, dx$$

2.1.20

Eq. 2.1.18 can also be written as

$$U = \tfrac{1}{2} GC \, (d\phi/dx)^2 L$$

or as

$$U = \frac{1}{2} GC \int_0^L \left(\frac{d\phi}{dx} \right)^2 dx$$

2.1.21

All the torsion formulas developed here are only valid for the Saint-Venant type of torsion in which the torque is constant along the bar and the sections of the bar are permitted to warp freely.   When the warping is restricted by geometric constraints such as rigid end fixation, or by variations in the applied torque along the bar, normal stresses arise in addition to the shear stresses, and they contribute an additional resistance to twisting.   This effect is completely absent from bars of circular cross section whose normal sections remain plane during torsion, but it is very important in thin-walled built-up and extruded sections whose resistance to Saint-Venant torsion is small because b in Eq. 2.1.19 is small. The strain energy associated with warping is

$$U = \frac{1}{2} E\Gamma \int_0^L \left(\frac{d^3\phi}{dx^3}\right)^2 dx \qquad\qquad 2.1.22$$

where ϕ is the angle of twist, x the distance along the axis of the bar, and $E\Gamma$ the warping rigidity.   For an I beam of uniform wall thickness t, height h, and total flange width b

$$\Gamma = \tfrac{1}{24} t b^3 h^2 \qquad\qquad 2.1.22a$$

These formulas are given here without proof because of their usefulness in applications.   For their derivation reference should be made to the publications listed in Appendix I.

2.2   Statement of the Principle of Virtual Displacements in Terms of Strain Energy

The Change in Internal Work during a Virtual Displacement

The curve in Fig. 2.2.1 represents the strain energy stored in the bar shown in Fig. 2.1.1 when it is subjected to a uniform normal stress in the axial direction.   If the strain energy $U(\epsilon_1)$ corresponding to a strain ϵ_1 is known, the strain energy $U(\epsilon_1 + \delta\epsilon)$, corresponding to the strain $\epsilon_1 + \delta\epsilon$, where $\delta\epsilon$ is a finite or infinitesimal increment of the strain, can be calculated from Taylor's theorem:

$$U(\epsilon_1 + \delta\epsilon) = U(\epsilon_1) + [dU(\epsilon_1)/d\epsilon]\delta\epsilon + (1/2!)[d^2U(\epsilon_1)/d\epsilon^2](\delta\epsilon)^2$$
$$+ (1/3!)[d^3U(\epsilon_1)/d\epsilon^3](\delta\epsilon)^3 + \cdots$$

The theorem is discussed in detail in textbooks on the calculus.   It means that in a first, rough approximation the value of the function U corresponding to $\epsilon_1 + \delta\epsilon$, written symbolically as $U(\epsilon_1 + \delta\epsilon)$, can be taken equal to its value corresponding to ϵ_1, denoted by the letters $U(\epsilon_1)$. A second and better approximation can be obtained if the second term of the Taylor expansion is added to the first.   This is the product of the slope of the tangent to the curve of Fig. 2.2.1 at point A, namely, $dU(\epsilon_1)/d\epsilon$, and the increment $\delta\epsilon$ of the independent variable ϵ. In other words, the actual ordinate FB of the U curve is replaced first by the height FC and in the second approximation by the height FD. The Taylor expansion states how the approximation can be further improved. We can add the rate of change of the slope $d^2U(\epsilon_1)/d\epsilon^2$ multiplied by the square of the increment in strain and by the factor $1/2$, then the rate of change of this quantity multiplied by the cube of the increment in strain and by $(1/3!) = (1/6)$, and continue further until the ordinate obtained is considered to be close enongh to the actual value FB of the U function. In

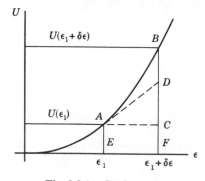

Fig. 2.2.1.   Strain energy.

general, the actual value is reached exactly if an indefinitely large number of terms is taken into account for an analytic function. Sometimes a finite number of terms suffices, namely, when the Taylor expansion is finite, that is, when all the derivatives of the function vanish beyond that of a certain finite order.

The increment δU of the strain energy during the displacement corresponding to the increment of strain $\delta\epsilon$ is

$$\delta U = U(\epsilon_1 + \delta\epsilon) - U(\epsilon_1) = \delta\epsilon\{[dU(\epsilon_1)/d\epsilon] + (1/2!)[d^2U(\epsilon_1)/d\epsilon^2]\delta\epsilon$$
$$+ (1/3!)[d^3U(\epsilon_1)/d\epsilon^3](\delta\epsilon)^2 + \cdots\} \quad (a)$$

The change in the strain energy is, by definition, equal to the work done by the external loads

$$\delta U = \delta W_e \qquad (b)$$

and consequently equal to (-1) times the work done by the stresses in the interior of the bar which oppose the deformations:

$$\delta U = -\delta W_i \qquad (c)$$

The external work corresponding to a variation $\delta\epsilon$ of the strain can also be obtained from Eq. 2.1.1:

$$\delta W_e = V\sigma_1(1 + k)\,\delta\epsilon \qquad (d)$$

where V is the volume of the bar.   Equations a, c, and d yield

$$-\delta W_i = V\sigma_1(1 + k)\delta\epsilon = \delta\epsilon\{[dU(\epsilon_1)/d\epsilon] + (1/2!)[d^2U(\epsilon_1)/d\epsilon^2]\delta\epsilon$$
$$+ (1/3!)[d^3U(\epsilon_1)/d\epsilon^3](\delta\epsilon)^2 + \cdots\} \quad (e)$$

As the stress-strain curves of structural materials do not have vertical tangents, k becomes negligibly small as compared to unity if the variation $\delta\epsilon$ in the strain is made sufficiently small.   At the same time inside the braces the terms multiplied by $\delta\epsilon$ or its powers can be disregarded because the term without $\delta\epsilon$ is much larger.   (The quantity $dU(\epsilon_1)/d\epsilon$ cannot vanish when the strain ϵ_1 is not zero.)   Thus Eq. e reduces to

$$-\delta W_i = V\sigma_1\delta\epsilon = [dU(\epsilon_1)/d\epsilon]\delta\epsilon \qquad 2.2.1$$

But Eq. e becomes Eq. 2.2.1 even for finite values of $\delta\epsilon$ if $L\,\delta\epsilon$ is a virtual displacement of the free end of the bar in Fig. 2.1.1 corresponding to a uniform variation $\delta\epsilon$ in the strain throughout the bar, and the change in the internal work in the bar during this virtual displacement is sought. It was discussed at length in Part 1 that the stresses (and the forces) must be considered constant when the principle of virtual displacements is applied.   With unchanging stresses, σ has the same value before and after the increment $\delta\epsilon$ in Fig. 2.1.1 and k is equal to zero.   Under these conditions the strain energy in Fig. 2.2.1 increases from point A at the constant rate represented by the tangent AD and not along the curve AB.   Hence the change in the strain energy is the expression in the right hand member of Eq. 2.2.1 and the other terms of the Taylor series of Eq. e are absent. These terms contain $\delta\epsilon$ or its powers and represent the effect of the change in stress during the displacement upon the strain energy.   As the stresses remain constant during a virtual displacement, these terms must be disregarded.

It can be concluded therefore that the change in the internal work during a virtual displacement can be calculated as the product of (-1) by the virtual change in the strain and by the first derivative of the strain energy U with respect to the strain.   Terms containing higher powers of $\delta\epsilon$, which are multiplied by derivatives higher than the first, must be disregarded.   Of course all the terms of the Taylor series are needed when the actual change in the strain energy is sought corresponding to a finite real displacement, but this quantity is of no interest in the statement of the principle of virtual displacements.

In the derivations just presented, the work done by the external force could be calculated with equal ease for infinitesimal or finite displacements because of the simple nature of the body and of the force acting on it.   In more complicated systems the simplicity of the analysis can be retained only if finite rotations of the body, or parts of it, are excluded from the displacement patterns.   For this reason virtual displacements are generally assumed to be very small.

General State of Stress and Strain

When the infinitesimal parallelepiped of Fig. 2.1.5 is subjected to the most general state of stress, the normal stress in any one direction is a function of the normal strains in all three directions.   Nevertheless the internal work done during a fictitious displacement corresponding to a variation $\delta\epsilon_x$ can be calculated as the negative of the product of $\delta\epsilon_x$ by the first partial derivative of the strain energy with respect to ϵ_x as long as the stresses are assumed unchanged by the fictitious displacement.

If such a general state of stress and strain is investigated, Fig. 2.2.1 can be thought of as representing the change in the strain energy U stored in the parallelepiped as affected by one single strain component ϵ while all the other strain components have fixed values.   Physically such a situation can be realized by attaching to rigid walls those four faces of the infinitesimal parallelepiped of Fig. 2.1.5 which are not perpendicular to the x direction.   When subsequently the distance between the remaining two free faces, namely, those perpendicular to x, is changed and the corresponding change in U is plotted against the independent variable ϵ_x, a curve similar to that in Fig. 2.2.1 is obtained.   Of course the ϵ axis should be shifted downward since the strain energy is not equal to zero when ϵ_x is zero, because the other strain components have fixed, and in general non-zero, values.   When the variation in U corresponding to $\delta\epsilon_x$ is expressed, an equation similar to Eq. a is obtained

$$\delta U = \delta\epsilon_x[(\partial U/\partial\epsilon_x) + (1/2!)(\partial^2 U/\partial\epsilon_x^2)\delta\epsilon_x + (1/3!)(\partial^3 U/\partial\epsilon_x^3)(\delta\epsilon_x)^2 + \cdots]$$

$$(a')$$

where U is a function of ϵ_x, ϵ_y, ϵ_z, γ_{yz}, γ_{zx}, and γ_{xy}.

During this variation in the strain only the stresses σ_x do work.   The stresses σ_y and σ_z are changed, but they cannot do work because the faces on which they act are held fixed.   The variation of σ_x with ϵ_x can again be represented by a curve similar to, but not identical with, that in Fig. 2.1.1b. The differences between the old and new curves are due to the differences in the boundary conditions.   In the case now under consideration four faces of the cube are fixed, while the elemental parallelepiped cut out of the tension bar of Fig. 2.1.1a is free to contract in the y and z directions.

Again the work done by the external forces can be represented by the equation

$$\delta W_e = V\sigma_{x_1}(1 + k)\delta\epsilon_x \qquad (d')$$

where V is the volume of the infinitesimal parallelepiped. The value of k is not the same as that in the earlier discussion of the uniaxial stress condition, since the new curve and that shown in Fig. 2.1.1b are not identical. When the work and strain-energy expressions are equated to each other, an equation analogous to Eq. e is obtained which reduces to

$$-\delta W_i = V\sigma_{x_1}\delta\epsilon_x = (\partial U/\partial\epsilon_x)\,\delta\epsilon_x \qquad 2.2.1a$$

where $\delta\epsilon_x$ is the change in strain corresponding to a virtual displacement.

This equation can be interpreted as follows: The work done by the internal forces prevailing when $\epsilon_x = \epsilon_{x_1}$, $\epsilon_y = \epsilon_{y_1}$, $\cdots$, $\gamma_{xy} = \gamma_{xy_1}$, during a virtual displacement corresponding to $\delta\epsilon_x$, can be calculated as the negative of the product of $\delta\epsilon_x$ by the first partial derivative of U with respect to ϵ_x.

Displacements Defined by Parameters

Usually the variation of the state of strain in a finite body is undertaken by varying some distortion parameter q rather than one rectangular component of strain. The parameter q can be suitably chosen to be the axial displacement of plane sections of the bar subjected to uniform tension in Fig. 2.1.1, or the relative angle of twist of two plane sections of the twisted tube in Fig. 2.1.2. For the beam of Fig. 2.1.3 the deflection η and for the plate in Fig. 2.1.6 the deflection w are parameters suitable for the variation of the state of strain.

When these deflection parameters are varied, several or all the strain components are likely to vary simultaneously in the elastic body. When the work done by the unchanged forces is sought, the contributions of the individual strain components to the total work $\delta W_e = \delta U = -\delta W_i$ are independent and can be added up according to the equation

$$\delta U = (dU/dq)\delta q = [(\partial U/\partial\epsilon_x)(\partial\epsilon_x/\partial q) + (\partial U/\partial\epsilon_y)(\partial\epsilon_y/\partial q) + \cdots$$
$$+ (\partial U/\partial\gamma_{xy})(\partial\gamma_{xy}/\partial q)]\delta q \qquad (f)$$

When the variation in the strain energy is required for a finite body in which the stress and the strain vary from point to point, the expressions are written first for the infinitesimal parallelepiped and are integrated subsequently over the entire volume of the body.

Because formulas have been developed that express the strain energy U stored in elastic bodies of various shapes subjected to different types of loading, the first derivative of the strain energy can be easily calculated in most cases. Consequently a convenient procedure is now available for the calculation of the internal work δW_i done during an imagined, fictitious variation δq of the state of strain. The procedure can be expressed by the equation

$$\delta W_i = -\delta U = -(\partial U/\partial q)\delta q \qquad 2.2.2$$

or by the statement:

> The work done during an imagined variation δq of the state of strain by the internal stresses prevailing in an elastic body is equal to the negative of the product of δq and the first derivative of the strain energy with respect to q.   **Theorem 11**

The Minimum of the Total Potential

It was shown in Arts. 1.2 to 1.4 that the principle of virtual displacements establishes the vanishing of the sum of the external and internal work as the necessary and sufficient condition of equilibrium of a system of mass points. Equation 2.2.2 gives the rule according to which the internal work δW_i can be calculated for a perfectly elastic body containing an indefinitely large number of mass points. The external work δW_e is $\sum P \, \delta p$ if the external loads P are all concentrated forces and the displacements δp are those of their points of attack in the directions of the forces corresponding to the variation of strain undertaken. When the elastic body is under the action of distributed loads, the external work can be calculated by integration rather than by summation. If for the sake of simplicity the principle of virtual displacements is written for the case of concentrated external loads, it becomes

$$\delta W_e + \delta W_i = \sum P \, \delta p - \delta U = 0 \qquad 2.2.3$$

Equation 2.2.3 is a necessary and sufficient condition of equilibrium, provided the variation sign δ is understood to imply any arbitrary displacement. In agreement with Art. 1.4 it suffices to investigate only those variations of the state of strain that do not infringe on the geometric restraints imposed on the elastic body. Such variations are denoted as virtual displacements. It may be stated therefore that

> A necessary and sufficient condition of the equilibrium of an elastic body is that the work done by the external loads less the change in the strain energy vanish for every virtual displacement.   **Theorem 12**

The change in the strain energy has to be calculated on the assumption that the forces remain unchanged during the variation of the state of strain. Hence, in accordance with Theorem 11, in the Taylor expansion of the strain-energy function, only the first derivative has to be considered.

It is convenient to define a potential V of the external forces in such a manner that the work done by the forces during a variation of the state of deformations be equal to $-\delta V$.   In the form of an equation

$$-\delta V = \sum P\,\delta p \qquad\qquad 2.2.4$$

when all the external loads are concentrated forces.   The summation must be replaced by a surface integral when the external loads are distributed over part or all of the surface of the elastic body.   Substitution of $-\delta V$ in Eq. 2.2.3 yields

$$-\delta U - \delta V = 0$$

or

$$\delta(U + V) = 0 \qquad\qquad 2.2.5$$

The expression $U + V$ is known as the total potential of the system. Consequently Eq. 2.2.5 can be stated as follows:

> The first-order change in the total potential must vanish **Theorem 13**
> for every virtual displacement when an elastic body is in
> equilibrium.

By first-order change is meant one in which only those terms are considered that contain the first power of the displacement quantities.   Terms containing higher powers of the virtual displacements are disregarded.

Theorem 13 is a concise statement of the principle of virtual displacements for a system composed of an elastic body and external loads.

Let it now be assumed that the total potential $U + V$ is a function of one single displacement parameter q.   Then the elastic body is in equilibrium if

$$\delta(U + V) = [d(U + V)/dq]\delta q = 0$$

that is, if

$$d(U + V)/dq = 0 \qquad\qquad (g)$$

since δq is not zero by assumption.   On the other hand, if the function $U + V$ is plotted against the independent variable q, Eq. (g) requires that the curve representing $U + V$ have a horizontal tangent.   This is so when $U + V$ is a maximum or a minimum, or when it has an inflection point with a horizontal tangent as shown in Fig. 2.2.2.   In each case the function is said to have a stationary value.

When $U + V$ is a function of two independent displacement parameters q_1 and q_2, it can be plotted as a surface over the plane defined by two mutually perpendicular axes representing q_1 and q_2.   In this case the requirement of the principle of virtual displacements is expressed by two equations,

$$\partial(U + V)/\partial q_1 = 0$$
$$\partial(U + V)/\partial q_2 = 0 \qquad\qquad (h)$$

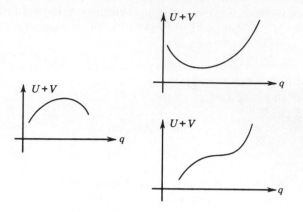

Fig. 2.2.2.   The stationary value of the total potential;
single degree of freedom.

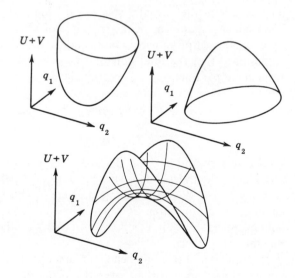

Fig. 2.2.3.   The stationary value of the total potential;
two degrees of freedom.

stating that the variation of the total potential must vanish for each of the two possible displacements.   Equations (h) define either a maximum or a minimum of the surface, or a saddle point with a horizontal tangent plane as shown in Fig. 2.2.3.   Again it can be said that the total potential has a stationary value.

This argument can be extended to cases in which the number of independent variables is 3, 4, or 5, or even when it is indefinitely large.   Although this is impossible to visualize, it is convenient to keep the phraseology and say that for equilibrium the total potential surface must have a horizontal tangent plane when it is plotted in an $(n + 1)$-dimensional *space* over an *n*-dimensional *plane*, provided the number of independent variables is *n*.

Theorem 13 can now be restated as follows:

The total potential has a stationary value when an   **Corollary† 1**
elastic body is in equilibrium.

It will be shown in Part 3 that the stationary value always corresponds to a minimum when the equilibrium is stable.   In investigations of stable equilibrium patterns Corollary 1 can therefore be stated in the form:

The total potential is a minimum when an elastic body   **Corollary 2**
is in equilibrium.

This minimal principle of the theory of elasticity is of great importance in structural analysis.   Methods of calculating stresses by its use are known as strain energy methods.   It is of interest to note that the validity of Hooke's law and of the principle of superpositions was not assumed in its derivation.

The Total Potential Surface When Hooke's Law Holds

When Hooke's law holds, the strain energy is a quadratic function of the strain, as was shown in Art. 2.1.   The strain in turn is a linear function of the displacements as long as the displacements are small.   Hence the strain energy is a quadratic function of the displacement parameters, and so is the total potential because the potential of the applied loads only contributes linear terms.   If the total potential is plotted against the displacement coordinates, a quadric surface is obtained.   The quadric cannot be closed because the strain energy must increase beyond all limits when the strain increases indefinitely.   Thus the quadric must be a paraboloid, a hyperboloid, or in the degenerate case a cylinder.   It cannot be a cone because a tangent to the surface must exist at every point of the surface.   The cylinder corresponds to the critical condition of stability, as will be

† A corollary is a consequence of a proved proposition.

shown in Part 3; it will not be discussed further in Part 2. The paraboloid and the hyperboloid have only a single point with a horizontal tangent plane, and that point represents the equilibrium conditions in consequence of the principle of the stationary value of the total potential. This means that under given loads there is only one set of displacements that satisfies all the conditions of equilibrium and continuity. This conclusion is known as Kirchhoff's uniqueness theorem.

Comparison with the Principle of the Conservation of Energy

It is important to realize that the principle of the minimum of the total potential has nothing in common with the law of the conservation of energy. The latter can be stated as

$$U - W_e = 0 \qquad\qquad 2.2.6$$

where U is again the strain energy stored in the system, but W_e is the actual work done by all the applied loads during the loading process, and not the virtual work. Consequently

$$W_e = \tfrac{1}{2}\sum Pp$$

when the loads are concentrated and force and displacement are linearly related while, in the principle of virtual displacements, as well as in its energy form, the principle of the minimum of the total potential,

$$\delta W_e = \sum P \, \delta p$$

The two expressions differ essentially by the factor 2. Moreover the minimal principle does not contain any statement about the total potential $(U + V)$ itself but only about its first derivative with respect to virtual displacements, while the conservation of energy law is concerned only with the energy and not with its derivatives. The two independent principles expressed in Eqs. 2.2.5 and 2.2.6 can be used to advantage in the solution of entirely different types of problems.

2.3   Application of the Minimal Principle to a System Having Two Degrees of Freedom of Motion

The principle of the minimum of the total potential, which is the strain-energy version of the principle of virtual displacements, was developed for use with systems having many, and in particular an infinite number of, degrees of freedom of motion. When the number of degrees of freedom is small, calculation of the equilibrium can be carried out with the aid of more elementary procedures based on the principle of virtual displacements, as was shown in Part 1. Although the minimal principle is too refined a tool for the analysis of a two-degrees-of-freedom system, it will be applied here in order to give some insight into the physical concepts involved in the use of the strain energy methods.

The system to be investigated consists of three straight bars attached to one another and to a rigid wall by means of ideal pin joints. The resulting ideal plane framework shown in Fig. 2.3.1 is statically indeterminate. The forces in the bars could be calculated easily by the method presented in Art. 1.8 but this will not be done here. It will be assumed that the apex of the system of bars, where the external load P is applied, is displaced a distance u in the x direction and a distance v in the y direction. The strain energy stored in the three bars will be calculated for the case when the

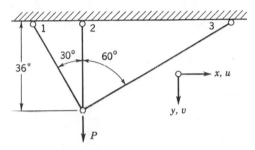

Fig. 2.3.1.   Three-bar system.

cross-sectional area of each bar is 0.1 sq in., the modulus of the material is 29×10^6 psi, and the load P is 1500 lb. In addition, the potential of this load will be determined and added to the expression obtained for the strain energy. The sum of the two is the total potential of the system.

Naturally the actual magnitudes of the displacements u and v are not known. They can be calculated from the minimal principle, which states that the change in the value of the total potential is zero during any virtual displacement when the system is in equilibrium. In other words, for arbitrary values of u and v an arbitrary variation δu and δv of the displacement quantities will lead to changes in the value of the total potential. When, however, u and v happen to be the correct displacement quantities corresponding to the load $P = 1500$ lb, the change in the total potential must vanish for any arbitrary choice of the variations δu and δv. This unique property of the total potential at the true values of the displacement quantities corresponding to equilibrium permits the calculation of the displacements. When the displacements are known, the forces in the individual bars are easy to obtain.

The first step in the calculations is the determination of the strain energy as a function of the elastic displacements u and v. If the free end point of the bar shown in Fig. 2.3.2 is displaced a distance u in the x direction and a distance v in the y direction, the x component of its length in the extended state will be $L_x + u$ and the y component of its extended length $L_y + v$. The total extended length can be calculated from the Pythagorean theorem:

$$(L + \Delta L)^2 = (L_x + u)^2 + (L_y + v)^2 \qquad 2.3.1$$

After squaring out, we obtain

$$L^2 + 2L\,\Delta L + \Delta L^2 = L_x{}^2 + 2L_x u + u^2 + L_y{}^2 + 2L_y v + v^2 \qquad 2.3.2$$

Division by L^2 yields

$$1 + 2(\Delta L/L) + (\Delta L/L)^2 = (L_x/L)^2 + 2(L_x/L)(u/L) + (u/L)^2$$
$$+ (L_y/L)^2 + 2(L_y/L)(v/L) + (v/L)^2 \qquad 2.3.3$$

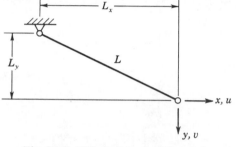

Fig. 2.3.2.   Displacements of free end.

Although in particular cases L_x or L_y might vanish, at least one of the two components of the length is always of the same order of magnitude as the length itself.   As a matter of fact either L_x/L or L_y/L must have a value ranging between 0.707 and 1.   On the other hand the quantities u/L and v/L, which might be termed strain components, cannot be much greater than the maximum elastic strain the material is capable of sustaining. For a steel bar the maximum working stress is of the order of magnitude of 30,000 psi.   With an elastic modulus of 29,000,000 psi Hooke's law gives a strain of about 0.001.   Hence the strain components cannot be much larger than this approximate value.   This consideration shows that the squares of the strain components $(u/L)^2$ and $(v/L)^2$ can be neglected in Eq. 2.3.3.   Similarly the square of $\Delta L/L$, where $\Delta L/L$ is the strain in the bar, can be omitted.   The values of these squares are about one millionth of the values of the other squared expressions in the equation, and about one thousandth of the two expressions which are multiplied by 2.   Since Eq. 2.3.2 and Eq. 2.3.3 are closely related, u^2, v^2, and ΔL^2 can be omitted from the former.   The resulting equation can be simplified if it is observed that from the Pythagorean theorem

$$L^2 = L_x{}^2 + L_y{}^2 \qquad 2.3.4$$

Subtraction of Eq. 2.3.4 from Eq. 2.3.2 yields, after omission of the terms which were found small,

$$2L\,\Delta L = 2L_x u + 2L_y v \qquad\qquad 2.3.5$$

Division by $2L^2$ gives the strain in the bar as a function of the displacements u and v:

$$\Delta L/L = (L_x/L)(u/L) + (L_y/L)(v/L) \qquad\qquad 2.3.6$$

It can be seen that the strain in the bar is the sum of the strain component u/L multiplied by the ratio of the x component of the length of the bar to the total length, and of the strain component v/L multiplied by the ratio of the y component of the length to the total length. With the aid of this equation the strains in the three bars can be easily determined if one observes that L_x and L_y are considered positive when a vector pointing from the fixed end toward the moving end of the bar has positive x and y components.

According to Fig. 2.3.1, the length of bar 2 is 36 in., that of bar 1 is $36/\cos 30° = 41.6$ in., and that of bar 3 is $36/\cos 60° = 72$ in. Consequently the strain in bar 1 is

$$\epsilon_1 = \tfrac{1}{2}(u/41.6) + 0.886(v/41.6) \qquad\qquad 2.3.7$$

The strains in bars 2 and 3 can be found in a similar manner:

$$\epsilon_2 = v/36$$
$$\qquad\qquad 2.3.8$$
$$\epsilon_3 = -0.866(u/72) + \tfrac{1}{2}(v/72)$$

Since the strains and the stresses are constant over the length of each individual bar, the strain energy U stored in each of them can be calculated from the formula

$$U = \tfrac{1}{6}E\epsilon^2 LA \qquad\qquad 2.3.9$$

Here L and A are the length and the cross-sectional area of the individual bar. Consequently the strain-energy quantities become

$$U_1 = \tfrac{1}{2}\,29 \times 10^6 \times 41.6 \times 10^{-1}\epsilon_1{}^2 = 6.03 \times 10^7\epsilon_1{}^2$$
$$U_2 = \tfrac{1}{2}\,29 \times 10^6 \times 36 \times 10^{-1}\epsilon_2{}^2 = 5.21 \times 10^7\epsilon_2{}^2 \qquad\qquad 2.3.10$$
$$U_3 = \tfrac{1}{2}\,29 \times 10^6 \times 72 \times 10^{-1}\epsilon_3{}^2 = 10.42 \times 10^7\epsilon_3{}^2$$

The square of the strain quantities listed in Eqs. 2.3.7 and 2.3.8 are

$$\epsilon_1{}^2 = 1.45 \times 10^{-4}u^2 + 5.1 \times 10^{-4}uv + 4.34 \times 10^{-4}v^2$$
$$\epsilon_2{}^2 = 7.72 \times 10^{-4}v^2 \qquad\qquad 2.3.11$$
$$\epsilon_3{}^2 = 1.45 \times 10^{-4}u^2 - 1.67 \times 10^{-4}uv + 0.47 \times 10^{-4}v^2$$

Substitution of the squares of the strain quantities in Eq. 2.3.10 and addition give

$$U = U_1 + U_2 + U_3 = 23{,}830u^2 + 13{,}400uv + 71{,}300v^2 \qquad 2.3.12$$

The potential V of the external load is defined as -1 times the work done by the load during the deformations. Since the load P is perpendicular to the displacement u, work is done only during displacements in the y direction. For this reason the potential is

$$V = -Pv \qquad 2.3.13$$

Addition of the quantities contained in Eqs. 2.3.12 and 2.3.13 results in an expression for the total potential of the system

$$U + V = 23{,}830u^2 + 13{,}400uv + 71{,}300v^2 - Pv \qquad 2.3.14$$

According to the minimal principle, the variation of the total potential vanishes when the system is in equilibrium. During the variation each displacement u is increased by an additional displacement δu, and each displacement v by the quantity δv. Thus u^2 becomes $(u + \delta u)^2$ which is equal to $u^2 + 2u\,\delta u + (\delta u)^2$. In the application of the minimal principle the external as well as the internal forces are considered constant during the variation of the state of deformation. It was shown earlier that for this reason expressions containing squares or products of the variations must be disregarded. Consequently u^2 becomes simply $u^2 + 2u\,\delta u$ when the displacement u is increased to $u + \delta u$. The increase in the quantity u^2 during the variation of the state of deformation therefore amounts to $2u\,\delta u$.

Naturally a similar argument would show that the increment in v^2 is $2v\,\delta v$ when the state of deformation is varied. The middle term in the right-hand member of Eq. 2.3.12 becomes $(u + \delta u)(v + \delta v) = uv + u\,\delta v + v\,\delta u + \delta u\,\delta v$. The last term in this expression is a product of two increments of the displacement quantities. It must be disregarded when the principle of virtual displacements is applied to the problem. The increment of the middle term is therefore $u\,\delta v + v\,\delta u$.

The change in the total potential of the system can now be written without difficulty:

$$\delta(U + V) = 47{,}660u\,\delta u + 13{,}400(u\,\delta v + v\,\delta u) + 142{,}600v\,\delta v - P\,\delta v = 0$$

$$2.3.15$$

The expression can be rearranged in the following manner:

$$\delta(U + V) = (47{,}600u + 13{,}400v)\delta u$$

$$+ (13{,}400u + 142{,}600v - P)\delta v = 0 \qquad 2.3.16$$

The minimal principle requires that the change in the total potential vanish for any arbitrary variation in the displacement quantities; consequently Eq. 2.3.16 must be valid for any arbitrary choice of δu and δv. This is possible only if the two expressions in parentheses vanish individually.   The proof of the statement can be given without difficulty.

One possible choice for the variations is obviously

$$\delta u = 1, \qquad \delta v = 0 \qquad\qquad 2.3.17$$

Because Eq. 2.3.16 must hold for any arbitrary choice of the variations in the displacement quantities, it must naturally hold for the particular choice represented by Eq. 2.3.17.   This means, however, that the first expression in parentheses must vanish:

$$47,660u + 13,400v = 0 \qquad\qquad 2.3.18$$

A second possible choice of the variations in the deformation quantities is represented by

$$\delta u = 0, \qquad \delta v = 1 \qquad\qquad 2.3.19$$

For this choice again the right-hand side of Eq. 2.3.16 must vanish. Consequently the second expression in parentheses is also zero:

$$13,400u + 142,600v - P = 0 \qquad\qquad 2.3.20$$

In this manner the minimal principle leads to two equations of equilibrium, namely, Eqs. 2.3.18 and 2.3.20, which must be satisfied simultaneously when the system consisting of the three bars and the load P is in equilibrium.   The two equations can be readily solved for the displacements u and v:

$$u = -0.00304, \qquad v = 0.0108 \qquad\qquad 2.3.21$$

when P is 1500 lb.

Under a load of 1500 lb the joint where the three bars are connected is therefore displaced 0.0108 in. downward and 0.00304 in. to the left.   These displacements can be substituted in Eqs. 2.3.7 and 2.3.8.   The following values are then obtained for the strains in the individual bars:

$$\epsilon_1 = 188.5 \times 10^{-6}$$
$$\epsilon_2 = 300 \times 10^{-6} \qquad\qquad 2.3.22$$
$$\epsilon_3 = 111.6 \times 10^{-6}$$

Multiplication of these values by the cross-sectional area and by Young's modulus yields the forces in the bars:

$$F_1 = 547 \text{ lb}$$
$$F_2 = 870 \text{ lb} \qquad\qquad 2.3.23$$
$$F_3 = 324 \text{ lb}$$

It is easy to check whether these forces and the applied load are in equilibrium.   When the sums of the x and y components of the forces are compared with the external load, deviations from perfect equilibrium are found which amount to less than one-half per cent of the applied load. This can be considered as a satisfactory check when all the computations are carried out with a slide rule.

Equation 2.3.14 can be used to plot values of the total potential $U + V$ as a function of the two displacement coordinates u and v.   The plot can be conveniently arranged in the form of a contour diagram in which u and v denote directions along two Cartesian coordinate axes.   The curves $U + V = $ constant appear then as constant elevation lines in a topographic map of a fictitious $U + V$ valley.   Points of such curves can be obtained from Eq. 2.3.14 by assuming fixed values for $U + V$ and for v, and solving the resulting quadratic equation for u.   The solution of the quadratic can be obviated if u and v are assumed, the value of $U + V$ calculated, and the constant total potential curves drawn by interpolation. Another procedure makes use of transformations of the coordinate axes.

First, a new set of coordinates u' and v' is introduced.   The new coordinates have the same directions as the old ones, but their origin is the point corresponding to the equilibrium of the system.   With the values given in Eq. 2.3.21

$$u' = u + 0.00304$$
$$v' = v - 0.0108$$

2.3.24

From Eqs. 2.3.24, u and v can be expressed as

$$u = u' - 0.00304$$
$$v = v' + 0.0108$$

2.3.25

Substitution of these expressions in Eq. 2.3.14 yields

$$U + V = 23{,}830u'^2 + 13{,}400u'v' + 71{,}300v'^2 - 8.11 \qquad 2.3.26$$

provided the load P is assumed to be 1500 lb.

This expression can be simplified further by introducing a rotation of the system of coordinates in such a manner as to eliminate the term containing $u'v'$.   It can be seen from Fig. 2.3.3 that the coordinates u' and v' can be expressed in terms of the coordinates u'' and v'' in the following way:

$$u' = u'' \cos \alpha - v'' \sin \alpha$$
$$v' = v'' \cos \alpha + u'' \sin \alpha$$

2.3.27

Substitution of these expressions in Eq. 2.3.26 yields the following new equation for the total potential of the system:

$$U + V = (23{,}830 \cos^2 \alpha + 13{,}400 \cos \alpha \sin \alpha + 71{,}300 \sin^2 \alpha)u''^2$$
$$+ (-23{,}830 \sin 2\alpha + 13{,}400 \cos 2\alpha + 71{,}300 \sin 2\alpha)u''v''$$
$$+ (23{,}830 \sin^2 \alpha - 13{,}400 \cos \alpha \sin \alpha + 71{,}300 \cos^2 \alpha)v''^2 - 8.11$$
$$\qquad\qquad 2.3.28$$

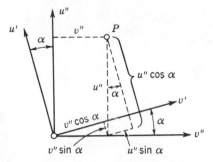

Fig. 2.3.3. Rotation of coordinate system.

The term containing $u''v''$ can be made to vanish if the angle of rotation α is so chosen that

$$-23{,}830 \sin 2\alpha + 13{,}400 \cos 2\alpha + 71{,}300 \sin 2\alpha = 0$$

This equation is equivalent to the requirement that

$$\tan 2\alpha = \frac{-13{,}400}{47{,}470} = -0.282 \qquad\qquad 2.3.29$$

Hence

$$2\alpha = -15°\,45' \qquad\qquad 2.3.30$$

With $\alpha = -7°\,53'$ and thus $\sin \alpha = -0.137$ and $\cos \alpha = 0.99$, substitutions in Eq. 2.3.28 yield the final expression for the total potential:

$$U + V = 22{,}730u''^2 + 72{,}160v''^2 - 8.11 \qquad\qquad 2.3.31$$

The curves $U + V = $ constant are obviously ellipses with the directions of their major and minor axes coinciding with the axes of the coordinates u'' and v''. The lengths of the semi-axes of the ellipses can be easily calculated and laid off along the coordinate axes, and the ellipses can be drawn according to the rules of plane geometry. As an example the lengths of the semi-axes are calculated for the case when $U + V = -2$ in.-lb. Substitution yields

$$22{,}750u''^2 + 72{,}160v''^2 = 6.11 \qquad\qquad 2.3.32$$

Division by 6.11 gives

$$3730u''^2 + 11{,}800v''^2 = 1 \qquad\qquad 2.3.33$$

This equation can also be written in the form

$$\frac{u''^2}{(\pm 0.0163)^2} + \frac{v''^2}{(\pm 0.0092)^2} = 1 \qquad\qquad 2.3.34$$

from which the end points of the semi-axes obviously have the coordinates

$$x'' = \pm 0.0163, \qquad y'' = 0$$
$$x'' = 0 \qquad\qquad y'' = \pm 0.0092 \qquad\qquad 2.3.35$$

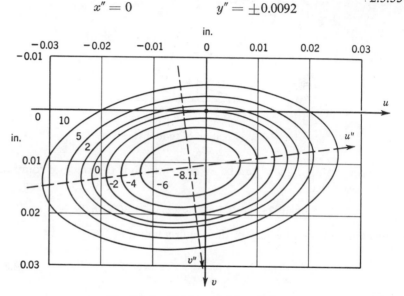

Fig. 2.3.4.   Total potential surface.

The contour map of the $U + V$ surface is shown in Fig. 2.3.4.   It can be seen that the surface represents a valley with its deepest point at $u = -0.00304$ in. and $v = 0.0108$ in.   The mountain sides become steeper as they rise from the bottom of the valley.   Plane sections through the surface perpendicular to the u or v axes are parabolas; this can be ascertained from Eq. 2.3.14 if either u or v is set equal to a constant.   The surface is therefore a paraboloid.

In agreement with the minimal principle, equilibrium corresponds to the minimum value of the total potential.   The change in the total potential during a virtual displacement must be calculated on the basis of the forces acting before the virtual displacements are undertaken.   It was shown in Art. 2.2 that for this reason terms containing the products or the squares of the variations in the displacements have to be disregarded. The geometric meaning of this requirement is that during a virtual displacement the change in the total potential takes place along the tangents of the total potential surface. Naturally this change vanishes for any virtual displacement when the tangent plane is horizontal.

It is of interest to note that the line $U + V = 0$ passes through the origin of the system of coordinates u, v. This must be so because the total potential is calculated from the undeformed position, and thus $U + V = 0$ when $u = 0$ and $v = 0$. This undeformed position, however, does not correspond to equilibrium if the external load is not zero. The origin of the true displacement coordinates u, v is not the minimum point of the surface.

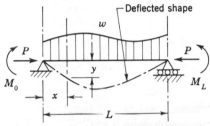

Fig. 2.4.1.   Beam column.

2.4   Derivation of the Beam-Column Differential Equation by the Minimal Principle

As a second example of the use of the minimal principle the differential equation governing the deflections of a beam column will be derived. Of course, the differential equation is known and can be developed on the basis of other, and even simpler, considerations. However, the present example shows how powerful a tool the minimal principle is in the analysis of elastic structures, demonstrates the technique employed in the solution of problems of equilibrium, and incidentally yields results that will be of use in later articles.

A beam column is a beam, as shown in Fig. 2.4.1, on which compressive end loads P act simultaneously with the transverse loading. In the problem to be analyzed, the beam column is simply supported, and its transverse loading consists of two concentrated end moments M_0 and M_L and a variable distributed load w. Forces, moments, and deflections are counted positive, as shown in the figure.

In the analysis it is assumed that the principle of the minimum of the total potential and the formula expressing the strain energy stored in a bent beam are the only known facts in the field of mechanics. They suffice for the derivation of the differential equation of the beam column.

The following procedure will be used in the investigation of the equilibrium of the beam column: A deflected shape $y = f(x)$ is assumed (see Fig. 2.4.2) without any further specification of the function. The actual function $f(x)$ differs from all other functions that might be chosen in that it makes the change in the total potential $(U + V)$ zero for any arbitrary variation δy of the deflections. In other words, if one happens to find the correct function $y = f(x)$ and increases the deflection y by an amount $\delta y = g(x)$ which may vary arbitrarily along the span, the change $\delta(U + V)$

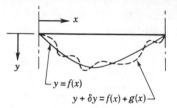

Fig. 2.4.2. Variation of deflected shape.

in the total potential vanishes whatever the choice of δy. On the other hand, if the function $y = f(x)$ is not the correct function, it is always possible to find variations δy which make $\delta(U + V)$ different from zero.

This procedure makes use of the knowledge that the average compressive strain is constant along the beam column under the action of the prescribed end loads P as long as the deflections y are small. Therefore it is not necessary to vary this state of constant average strain when the conditions of equilibrium are established by the minimal principle. It is permissible then to consider the uniformly compressed state as the zero level from which the strain energy is counted. The validity of these statements will be re-examined at the end of this article for the benefit of those who wish to have a more compelling proof than the reasoning just given.

In agreement with Eq. 2.1.15 the strain energy of bending is

$$U = \frac{1}{2} \int_0^L EI \left(\frac{d^2y}{dx^2}\right)^2 dx \qquad\qquad 2.4.1$$

where y replaces η as the deflection. The change δU in the strain energy corresponding to a variation δy of the assumed deflection y can be calculated by subtracting the strain energy corresponding to the deflection y from that corresponding to the deflection $y + \delta y$:[†]

$$U = \frac{1}{2} \int_0^L EI \left[\frac{d^2(y + \delta y)}{dx^2}\right]^2 dx - \frac{1}{2} \int_0^L EI \left(\frac{d^2y}{dx^2}\right)^2 dx \qquad (a)$$

[†] The attention of the reader is called to the similarity between the variations calculated here and the differentials of the calculus.

Since

$$\int_0^L \left[\frac{d^2(y + \delta y)}{dx^2} \right]^2 dx = \int_0^L \left(\frac{d^2y}{dx^2} + \frac{d^2 \delta y}{dx^2} \right)^2 dx$$

$$= \int_0^L \left[\left(\frac{d^2y}{dx^2} \right)^2 + 2 \left(\frac{d^2y}{dx^2} \right) \left(\frac{d^2 \delta y}{dx^2} \right) + \left(\frac{{}^2 \delta y}{dx^2} \right)^2 \right] dx$$

and, because terms containing the square of the virtual displacement δy or of its derivatives need not be considered when the minimal principle is used, as was stated in Art. 2.2, Eq. a reduces to

$$\delta U = \frac{1}{2} \int_0^L EI\, 2 \left(\frac{d^2y}{dx^2} \right) \left(\frac{d^2 \delta y}{dx^2} \right) dx \qquad (b)$$

The potential of the distributed load w is

$$V_w = - \int_0^L wy\, dx$$

This is in agreement with the definition of the potential in Eq. 2.2.4 since the change in this potential

$$\delta V_w = - \int_0^L w(y + \delta y)\, dx + \int_0^L wy\, dx$$

is equal to -1 times the work done by the load w during the displacements δy:

$$\delta V_w = - \int_0^L w\, \delta y\, dx \qquad (c)$$

The potential of the moment M_0 is

$$V_{M_0} = M_0(dy/dx)_{x=0}$$

since then

$$\delta V_{M_0} = -M_0[d(y + \delta y)/dx]_{x=0} + M_0(dy/dx)_{x=0}$$

$$= -M_0(d\delta y/dx)_{x=0} \qquad (d)$$

is the work done by the end moment M_0 during the rotation $(d\delta y/dx)_{x=0}$ of the end tangent caused by the variation δy of the deflections. Similar expressions hold for the work and the potential of the end moment M_L, but the sign is opposed since a positive slope dy/dx at $x = L$ is opposed to the positive sense of M_L, while at the left end the positive slope and moment have the same sense. Consequently

$$V_{M_L} = M_L(dy/dx)_{x=L}$$

$$\delta V_{M_L} = M_L(d\delta y/dx)_{x=L} \qquad (e)$$

The potential of the compressive end loads P is

$$V_P = -P\,\Delta L$$

where ΔL is the difference between the distance of the two supports before and after the deformations $y = f(x)$. When y is changed to $y + \delta y$, the difference also changes, and the change in the difference may be denoted by $\delta\Delta L$. Hence

$$\delta V_P = -P\delta\,\Delta L \qquad\qquad (f)$$

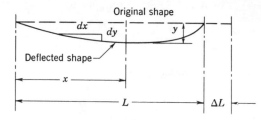

Fig. 2.4.3.   Shortening of distance between end points of beam.

The value of ΔL can be calculated with the aid of Fig. 2.4.3. According to the Pythagorean theorem the length of an infinitesimal element of arc ds is

$$ds = \sqrt{dx^2 + dy^2} = dx\sqrt{1 + (dy/dx)^2}$$

Since the actual deflected shape differs little from the original straight line before loading, $(dy/dx)^2$ is certainly a small quantity as compared to unity. Hence in good approximation

$$ds = dx[1 + \tfrac{1}{2}(dy/dx)^2]$$

as may be ascertained by squaring the expression in the brackets:

$$[1 + \tfrac{1}{2}(dy/dx)^2]^2 = 1 + (dy/dx)^2 + \tfrac{1}{4}(dy/dx)^4$$

The third term in the right-hand member of this equation is a negligibly small quantity.

The total length of the arc can be obtained by integration:

$$L + \Delta L = \int_0^L \left[1 + \frac{1}{2}\left(\frac{dy}{dx}\right)^2\right] dx = L + \frac{1}{2}\int_0^L \left(\frac{dy}{dx}\right)^2 dx$$

Hence the difference between the length of the arc and its horizontal projection is

$$\Delta L = \frac{1}{2}\int_0^L \left(\frac{dy}{dx}\right)^2 dx \qquad\qquad 2.4.2$$

When the deflections y are increased to $y + \delta y$, the value of the shortening ΔL of the distance between supports changes. The change $\delta \Delta L$ can be calculated in the following manner:

$$\delta \Delta L = \frac{1}{2} \int_0^L \left[\frac{d(y + \delta y)}{dx} \right]^2 dx - \frac{1}{2} \int_0^L \left(\frac{dy}{dx} \right)^2 dx$$

$$= \frac{1}{2} \int_0^L \left(\frac{dy}{dx} + \frac{d\delta y}{dx} \right)^2 dx - \frac{1}{2} \int_0^L \left(\frac{dy}{dx} \right)^2 dx$$

$$= \frac{1}{2} \int_0^L \left[\left(\frac{dy}{dx} \right)^2 + 2 \left(\frac{dy}{dx} \right) \left(\frac{d\delta y}{dx} \right) + \left(\frac{d\delta y}{dx} \right)^2 \right] dx - \frac{1}{2} \int_0^L \left(\frac{dy}{dx} \right)^2 dx$$

If the second-order small quantity $(d\delta y / dx)^2$ is neglected, one obtains

$$\delta \Delta L = \int_0^L \left(\frac{dy}{dx} \right) \left(\frac{d\delta y}{dx} \right) dx$$

With this value Eq. f becomes

$$\delta V_P = -P \int_0^L \left(\frac{dy}{dx} \right) \left(\frac{d\delta y}{dx} \right) dx \qquad (g)$$

The change in the total potential is the sum of the quantities given in Eqs. b, c, d, e, and g:

$$\delta(U + V) = EI \int_0^L \left(\frac{d^2y}{dx^2} \right) \left(\frac{d^2\delta y}{dx^2} \right) dx - \int_0^L w \, \delta y \, dx$$

$$-M_0 \left(\frac{d\delta y}{dx} \right)_{x=0} + M_L \left(\frac{d\delta y}{dx} \right)_{x=L} - P \int_0^L \left(\frac{dy}{dx} \right) \left(\frac{d\delta y}{dx} \right) dx$$

$$2.4.3$$

The Differential Equation and the Boundary Conditions

In accordance with the minimal principle, $\delta(U + V)$ must vanish for any variation δy of the displacements if y corresponds to the actual deflected shape, and thus the beam column is in equilibrium under the loads. In an attempt to find the correct deflection function $y = f(x)$ it is advisable to eliminate the derivatives of δy through integrations by parts. If the first term in the right-hand member of Eq. 2.4.3

$$\int_0^L \left(\frac{d^2y}{dx^2} \right) \left(\frac{d^2\delta y}{dx^2} \right) dx$$

is considered first, and the substitutions are made

$$u = (d^2y/dx^2), \qquad dv = (d^2\delta y/dx^2) \, dx$$

then from the formula for integration by parts

$$\int_0^L u \, dv = uv \Big|_0^L - \int_0^L v \, du$$

one obtains

$$\int_0^L \left(\frac{d^2y}{dx^2}\right)\left(\frac{d^2\delta y}{dx^2}\right) dx = \left[\left(\frac{d^2y}{dx^2}\right)\left(\frac{d\delta y}{dx}\right)\right]_{x=L}$$

$$- \left[\left(\frac{d^2y}{dx^2}\right)\left(\frac{d\delta y}{dx}\right)\right]_{x=0} - \int_0^L \left(\frac{d^3y}{dx^3}\right)\left(\frac{d\delta y}{dx}\right) dx$$

The last term in the right-hand member can be reduced further by one more integration by parts in which

$$u = (d^3y/dx^3), \qquad dv = (d\delta y/dx) \, dx$$

Hence

$$-\int_0^L \left(\frac{d^3y}{dx^3}\right)\left(\frac{d\delta y}{dx}\right) dx = -\left(\frac{d^3y}{dx^3}\,\delta y\right)_{x=L} + \left(\frac{d^3y}{dx^3}\,\delta y\right)_{x=0} + \int_0^L \frac{d^4y}{dx^4}\,\delta y\, dx$$

However, only virtual displacements need be considered in the investigation, and consequently the variation δy can be set equal to zero at the two supports. Vertical displacements at the rigid supports would infringe upon the geometric constraints of the beam column. With $\delta y_{x=L} = \delta y_{x=0} = 0$, two terms in the right-hand member of the last equation vanish. Consequently the entire process yields

$$\int_0^L \left(\frac{d^2y}{dx^2}\right)\left(\frac{d^2\delta y}{dx^2}\right) dx = \int_0^L \frac{d^4y}{dx^4}\,\delta y\, dx$$

$$+ \left[\left(\frac{d^2y}{dx^2}\right)\left(\frac{d\delta y}{dx}\right)\right]_{x=L} - \left[\left(\frac{d^2y}{dx^2}\right)\left(\frac{d\delta y}{dx}\right)\right]_{x=0} \qquad (h)$$

A similar integration by parts of the last term in the right-hand member of Eq. 2.4.3 gives

$$\int_0^L \left(\frac{dy}{dx}\right)\left(\frac{d\delta y}{dx}\right) dx = \left(\frac{dy}{dx}\,\delta y\right)_{x=L} - \left(\frac{dy}{dx}\,\delta y\right)_{x=0} - \int_0^L \frac{d^2y}{dx^2}\,\delta y\, dx \qquad (i)$$

Here again two terms of the right-hand member of the equation vanish because δy is zero when $x = 0$ and when $x = L$. After substitutions from Eqs. h and i and rearrangement of the terms, Eq. 2.4.3 becomes

$$\delta(U + V) = \int_0^L \left(EI\frac{d^4y}{dx^4} + P\frac{d^2y}{dx^2} - w\right)\delta y\, dx$$

$$+ \left[\left(EI\frac{d^2y}{dx^2} + M\right)\frac{d\delta y}{dx}\right]_{x=L}$$

$$- \left[\left(EI\frac{d^2y}{dx^2} + M\right)\frac{d\delta y}{dx}\right]_{x=0} = 0 \qquad 2.4.4$$

Equation 2.4.4 is an identity since it must be satisfied for any arbitrary choice of the variation δy of the deflected shape.   If δy is chosen such that $d\delta y/dx$ is zero at both ends of the beam, the second and third expressions in the right-hand member of Eq. 2.4.4 vanish. The equation reduces to

$$\int_0^L \left(EI\frac{d^4y}{dx^4} + P\frac{d^2y}{dx^2} - w \right) \delta y \, dx = 0 \qquad (j)$$

In Eq. j δy is still an arbitrary function of x except for the requirement regarding the end tangent, and the equation must be satisfied for any arbitrary choice of $\delta y = g(x)$.   It is easy to prove that this is possible only if

$$h(x) = EI(d^4y/dx^4) + P(d^2y/dx^2) - w = 0 \qquad 2.4.5$$

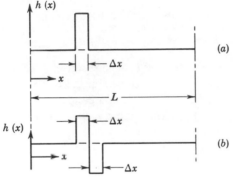

Fig. 2.4.4.   Two choices for $h(x)$.

For the proof let it be assumed that $h(x)$ is everywhere zero except for a short interval Δx, as shown in Fig. 2.4.4a.   Since δy is arbitrary, it is permissible to assume that it is represented by the same curve as $h(x)$.   If the product $h(x) \, \delta y$ is now plotted and the integral of the product is calculated by determining the area under the product curve, it will be found to be different from zero.   The same reasoning shows that the value of $\int h(x) \, \delta x \, dx$ can always be made positive, even if the shape of the $h(x)$ curve is like that in Fig. 2.4.4b, by a suitable choice of δy, for instance, by choosing it to be identical with $h(x)$.   The only exception to this statement occurs when $h(x)$ is identically equal to zero.   Hence the only possibility for fulfilling the requirement of the minimal principle expressed by the identity Eq. j is to set

$$EI(d^4y/dx^4) + P(d^2y/dx^2) - w = 0 \qquad 2.4.6$$

The beam column is in equilibrium only if Eq. 2.4.6 is satisfied. Consequently Eq. 2.4.4 reduces to

$$\{[EI(d^2y/dx^2) + M](d\delta y/dx)\}_{x=L} - \{[EI(d^2y/dx^2) + M](d\delta y/dx)\}_{x=0} = 0 \quad (k)$$

Equation k must be satisfied for any choice of the $\delta y = g(x)$ function because the system is in equilibrium only if the virtual work is equal to zero for any arbitrary displacement. Hence it is now permissible to assume that the slope $d\,\delta y/dx$ is zero when $x = L$ and that it is not equal to zero when $x = 0$. Then Eq. k can be satisfied only by setting

$$EI(d^2y/dx^2) + M = 0 \quad \text{when} \quad x = 0$$

from which it follows that

$$M_0 = -EI(d^2y/dx^2)_{x=0} \qquad\qquad 2.4.7a$$

A similar argument yields the third requirement of equilibrium:

$$M_L = -EI(d^2y/dx^2)_{x=L} \qquad\qquad 2.4.7b$$

The statement of the minimal principle contained in Eq. 2.4.4 is now expressed by a differential equation, Eq. 2.4.6, and two end conditions, Eqs. 2.4.7a and 2.4.7b. Two more end conditions were mentioned in the development of these equations, namely, that the deflection y must vanish at the two supports. Since the differential equation is of the fourth order, and consequently its integral will contain four arbitrary constants of integration, four end conditions are needed for determining the deflections y completely. The problem is therefore completely stated because four conditions are available.

In the strength of materials, the differential equation of the bent beam is usually given in the form

$$EI(d^2y/dx^2) = -M$$

If the moment M consists of a moment M_w caused by the transverse loads and of a moment Py due to the end loads, the equation can be written as

$$EI(d^2y/dx^2) + Py + M_w = 0$$

This equation will now be differentiated twice with respect to x, and it will be observed that $dM_w/dx = V$, where V is the shear force in a normal section of the beam, and $dV/dx = -w$, the distributed load. When the bending rigidity is constant, one obtains

$$EI(d^4y/dx^4) + P(d^2y/dx^2) - w = 0$$

which is identical with Eq. 2.4.6.

Integration of the Differential Equation

With the derivation of the differential equation and the boundary conditions, the problem of establishing the conditions of equilibrium of the beam column is solved.   For practical purposes, of course, there remains the task of finding explicit expressions for deflections and bending moments when the loading is prescribed.   This task can be accomplished by means of the well-known theory of linear differential equations with constant coefficients.

First Eq. 2.4.6 will be written in the condensed form

$$y^{iv} + k^2 y'' = w/EI \qquad\qquad 2.4.8$$

where

$$k^2 = P/EI \qquad\qquad 2.4.8a$$

and y^{iv} is the fourth and y'' the second derivative of y with respect to x. Introduction of a new variable

$$z = y'' \qquad\qquad (l)$$

results in

$$z'' + k^2 z = w/EI \qquad\qquad (m)$$

This is a second-order linear differential equation in z with constant coefficients.   Its solution can be found most readily in two steps.   First the so-called homogeneous equation is solved which is obtained from Eq. m by replacing its right-hand member with zero:

$$z''_c + k^2 z_c = 0 \qquad\qquad (n)$$

The solution of the homogeneous equation is known as the complementary solution, and the subscript c is added to the dependent variable to indicate "complementary."   The general solution of Eq. n which is also the equation of harmonic vibrations, is well known:

$$z_c = A' \cos kz + B' \sin kz \qquad\qquad (o)$$

where A' and B' are arbitrary constants of integration.   That z_c in Eq. o is indeed a solution of Eq. n is easily ascertained: When it is substituted in Eq. n, an identity, valid for any value of z_c, results.   It is also the most general solution because it contains two arbitrary constants of integration.

In the second step any particular solution z_p of Eq. m must be found, and this solution, of course, depends on the distribution of the load w.   In Fig. 2.4.1 w is shown as an arbitrary function of x.   In the following calculations, however, w will be assumed to be constant.   Then it is easy to see that $z_p = $ constant is a particular solution of Eq. m, since the first term in the left-hand member of the equation vanishes and the equation reduces to an equation between constants. If the unknown constant is denoted by C',

$$z_p = C'$$

substitution in Eq. m yields

$$k^2 C' = w/EI$$

Consequently

$$z_p = C' = w/k^2 EI \tag{p}$$

Since the substitution of z_c in Eq. m makes the left-hand member of the equation vanish, while substitution of z_p results in the left-hand side becoming w/EI, obviously the sum of the complementary and particular solutions is the most general solution of the complete equation:

$$z = z_c + z_p = A' \cos kx + B' \sin kx + (w/k^2 EI) \tag{q}$$

If there is any doubt regarding the correctness of the solution given by Eq. q, substitution of z in Eq. m will dispel it since an identity will result.

The justification of the procedure adopted for solving the differential equation lies in the fact that it is easy to find the general solution of the homogeneous equation and a particular solution of the complete equation, while it would be difficult to find directly the general solution containing two arbitrary constants of the complete, non-homogeneous differential equation, Eq. m. The linearity of the differential equation permits the addition of the two easily obtainable solutions to yield the complete solution.

The solution y of Eq. 2.4.8 can now be found with the aid of Eq. l. Two successive integrations of z yield y:

$$y = -(A'/k^2) \cos kx - (B'/k^2) \sin kx + (wx^2/2k^2 EI) + Cx + D$$

Since A' and B' are arbitrary $-A'/k^2$ may be replaced by A and $-B'/k^2$ by B where A and B are two other arbitrary constants of integration. Consequently

$$y = A \cos kx + B \sin kx + (wx^2/2k^2 EI) + Cx + D \tag{2.4.9}$$

The constants of integration must now be determined from the four end conditions.

One of them requires that the deflection be zero at the first support; substitution of $x = 0$ in Eq. 2.4.9 yields

$$y = 0 = A + D$$

Hence

$$D = -A \tag{r}$$

The second derivative of y in Eq. 2.4.9,

$$y'' = -k^2 A \cos kx - k^2 B \sin kx + (w/k^2 EI)$$

must be equal to $-M_0/EI$ in accordance with Eq. 2.4.7a when $x = 0$. Thus

$$y''_{x=0} = -k^2 A + (w/k^2 EI) = -M_0/EI$$

Solution for A gives

$$A = (1/k^2 EI)[M_0 + (w/k^2)] \tag{s}$$

Substitution of A and D in Eq. 2.4.9 yields

$$y = \frac{1}{k^2 EI}\left(M_0 + \frac{w}{k^2}\right)(\cos kx - 1) + B \sin kx + \frac{wx^2}{2k^2 EI} + Cx$$

The second derivative of y is

$$y'' = \frac{-1}{EI}\left(M_0 + \frac{w}{k^2}\right)\cos kx - Bk^2 \sin kx + \frac{w}{k^2 EI}$$

When $x = L$, the second derivative of y must be equal to $-M_L/EI$ in agreement with Eq. 2.4.7b. Hence

$$-(1/EI)[M_0 + (w/k^2)]\cos kL - Bk^2 \sin kL + (w/k^2 EI) = -M_L/EI$$

Solution for B yields

$$B = \frac{1}{k^2 EI \sin kL}\left[\frac{w}{k^2} + M_L - \left(\frac{w}{k^2} + M_0\right)\cos kL\right] \tag{t}$$

The deflection must vanish when $x = L$:

$$y = 0 = \frac{1}{k^2 EI}\left(M_0 + \frac{w}{k^2}\right)(\cos kL - 1) + B \sin kL + \frac{wL^2}{2k^2 EI} + CL$$

Substitution of B and solution for C result in

$$C = \frac{-1}{k^2 LEI}\left(M_L - M_0 + \frac{wL^2}{2}\right) \tag{u}$$

after some manipulations. Because of Eqs. r, s, t, and u, Eq. 2.4.9 becomes

$$y = \frac{1}{k^2 EI}\left\{\left(M_0 + \frac{w}{k^2}\right)(\cos kx - 1) + \left[M_L + \frac{w}{k^2} - \left(M_0 + \frac{w}{k^2}\right)\cos kL\right]\right.$$

$$\left. \times \frac{\sin kx}{\sin kL} - \left[M_L - M_0 + \frac{wL}{2}(L - x)\right]\frac{x}{L}\right\} \qquad 2.4.10$$

The Stiffness Coefficient of a Beam Column

Equation 2.4.10 represents the final solution of the problem of the deflections of a simply supported beam column under the action of a uniformly distributed transverse load and two end moments. Two special cases of loading are of particular interest. One is the beam column, as shown in Fig. 2.4.5, the left end of which is rigidly fixed and on whose simply supported right end the moment M_L is acting. The solution of this

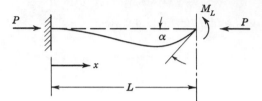

Fig. 2.4.5.   Beam column with one fixed and one supported end.

problem is included in Eq. 2.4.10. When $w = 0$, algebraic and trigonometric manipulations yield

$$y = \frac{1}{k^2 EI} \left\{ \left[\frac{\sin k(L-x)}{\sin kL} - \frac{L-x}{L} \right] M_0 + \left(\frac{\sin kx}{\sin kL} - \frac{x}{L} \right) M_L \right\} \quad 2.4.11$$

The slope is

$$y' = \frac{1}{k^2 LEI} \left\{ \left[\frac{-kL \cos k(L-x)}{\sin kL} + 1 \right] M_0 + \left(\frac{kL \cos kx}{\sin kL} - 1 \right) M_L \right\} \quad (v)$$

The requirement that y' vanish when $x = 0$ yields a condition that can be used to express M_0 in terms of M_L:

$$M_0 = \frac{\sin kL - kL}{\sin kL - kL \cos kL} M_L \quad 2.4.12$$

Substitution in Eq. v of $x = L$ and of the value of M_0 given in Eq. 2.4.12 yields

$$\alpha = -y'_{x=L} = \frac{M_L L}{EI} \frac{1}{kL} \frac{2 - 2 \cos kL - kL \sin kL}{\sin kL - kL \cos kL} \quad 2.4.13$$

In this equation $\alpha = -y'$ at $x = L$ since the angle α is counted positive as shown in Fig. 2.4.5, while $y' = dy/dx$ is negative; the deflection decreases with increasing x in the neighborhood of the right-hand support.

In Art. 1.17 the stiffness coefficient of a beam was defined as the moment required at the simply supported end of a beam in order to cause a unit angle of rotation there when the other end of the beam is rigidly fixed.

This quantity is of importance in calculations carried out by means of the Hardy Cross moment-distribution method. Equation 2.4.13 provides the answer to the question of how the stiffness coefficient of a beam changes because of a compressive end load. If α is set equal to unity and Eq. 2.4.13 is solved for M_L, which here is equal to the stiffness coefficient S by definition, the solution is

$$S = \frac{EI}{L} kL \frac{\sin kL - kL \cos kL}{2 - 2 \cos kL - kL \sin kL} \qquad 2.4.14$$

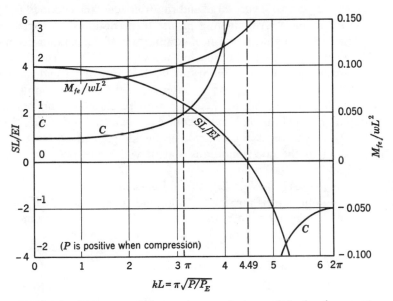

Fig. 2.4.6.   Stiffness coefficient, carry-over factor, and fixed-end moment.

The value of S calculated from Eq. 2.4.14 is plotted in Fig. 2.4.6 against the parameter kL. The physical significance of this parameter can be understood if the definition of k given in Eq. 2.4.8a is recalled. Simple manipulations yield

$$kL = L\sqrt{P/EI} = \pi[P/(\pi^2 EI/L^2)]^{1/2}$$

$$kL = \pi\sqrt{P/P_E} \qquad 2.4.15$$

where P_E is Euler's buckling load of a column as given in all textbooks on strength of materials:

$$P_E = \pi^2 EI/L^2 \qquad 2.4.16$$

According to Fig. 2.4.6 the stiffness coefficient S is $4EI/L$ when the end load $P = 0$. This is in agreement with the value found earlier in Art. 1.17

and given in Eq. 1.17.2. With increasing kL, that is, with increasing end load P, the stiffness coefficient decreases. Consequently, when an end load P is acting on the beam, a smaller end moment is required to cause a unit rotation than in the absence of such an end load. This could be anticipated qualitatively since the bending moment Py increases the deflections caused by M_L alone.

When with increasing kL the end load P reaches the value P_E the stiffness coefficient is only 62 per cent of its original value, and S becomes zero when $P = 2.04P_E$, that is, when $kL = 4.49$. Consequently a very small end moment M_L causes a very large end rotation when the end load is 2.04 times the Euler load. Anticipating results to be obtained in Part 3, we may say that the straight-line form of equilibrium of the column becomes unstable when the end load approaches the value $2.04P_E$, since then the slightest disturbance of the equilibrium in the form of an end moment causes very large deflections. As a consequence the bending moment Py attains such high values that the bending stress reaches the yield point of the material and the beam column becomes permanently deformed. The load $2.04P_E$ is known as the critical, or buckling, load of the column corresponding to the end conditions stated, which are one end simply supported and the other rigidly fixed.

In the region $4.49 < kL < 2\pi$, the stiffness factor is negative. This means that equilibrium is possible with an end rotation α only if an end moment is available that opposes the rotation. Such a restraining moment can be realized if a spring is attached to the end of the beam.

Carry-Over Factor and Fixed-End Moment

The second quantity of importance in the Hardy Cross moment-distribution method is the carry-over factor C defined as the ratio of the moment M_0, caused at the fixed end by the application of a moment M_L at the simply supported end, to the moment M_L. This quantity is easily obtained from Eq. 2.4.12. However, in the differential equation, the beam convention of signs was used, while in Art. 1.17 the rigid-frame convention was adopted. According to the former, a positive moment is clockwise at $x = 0$ in Fig. 2.4.5 and counterclockwise at $x = L$, while according to the latter the positive sense is the same at both ends of the beam. Consequently the ratio M_0/M_L must be negative according to the beam convention if it is positive according to the rigid-frame convention. On the basis of this argument and Eq. 2.4.12, the carry-over factor according to the frame convention is

$$C = \frac{kL - \sin kL}{\sin kL - kL \cos kL} \qquad\qquad 2.4.17$$

The carry-over factor C is also plotted in Fig. 2.4.6.   It has the value 1/2 when the end load is zero in agreement with the results obtained in Art. 1.17, and it becomes infinite when kL approaches the value 4.49.   Between 4.49 and 2π the carry-over factor is negative.

Because of its importance in the generalized Hardy Cross moment-distribution method the next quantity to be calculated here is the fixed-end moment defined in Art. 1.17.   It is the end moment caused at the supports by the transverse loading when the ends of the beam are assumed to be fixed rigidly against rotation.   When the transverse loading w is distributed

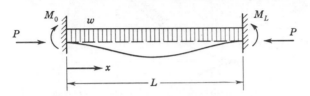

Fig. 2.4.7.   Beam column with both ends fixed.

uniformly $M_0 = M_L$ because of the symmetry, as may be seen from Fig. 2.4.7.   Consequently Eq. 2.4.10 reduces to

$$y = \frac{w}{k^2 EI}\left[\frac{1}{k^2}(1 - \cos kL)\frac{\sin kx}{\sin kL} - \frac{1}{k^2}(1 - \cos kx) - \frac{x}{2}(L - x)\right]$$

$$+ \frac{M_{\text{fe}}}{k^2 EI}\left[(1 - \cos kL)\frac{\sin kx}{\sin kL} - (1 - \cos kx)\right] \tag{v}$$

if the notation is introduced

$$M_0 = M_L = M_{\text{fe}} \tag{x}$$

The requirement that the slope y' vanish when $x = 0$ yields a condition that can be used for expressing the fixed-end moment M_{fe}.   One obtains

$$M_{\text{fe}} = wL^2 \frac{1}{(kL)^2}\frac{2 - 2\cos kL - kL\sin kL}{2 - 2\cos kL} \tag{2.4.18}$$

Values of M_{fe} are also plotted in Fig. 2.4.6.   The fixed-end moment is $\frac{1}{12}wL^2$ when there is no end load acting, as was given in Eq. 1.17.1, and increases beyond all limits as P approaches the value $4P_E$.

Other End Conditions

Occasionally the end of a bar is attached to the rest of the structure by an actual hinge, and more often the analyst assumes the existence of such a hinge in order to simplify his calculations.   For this reason it is worth while to calculate the stiffness and carry-over factors for bars whose far end is hinged. If in Eqs. 2.4.11 and v one sets $x = 0$ and $M_L = 0$, the slope becomes

$$y' = \frac{M_0 L}{EI(kL)^2} \frac{\sin kL - kL \cos kL}{\sin kL}$$

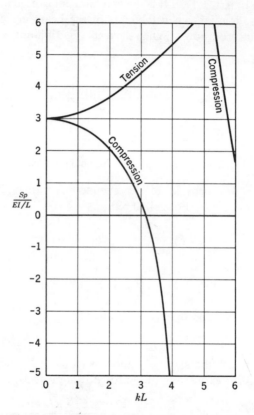

Fig. 2.4.8. Stiffness coefficients for pin-ended bars.

The stiffness S_p of the pin-ended bar is the value of M_0 when the slope y' is unity. Consequently

$$S_p = \frac{EI}{L} \frac{(kL)^2 \sin kL}{\sin kL - kL \cos kL} \qquad 2.4.19$$

Values of S_p are plotted against kL in Fig. 2.4.8. The carry-over factor is zero since the far-end moment must always be zero on account of the hinge.

When the bar is part of a welded or riveted framework, the ends of the bar are reinforced by gusset plates.   The effect of the gusset plates can be taken into account with the aid of simplifying assumptions.   Values of the

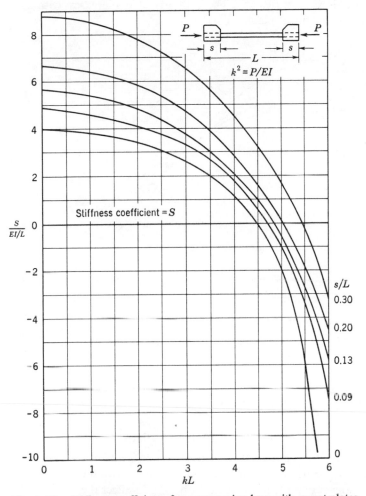

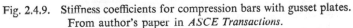

Fig. 2.4.9.   Stiffness coefficients for compression bars with gusset plates. From author's paper in *ASCE Transactions*.

stiffness coefficient and carry-over factor have been calculated both for bars subjected to compressive end loads and for bars on which tensile end loads are acting, but details of the calculations are not given here. The values are presented in Figs. 2.4.9 to 2.4.12.   In the diagrams L is the length of the bar between the mathematical end points and s the length of the gusset plate, also measured from the mathematical end point of the bar. The deflections of tension bars are governed by a differential equation obtainable from Eq. 2.4.6 through substitution of $-P$ for P. It can be seen from the figures that tension increases the stiffness of a bar while compression decreases it. The stiffness also increases with increasing length of the gusset plate.

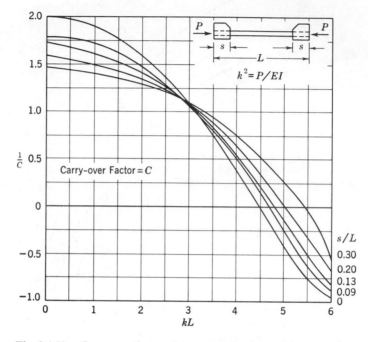

Fig. 2.4.10.   Carry-over factors for compression bars with gusset plates. From author's paper in *ASCE Transactions*

More General Variation of the Total Potential[†]

When the total potential was calculated and varied, the assumption was made that the normal strain in the beam column would remain unchanged during the variations of the deflections y.   In effect the beam column was considered inextensible when the shape of its deflected center line was varied.   Moreover, the strain energy of uniform compression was omitted from the expression representing the total potential of the system.   It will now be shown by a more general variation of the state of deformation that this procedure is permissible.

† This section can be omitted by readers more interested in applications than in the development of the theory.

In the proof every cross section of the beam column will be given a transverse displacement y and an axial displacement u. The extended length of an element has then the y component dy and the x component $dx + (du/dx)dx$.‡ From the Pythagorean theorem

$$ds = \{(dy)^2 + [1 + (du/dx)]^2\, dx^2\}^{1/2} \qquad (y)$$

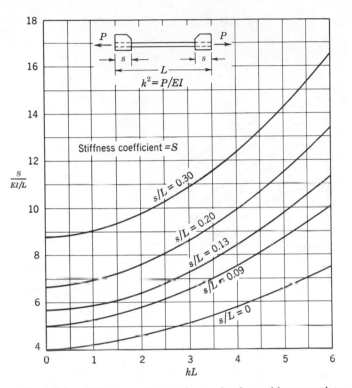

Fig. 2.4.11. Stiffness coefficients for tension bars with gusset plates. From author's paper in *ASCE Transactions*

which can also be written as

$$ds = \{1 + 2(du/dx) + (du/dx)^2 + (dy/dx)^2\}^{1/2}dx \qquad 2.4.20$$

According to the binomial theorem

$$(1 + p)^{1/2} = 1 + \tfrac{1}{2}p - \tfrac{1}{8}p^2 + \cdots$$

With

$$p = 2(du/dx) + (du/dx)^2 + (dy/dx)^2$$

‡ This is discussed in more detail in Art. 2.6.

one obtains

$$(1 + p)^{1/2} = 1 + (du/dx) + \tfrac{1}{2}(du/dx)^2 + \tfrac{1}{2}(dy/dx)^2 - \tfrac{1}{2}(du/dx)^2 + R$$

where the symbol R represents terms containing (du/dx) multiplied by $(dy/dx)^2$, the square of (du/dx) multiplied by $(dy/dx)^2$, as well as higher powers of these small quantities. If R is neglected, Eq. 2.4.20 becomes

$$ds = dx[1 + du/dx + \tfrac{1}{2}(dy/dx)^2] \tag{z}$$

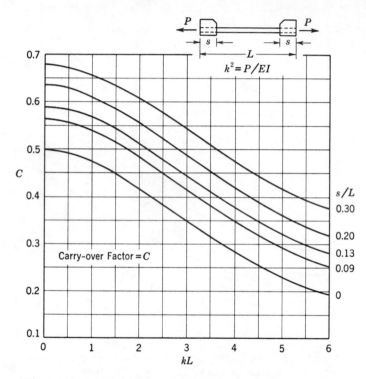

Fig. 2.4.12.   Carry-over factors for tension bars with gusset plates. From author's paper in *ASCE Transactions*.

As the length of the element was dx before the displacements, the average tensile strain ϵ is

$$\epsilon = (ds - dx)/dx = du/dx + \tfrac{1}{2}(dy/dx)^2 \tag{A}$$

The strain energy stored in the bar because of the extension is

$$U_e = \frac{1}{2}\, EA \int_0^L \left[\frac{du}{dx} + \frac{1}{2}\left(\frac{dy}{dx}\right)^2\right]^2 dx \tag{2.4.21}$$

and the change in this quantity when both u and y are varied is

$$\delta U_e = \frac{1}{2} EA \int_0^L \left\{ \frac{d(u + \delta u)}{dx} + \frac{1}{2} \left[\frac{d(y + \delta y)}{dx} \right]^2 \right\}^2 dx$$

$$- \frac{1}{2} EA \int_0^L \left[\frac{du}{dx} + \frac{1}{2} \left(\frac{dy}{dx} \right)^2 \right]^2 dx$$

$$= EA \int_0^L \left[\frac{du}{dx} + \frac{1}{2} \left(\frac{dy}{dx} \right)^2 \right] \left[\frac{d\, \delta u}{dx} + \left(\frac{dy}{dx} \right) \left(\frac{d\, \delta y}{dx} \right) \right] dx \qquad (B)$$

This expression can be integrated by parts.   One obtains

$$\delta U_e = EA \left[\frac{du}{dx} + \frac{1}{2} \left(\frac{dy}{dx} \right)^2 \right] \delta u \Bigg|_0^L$$

$$+ EA \left[\left(\frac{du}{dx} \right) \left(\frac{dy}{dx} \right) + \frac{1}{2} \left(\frac{dy}{dx} \right)^3 \right] \delta y \Bigg|_0^L$$

$$- EA \int_0^L \frac{d}{dx} \left[\frac{du}{dx} + \frac{1}{2} \left(\frac{dy}{dx} \right)^2 \right] \delta u \, dx$$

$$- EA \int_0^L \frac{d}{dx} \left[\left(\frac{du}{dx} \right) \left(\frac{dy}{dx} \right) + \frac{1}{2} \left(\frac{dy}{dx} \right)^3 \right] \delta y \, dx \qquad (C)$$

where the symbol d/dx indicates a differentiation that should be performed on the expression in square brackets. As δu and δy represent virtual displacements, they must vanish at $x = 0$. At the other end, $x = L$, the roller support permits any arbitrary displacement u, but it prevents displacements in the y direction. For these reasons the change in the extensional strain energy is

$$\delta U_e = \left\{ EA \left[\frac{du}{dx} + \frac{1}{2} \left(\frac{dy}{dx} \right)^2 \right] \delta u \right\}_{x=L}$$

$$- EA \int_0^L \frac{d}{dx} \left[\frac{du}{dx} + \frac{1}{2} \left(\frac{dy}{dx} \right)^2 \right] \delta u \, dx$$

$$- EA \int_0^L \frac{d}{dx} \left[\left(\frac{du}{dx} \right) \left(\frac{dy}{dx} \right) + \frac{1}{2} \left(\frac{dy}{dx} \right)^3 \right] \delta y \, dx \qquad 2.4.22$$

In the present approach the strain energy of bending and its change because of the variation of y are again represented by Eqs. 2.4.1, b, and h; similarly the potential of the distributed transverse load and the end moments, as well as its change due to δy, remains the same. When u is varied, these quantities are not affected since they depend only on y. On the other hand, the potential of the end load P must be represented by a new expression. The axial displacement at $x = L$ is now simply the value of u at that point. Thus

$$V_P = +Pu_{x=L} \tag{D}$$

The positive sign is appropriate here because the positive directions of u and P are opposed. The value of V_P does not change when y is varied. When u is varied, one has

$$\delta V_P = P\, \delta u_{x=L} \tag{2.4.23}$$

In the original derivation the total change in the total potential was given by Eq. 2.4.4. Because of the differences in the approach, the second term under the integral sign must be canceled and replaced by the expression given for δV_P in Eq. 2.4.23. Moreover the change in the strain energy of extension as presented in Eq. 2.4.22 must be added. The sum total of these quantities must vanish for every variation δu and δy when the system is in equilibrium. This implies that the expressions multiplied by δu must vanish:

$$\left\{ EA \left[\frac{du}{dx} + \frac{1}{2}\left(\frac{dy}{dx} \right)^2 \right] \delta u \right\}_{x=L}$$

$$- EA \int_0^L \frac{d}{dx}\left[\frac{du}{dx} + \frac{1}{2}\left(\frac{dy}{dx} \right)^2 \right] \delta u\, dx + P\, \delta u_{x=L} = 0 \tag{E}$$

The left-hand member of this equation vanishes identically if the expression under the integral sign and the sum of the two terms referring to the end point $x = L$ vanish individually. The first condition reduces to the statement that

$$(du/dx) + \tfrac{1}{2}(dy/dx)^2 = \text{constant} \tag{F}$$

If this equation is compared with Eq. A, one finds that the average strain ϵ in the beam column must be constant. The second condition derived from Eq. E is

$$[(du/dx) + \tfrac{1}{2}(dy/dx)^2]_{x=L} = -P/EA \tag{G}$$

Hence the average strain at $x = L$ is simply $-P/EA$; but from the first condition the same average strain must prevail over the entire length of the beam column, and thus

$$\epsilon = -P/EA \qquad \text{for all } x \tag{2.4.24}$$

The last two terms in Eq. 2.4.4 again imply the end conditions Eqs. 2.4.7a and b. There remains then the requirement that

$$\int_0^L \left\{ EI \frac{d^4y}{dx^4} - w - EA \frac{d}{dx} \left[\left(\frac{du}{dx}\right)\left(\frac{dy}{dx}\right) + \frac{1}{2}\left(\frac{dy}{dx}\right)^3 \right] \right\} \delta y \, dx = 0 \quad (H)$$

The expression in brackets can also be written as

$$(dy/dx)[(du/dx) + \tfrac{1}{2}(dy/dx)^2]$$

or, in view of Eqs. G and 2.4.24, as

$$-(P/EA)(dy/dx)$$

If this expression is differentiated with respect to x, one obtains $-(P/EA)(d^2y/dx^2)$. Substitution in Eq. H yields

$$\int_0^L \left(EI \frac{d^4y}{dx^4} - w + P \frac{d^2y}{dx^2} \right) \delta y \, dx = 0$$

from which it follows that one of the requirements of equilibrium is the differential equation

$$EI(d^4y/dx^4) + P(d^2y/dx^2) - w = 0$$

which is identical with Eq. 2.4.6.

This more rigorous analysis of the conditions of equilibrium justifies therefore the original approach. The beam column can be treated as if its axis were flexible and inextensible, and the strain energy stored in it can be calculated as if it were zero in the compressed state under the constant load P before the transverse displacements take place.

2.5 Large Deflections of a Flexible Cable

In all the problems discussed heretofore the displacements corresponding to equilibrium were considered very small. Under such conditions the curvature of a beam is proportional to the second derivative of the deflections, and the more complex non-linear curvature expression preceding Eq. 2.1.14 reduces to the usual linear one presented as Eq. 2.1.14. When the deflections are large, non-linear expressions of various kinds appear, and the mathematical difficulties of solving the problem increase greatly. The minimal principle, however, retains its validity and in some cases can still lead to comparatively simple solutions.

To illustrate this statement, the shape assumed by a freely hanging cable, rope, or chain will now be calculated under two different loading conditions. It will be assumed that the cable offers no resistance to flexure.

Flexible cables constitute the most important structural element of the suspension bridge. They are also widely used in electric transmission lines.

Figure 2.5.1 is the schematic representation of a cable subjected to loads uniformly distributed over the horizontal projection of the cable.

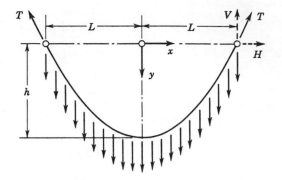

Fig. 2.5.1.   Flexible cable.

The vertical downward load is $w\,dx$ and not $w\,ds$, where w is a constant. Loads proportional to the horizontal projection of the cable may be uniform loads suspended from the cable or the weight of snow settled on the cable. The weight of the cable naturally is proportional to its length ds. This type of loading will be discussed later.

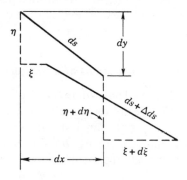

Fig. 2.5.2.   Element of cable.

Figure 2.5.2 shows an element ds of the cable in its unknown actual position as well as virtual displacements ξ and η of its end points. Before the virtual displacements are undertaken, the length of the element is ds. The new length $ds + \Delta ds$ after displacements can be calculated from the

Pythagorean theorem:

$$(ds + \Delta ds)^2 = (dx + d\xi)^2 + (dy + d\eta)^2 \qquad 2.5.1$$

Consequently

$$ds + \Delta ds = [dx^2 + dy^2 + 2dx\, d\xi + 2dy\, d\eta + d\xi^2 + d\eta^2]^{1/2}$$

Since $ds^2 = dx^2 + dy^2$, it follows that

$$ds + \Delta ds = ds[1 + 2(dx/ds)(d\xi/ds) + 2(dy/ds)(d\eta/ds)$$
$$+ (d\xi/ds)^2 + (d\eta/ds)^2]^{1/2}$$

The strain $\epsilon_{\xi,\eta}$ caused in the cable by the virtual displacements is $\Delta ds/ds$, that is,

$$\epsilon_{\xi,\eta} = [1 + 2(dx/ds)(d\xi/ds) + 2(dy/ds)(d\eta/ds) + (d\xi/ds)^2$$
$$+ (d\eta/ds)^2]^{1/2} - 1 \qquad 2.5.2$$

If the cable is made of steel, the order of magnitude of the strain must be smaller than 0.01; otherwise the cable would break. This means that the sum of the four terms added to unity in the brackets must be small as compared to unity. If the sum of the four terms is designated by a, we can write

$$(1 + a)^{1/2} = 1 + (a/2)$$

in good approximation, because the square of the right-hand member is $1 + a + (a/2)^2$, which differs from $1 + a$ only by the term $(a/2)^2$. This term, however, is smaller than 0.0001 when a is smaller than 0.01, and can be neglected in the calculations. Consequently Eq. 2.5.2 becomes

$$\epsilon_{\xi,\eta} = (dx/ds)(d\xi/ds) + (dy/ds)(d\eta/ds) + \tfrac{1}{2}(d\xi/ds)^2 + \tfrac{1}{2}(d\eta/ds)^2 \qquad 2.5.3$$

The tension prevailing in the cable can be denoted by T. It varies with x. The sum of the initial strain and that caused by the virtual displacements is then

$$\epsilon = T/EA + \epsilon_{\xi,\eta} \qquad 2.5.4$$

where A is the cross-sectional area of the cable and E is Young's modulus of its material. The strain energy stored in the cable before the virtual displacements is

$$U_0 = \frac{EA}{2} \int_l \left(\frac{T}{EA}\right)^2 ds \qquad 2.5.5$$

where the symbol l indicates that the integration must be extended over the entire length of the cable. After the virtual displacements the strain energy U is

$$U = \frac{EA}{2} \int_l \left[\left(\frac{T}{EA}\right)^2 + 2\frac{T}{EA}\,\epsilon_{\xi,\eta} + \epsilon^2_{\xi,\eta} \right] ds \qquad 2.5.6$$

The change δu in the strain energy during the virtual displacements is

$$\delta U = U - U_0 = \frac{EA}{2} \int_l \left(2\frac{T}{EA}\,\epsilon_{\xi,\eta} + \epsilon^2_{\xi,\eta} \right) ds \qquad 2.5.7$$

In the application of the principle of virtual displacements, terms containing products or powers of the virtual displacements must be disregarded. If $\epsilon_{\xi,\eta}$ is substituted from Eq. 2.5.3 in Eq. 2.5.7, and terms containing only the first power of ξ, η and their derivatives are retained, we obtain for the change in the strain energy the expression

$$\delta U = \int_l \left[T\left(\frac{dx}{ds}\right)\left(\frac{d\xi}{ds}\right) + T\left(\frac{dy}{ds}\right)\left(\frac{d\eta}{ds}\right) \right] ds \qquad 2.5.8$$

The horizontal component H of the tensile force T is connected with T by the equation

$$T = H(ds/dx) \qquad 2.5.9$$

Hence Eq. 2.5.8 can be written in the form

$$\delta U = \int_{-L}^{L} \left[H\frac{d\xi}{dx} + H\left(\frac{dy}{dx}\right)\left(\frac{d\eta}{dx}\right) \right] dx \qquad 2.5.10$$

where the limits $-L$ and L replace the symbol l because the last integrand contains dx rather than ds.

There is no strain energy of bending stored in the cable because, according to the assumption made, the cable does not resist bending. The change in the potential of the loads during the virtual displacements is equal to the negative of the product of the load $w\,dx$ acting on an element whose horizontal projection is dx and the displacement η, integrated over the total projected length of the cable $2L$:

$$\delta V = - \int_{-L}^{L} w\eta\,dx \qquad 2.5.11$$

In accordance with the principle of the minimum of the total potential the actual deflected shape $y = f(x)$ is such that, if any virtual displacements ξ and η are undertaken from it, the change in the total potential is zero. Thus the condition of equilibrium in the unknown deflected position $y = f(x)$ is

$$\delta(U + V) = 0 = \int_{-L}^{L} \left[H\frac{d\xi}{dx} + H\left(\frac{dy}{dx}\right)\left(\frac{d\eta}{dx}\right) - w\eta \right] dx \qquad 2.5.12$$

The term containing $d\xi/dx$ can be integrated by parts:

$$\int_{-L}^{L} H\frac{d\xi}{dx}\,dx = H\xi \Big|_{-L}^{L} - \int_{-L}^{L} \frac{dH}{dx}\,\xi\,dx$$

The first term in the right-hand member of this equation vanishes because the virtual displacement ξ must be zero at the fixed-end points of the cable $x = \pm L$. Integration by parts of the second term in Eq. 2.5.12 yields

$$\int_{-L}^{L} H\left(\frac{dy}{dx}\right)\left(\frac{d\eta}{dx}\right)\,dx = H\left(\frac{dy}{dx}\right)\eta \Big|_{-L}^{L} - \int_{-L}^{L}\left[\left(\frac{dH}{dx}\right)\left(\frac{dy}{dx}\right) + H\frac{d^2y}{dx^2}\right]\eta\,dx$$

Here again the first term in the right-hand member vanishes since no η displacements need be assumed at the fixed-end points of the cable. Consequently Eq. 2.5.12 reduces to

$$\delta(U+V) = -\int_{-L}^{L}\frac{dH}{dx}\,\xi\,dx - \int_{-L}^{L}\left[\left(\frac{dH}{dx}\right)\left(\frac{dy}{dx}\right) + H\frac{d^2y}{dx^2} + w\right]\eta\,dx$$

$$2.5.13$$

In this equation ξ and η are perfectly arbitrary virtual displacements. If η is chosen as zero over the entire length $2L$, Eq. 2.5.13 becomes

$$\int_{-L}^{L}\frac{dH}{dx}\,\xi\,dx = 0$$

This equation can be satisfied for arbitrary choices of the function ξ only if

$$dH/dx = 0 \qquad\qquad 2.5.14$$

This means that the horizontal component of the tension must be constant. Introduction of zero for dH/dx in Eq. 2.5.13 yields

$$\int_{-L}^{L}\left(H\frac{d^2y}{dx^2} + w\right)\eta\,dx = 0 \qquad\qquad 2.5.15$$

which can be satisfied for arbitrary η functions only if

$$H(d^2y/dx^2) + w = 0$$

or

$$d^2y/dx^2 = -w/H \qquad\qquad 2.5.16$$

with H a constant. This is the differential equation of the deflected shape of the cable. It can be readily integrated by separating the variables. We obtain

$$y = -(w/H)(x^2/2) + Cx + D$$

The constants of integration must be determined from the end conditions.

At $x = 0$ the cable has a horizontal tangent; that is (dy/dx) vanishes, and the ordinate y has the value h. This means that

$$C = 0 \quad \text{and} \quad D = h$$

The deflected shape is then characterized by

$$y = h - (w/H)(x^2/2) \qquad 2.5.17$$

Moreover the deflection y must vanish at the end points, $x = \pm L$. This condition yields the equation

$$h = (w/H)(L^2/2)$$

or

$$H = \tfrac{1}{2}L^2(w/h) \qquad 2.5.18$$

The equation gives the value of the horizontal tension H if the sag h is known. Substitution of H from Eq. 2.5.18 in Eq. 2.5.17 yields

$$y = h[1 - (x/L)^2] \qquad 2.5.19$$

The deflected shape of a cable on which loads proportional to the horizontal projection of the cable are acting is a parabola.

When the load is the weight of the cable, the equations just derived must be modified. δU remains unchanged but δV becomes

$$\delta V = - \int_l w\eta \, ds \qquad 2.5.20$$

Since

$$ds = [1 + (dy/dx)^2]^{1/2} \, dx$$

the change in the total potential is

$$\delta(U + V) = - \int_{-L}^{L} \frac{dH}{dx} \xi \, dx - \int_{-L}^{L} \left\{ \left(\frac{dH}{dx} \right) \left(\frac{dy}{dx} \right) + H \frac{d^2y}{dx^2} \right.$$
$$\left. + w \left[1 + \left(\frac{dy}{dx} \right)^2 \right]^{1/2} \right\} \eta \, dx \qquad 2.5.21$$

The same manipulations as those undertaken earlier result in

$$H = \text{constant} \qquad 2.5.22a$$
$$d^2y/dx^2 = -(w/H)[1 + (dy/dx)^2]^{1/2} \qquad 2.5.22b$$

This is a non-linear differential equation, but its solution does not present any difficulties. We can introduce the notation

$$z = dy/dx$$

which reduces Eq. 2.5.22b to

$$dz/dx = -(w/H)(1 + z^2)^{1/2}$$

From this we obtain

$$-(w/H)\,dx = dz/(1+z^2)^{1/2}$$

Use can be made of the substitution

$$z = \sinh v$$

from which it follows that

$$1 + z^2 = 1 + \sinh^2 v = \cosh^2 v$$

$$dz = \cosh v\,dv$$

Hence

$$-(w/H)\,dx = \cosh v\,dv/\cosh v = dv$$

Integration gives

$$-(w/H)x = v + C = \sinh^{-1} z + C$$

or

$$\sinh^{-1} z = -(wx/H) - C$$

which is equivalent to

$$z = -\sinh\left[(wx/H) + C\right] \qquad 2.5.23$$

When $x = 0$, the slope of the curve must vanish because of the symmetry. But the slope dy/dx is z.   Substitution of these values in Eq. 2.5.23 yields

$$C = 0$$

and thus

$$z = dy/dx = -\sinh\,(wx/H) \qquad 2.5.24$$

Separation of the variables and integration give

$$y = -(H/w)\cosh\,(wx/H) + D \qquad 2.5.25$$

The deflection is h when x is zero.   Thus

$$h = -(H/w) + D$$

Solution of the equation for D and substitution in Eq. 2.5.25 result in

$$y = h - (H/w)[\cosh\,(wx/H) - 1] \qquad 2.5.26$$

At $x = L$ the deflection must vanish

$$0 = h - (H/w)[\cosh\,(wL/H) - 1]$$

This equation can be solved for h:

$$h = (H/w)[\cosh\,(wL/H) - 1] \qquad 2.5.27$$

If the only load is its own weight, the cable assumes a hyperbolic cosine type of deflected shape.   This curve is known as the catenary.   The connection between span, sag, weight per unit length, and the horizontal component of the tension in the cable is given in Eq. 2.5.27.   If the only unknown is the constant horizontal pull, the equation can be solved easily by a trial-and-error method.   With H known, Eq. 2.5.26 is an explicit representation of the deflected shape.

2.6   Deflections of a Sandwich-Type Beam

Sandwich Construction

In the preceding article the principle of the minimum of the total potential was used to find the solutions of problems which could also have been obtained by means of more elementary considerations.   The calculations were meant as examples and did not yield new or unexpected results. In many other more complex problems application of the minimal principle yields the easiest, and sometimes the only, solution.   In such cases the principle proves itself a most valuable tool in the equipment of the stress analyst.   In the present article the differential equations of the bending of a sandwich beam will be established through the application of the minimal principle.   These differential equations cannot be developed in any simple manner from elementary considerations.

A sandwich beam, such as the one shown in Figs. 2.6.1a and b, consists of two comparatively thin faces of a strong material between which a thick layer of very light weight and comparatively weak core is sandwiched. The obvious advantage of this construction is the large moment of inertia of the section obtained by spacing far apart the main carrying elements, namely, the faces.   The weight of the structure is small because of the low density of the core.

During World War II sandwich structural elements attained some importance in airplane construction, where rigidity and light weight were the primary requirements.   For the faces, plywood, Papreg, aluminum alloy, stainless steel, or Fiberglas textile were used; for the core, balsa wood, expanded synthetic plastics, and built-up grids were available. Faces and core were cemented together by means of synthetic glues. Typical mechanical properties of the core material can be quoted as follows:

Specific gravity	5 lb per cu ft
Compressive strength	300 psi
Modulus of elasticity	5000 psi
Shear modulus	2500 psi

If under the action of a load W, a sandwich beam deflected substantially in bending, with plane sections perpendicular to the axis remaining plane and perpendicular to the axis after bending, its rigidity would be extremely high for its weight.   However, an ultralightweight core cannot be incorporated in a beam without some penalty.   The small shearing rigidity of the core gives rise to very substantial shearing deformations.

The equilibrium conditions of the sandwich beam will now be established with the aid of the principle of the minimum of the total potential.   It will be assumed that the deformations can be described with sufficient accuracy by superimposing in a suitable ratio the two types of displacements shown

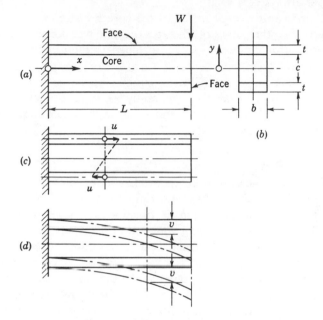

Fig. 2.6.1.   Cantilever sandwich beam.   (From author's paper in
J. Aero Sciences)

in Figs. 2.6.1c and d.   The former is basically an extensional and the latter a shearing pattern.

To arrive at sufficiently simple results for practical calculations, only the essential parts of the strain energy will be taken into account.   These are the extensional and the bending strain energy stored in the faces and the shear strain energy stored in the core.   The contribution of the core to the bending-moment equilibrium is neglected.   Consequently it is assumed that the layer of the core attached to the face is extended together with the face without any normal stresses of appreciable magnitude being set up in the core.   Such an assumption is justifiable because of the very low value of the elastic modulus of the core.   Whether the assumption leads to satisfactory results can best be decided from a comparison of the deflections predicted by this theory and those measured in experiment.   It might be mentioned here that such experiments have been carried out and that they have corroborated the theory.

Another quantity neglected in the calculations is the shear strain energy stored in the faces. The justification for disregarding this quantity lies in the known fact that the shear strain energy is very small in metal beams of rectangular section if their length is great compared to their depth. This requirement is always fulfilled by the faces of sandwich beams since the ratio t/L is always very small.

Finally the strain energy stored in the core because of normal stresses perpendicular to the planes of the faces must be mentioned. Its effect on the deflections of the beam can be included in the calculations without difficulty. However, the results of such a more comprehensive theory prove the effect to be negligible in straight beams and to be of some importance in curved beams. Apparently the distance between the faces remains unchanged in the former and varies in the latter under the action of the bending moments.

The Total Potential

For the reasons mentioned the theory has to include only the stretching and bending of the faces and the shearing of the core. The normal strain corresponding to the displacement function $u = u(x)$, where the x in parentheses indicates that u is a function of x, can be determined from Fig. 2.6.2a. In the figure the dashed lines show a small element of the face before deformations, and the full lines the position and shape taken by the element after the load is applied to the sandwich beam. If the horizontal displacement at x is denoted by u, the slightly different displacement at $x + dx$ can be designated as $u + du$. The change du in the magnitude of the displacement can be calculated from a Taylor expansion of the $u = u(x)$ function:

$$u(x + dx) = u(x) + \frac{du(x)}{dx}\, dx + \frac{1}{2}\frac{d^2u(x)}{dx^2}(dx)^2 + \frac{1}{3!}\frac{d^3u(x)}{dx^3}(dx)^3 + \cdots$$

The increment of the length dx of the element is

$$(u + du) - u = u(x + dx) - u(x)$$

and the strain ϵ_x is obtained by dividing this increment by the original length dx

$$\epsilon_x = \frac{u(x + dx) - u(x)}{dx} = \frac{du(x)}{dx} + \frac{1}{2}\frac{d^2u(x)}{dx^2}\, dx + \frac{1}{3!}\frac{d^3u(x)}{dx^3}(dx)^2 + \cdots$$

and by letting dx decrease indefinitely.   In the limit, when dx is very small

$$\epsilon_x = \frac{du(x)}{dx} = u'$$ 2.6.1

where the prime denotes differentiation of the function with respect to x.

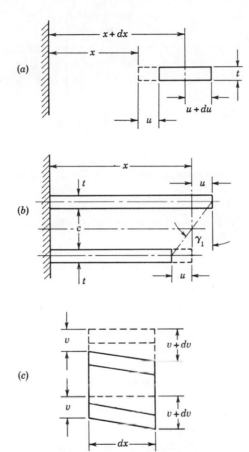

Fig. 2.6.2.   Strains caused by displacements.

The strain energy stored in one face is, in agreement with Eq. 2.1.8,

$$U = \frac{1}{2} E \int_V \epsilon_x^2 \, dV = \frac{1}{2} Etb \int_0^L u'^2 \, dx$$

According to Fig. 2.6.1c, a u displacement to the right in the upper face occurs simultaneously with a u displacement of equal magnitude to the left in the lower face. Hence a tensile strain in the upper face is accompanied by a compressive strain of equal magnitude in the lower face. Consequently the strain energy stored in the lower face is equal to that in the upper, and the total strain energy due to extensional deformations of the faces is

$$U_f = Etb \int_0^L u'^2 \, dx \qquad\qquad 2.6.2$$

The faces remain straight during the u deformations and are bent when the v deformations take place. The strain energy due to the bending of one face can be calculated from Eq. 2.1.15 in which the deflection η must be replaced by the symbol v:

$$U = \frac{1}{2} \int_0^L EI \left(\frac{d^2v}{dx^2}\right)^2 dx$$

If the shorter notation v'' is introduced for the second derivative of v with respect to x and the constant bending rigidity of one face

$$EI = E(bt^3/12)$$

is written before the integral sign, the total strain energy of bending stored in the two faces becomes

$$U_b = E \frac{bt^3}{12} \int_0^L v''^2 \, dx \qquad\qquad 2.6.3$$

since the two faces undergo identical v deformations.

The shear strain in the core consists of two parts. The first is caused by the u displacements as shown in Fig. 2.6.2b. When the upper face is displaced to the right and the lower to the left through a distance u, the original right angle between the axis of the beam and the cross section $x = $ constant is decreased by the angle γ_1 whose tangent is

$$\tan \gamma_1 = 2u/(c + t)$$

provided the displacement u is assumed to prevail at the middle of the faces. Because the elastic deformations considered are always small, the tangent is equal to the angle itself, when measured in radians, in very good approximation. Thus we may write

$$\gamma_1 = 2u/(c + t)$$

The v displacements increase the originally right angle by the angle γ_2 as may be seen from Fig. 2.6.2c. The tangent of this angle is

$$\tan \gamma_2 = \frac{(v + dv) - v}{dx} = v'$$

and because of the smallness of the angle we have

$$\gamma_2 = v'$$

in good approximation. The total decrement of the originally right angle due to both the u and v deformations is

$$\gamma = \gamma_1 - \gamma_2 = [2u/(c + t)] - v'$$

The shear strain energy stored in the core can be calculated from Eq. 2.1.9:

$$U_s = \frac{1}{2} G \int_V \gamma^2 \, dV = \frac{1}{2} bcG \int_0^L \left(\frac{2u}{c + t} - v' \right)^2 dx \qquad 2.6.4$$

The only external load is W. Its potential is

$$V = -Wv_{x=L} \qquad 2.6.5$$

since the work done by W is $W \, \delta v_{x=L}$ during a virtual displacement δv and zero during a virtual displacement δu. The subscript $x = L$ signifies that the value of v must be taken at $x = L$.

The expression for the total potential can now be written as

$$U + V = Etb \int_0^L u'^2 \, dx + E \frac{bt^3}{12} \int_0^L v''^2 \, dx$$

$$+ \frac{1}{2} Gbc \int_0^L \left(\frac{2u}{c + t} - v' \right)^2 dx - Wv_{x=L} \qquad 2.6.6$$

Minimization of the Total Potential

The physical model of the sandwich structural element whose total potential is given in Eq. 2.6.6 is a beam, having two thin but not infinitesimal faces, and a core attached to the median lines of the faces. The extensional rigidity of the core is very small compared to that of the faces, and its portion extending between $y = c/2$ and $y = (c + t)/2$ does not store strain energy. That this idealized model was suitable to serve as the basis for calculations was proved by the agreement between experiment and calculation.

According to the principle of the minimum of the total potential, the actual deflected shape characterized by the functions u and v has the property that the total potential corresponding to it has a stationary value. Consequently the change in the total potential is zero for any virtual variation of the deflection functions u and v. When u is increased to $u + \delta u$ and v to $v + \delta v$, the individual terms in the expression for the total potential become after manipulations similar to those carried out in the preceding article:

$$\delta \int_0^L u'^2 \, dx = \int_0^L (u + \delta u)'^2 \, dx - \int_0^L u'^2 \, dx = \int_0^L (u' + \delta u')^2 \, dx - \int_0^L u'^2 \, dx$$

$$= \int_0^L 2u' \, \delta u' \, dx$$

$$\delta \int_0^L v'' \, dx = \int_0^L (v + \delta v)''^2 \, dx - \int_0^L v''^2 \, dx = \int_0^L (v'' + \delta v'')^2 \, dx - \int_0^L v''^2 \, dx$$

$$= \int_0^L 2v'' \, \delta v'' \, dx$$

$$\delta \int_0^L \left(\frac{2u}{c+t} - v' \right)^2 dx = \int_0^L \left[\frac{2u + \delta u}{c+t} - (v + \delta v)' \right]^2 dx - \int_0^L \left(\frac{2u}{c+t} - v' \right)^2 dx$$

$$= \int_0^L \left[\frac{2u + \delta u}{c+t} - (v' + \delta v') \right]^2 dx - \int_0^L \left(\frac{2u}{c+t} - v' \right)^2 dx$$

$$\int_0^L \left[\left(\frac{2u}{c+t} - v' \right) + \left(\frac{2\delta u}{c+t} - \delta v' \right) \right]^2 dx - \int_0^L \left(\frac{2u}{c+t} - v' \right)^2 dx$$

$$= \int_0^L 2 \left(\frac{2u}{c+t} - v' \right) \left(\frac{2\delta u}{c+t} - \delta v' \right) dx$$

$$\delta W v_{x=L} = W \, \delta v_{x=L}$$

Hence

$$\delta (U + V) = 2Ebt \int_0^L u' \, \delta u' \, dx + E \frac{bt^3}{6} \int_0^L v'' \, \delta v'' \, dx$$

$$+ \, Gbc \int_0^L \left(\frac{2u}{c+t} - v' \right) \left(\frac{2\delta u}{c+t} - \delta v' \right) dx - W \, \delta v_{x=L} \quad 2.6.7$$

Next the derivatives of the variations of the deflection quantities will be eliminated from the integrals by integrating by parts. We obtain

$$\int_0^L u' \, \delta u' \, dx = u' \, \delta u \, \Big|_0^L - \int_0^L u'' \, \delta u \, dx$$

At the fixed end, $x = 0$, a u displacement would infringe on the geometric constraint, while at the free end, $x = L$, the tensile strain in the upper face and the corresponding compressive strain in the lower face, that is, $\epsilon_x = u'$ must be zero. Hence the first term in the right-hand member of the last equation vanishes:

$$\int_0^L u'\,\delta u'\,dx = -\int_0^L u''\,\delta u\,dx \qquad (a)$$

Similarly

$$\int_0^L v''\,\delta v''\,dx = v''\,\delta v' \Big|_0^L - \int_0^L v'''\,\delta v'\,dx$$

Here again the first term in the right-hand member of the equation vanishes, because at the fixed end the slope of the faces cannot change, and thus v' is zero, while at the free end the bending moment in the faces must vanish with the result that $v'' = 0$. Consequently

$$\int_0^L v''\,\delta v''\,dx = -\int_0^L v'''\,\delta v'\,dx = -v'''\,\delta v \Big|_0^L + \int_0^L v^{iv}\,\delta v\,dx$$

At the fixed end, $x = 0$, obviously $v = 0$, but at the free end nothing can be said of either v''' or δv. Hence

$$\int_0^L v''\,\delta v''\,dx = -(v'''\,\delta v)_{x=L} + \int_0^L v^{iv}\,\delta v\,dx \qquad (b)$$

Part of the expression representing the change in the shear strain energy

$$\int_0^L \left(\frac{2u}{c+t} - v'\right)\left(\frac{2\delta u}{c+t}\right)dx$$

remains unaltered. The rest is manipulated as follows:

$$-\int_0^L \left(\frac{2u}{c+t} - v'\right)\delta v'\,dx = -\left(\frac{2u}{c+t} - v'\right)\delta v \Big|_0^L + \int_0^L \left(\frac{2u'}{c+t} - v''\right)\delta v\,dx$$

In the first term of the right-hand member δv vanishes at $x = 0$, but at $x = L$ both its factors have non-zero values. Consequently

$$\int_0^L \left(\frac{2u}{c+t} - v'\right)\left(\frac{2\delta u}{c+t} - \delta v'\right)dx$$

$$= \int_0^L \left(\frac{2u}{c+t} - v'\right)\left(\frac{2\delta u}{c+t}\right)dx + \int_0^L \left(\frac{2u'}{c+t} - v''\right)dv\,dx - \left(\frac{2u}{c+t} - v'\right)\delta v_{x=L}$$

$$(c)$$

If expressions a, b, and c are substituted in Eq. 2.6.7, the terms are re-arranged, and $\delta(U + V)$ is set equal to zero, it follows that

$$
\begin{aligned}
\delta(U + V) = {}& -2Ebt \int_0^L u'' \, \delta u \, dx + Gbc \int_0^L \left(\frac{2u}{c+t} - v' \right) \frac{2\delta u}{c+t} \, dx \\
& + \frac{Ebt^3}{6} \int_0^L v^{\mathrm{iv}} \, \delta v \, dx + Gbc \int_0^L \left(\frac{2u'}{c+t} - v'' \right) \delta v \, dx \\
& - \frac{Ebt^3}{6} (v''' \, \delta v)_{x=L} - Gbc \left(\frac{2u}{c+t} - v' \right) \delta v_{x=L} - W \delta v_{x=L} \\
= {}& \int_0^L \left[-2Ebtu'' + \frac{2Gbc}{c+t} \left(\frac{2u}{c+t} - v' \right) \right] \delta u \, dx \\
& + \int_0^L \left[\frac{Ebt^3}{6} v^{\mathrm{iv}} + Gbc \left(\frac{2u'}{c+t} - v'' \right) \right] \delta v \, dx \\
& - \left[\frac{Ebt^3}{6} v''' + Gbc \left(\frac{2u}{c+t} - v' \right) + W \right] \delta v_{x=L} = 0
\end{aligned}
$$

In accordance with the principle of the minimum of the total potential this equation must be satisfied identically for any arbitrary choice of the variations δu and δv of the displacement functions u and v. An argument paralleling that given in Art. 2.4 would show that this is possible only if

$$
-2Ebtu'' + \frac{2Gbc}{c+t} \left(\frac{2u}{c+t} - v' \right) = 0 \qquad\qquad 2.6.8
$$

$$
\frac{Ebt^3}{6} v^{\mathrm{iv}} + Gbc \left(\frac{2u'}{c+t} - v'' \right) = 0 \qquad\qquad 2.6.9
$$

$$
\left[\frac{Ebt^3}{6} v''' + Gbc \left(\frac{2u}{c+t} - v' \right) + W \right]_{x=L} = 0 \qquad\qquad 2.6.10
$$

Equations 2.6.8 and 2.6.9, boundary condition Eq. 2.6.10, and the boundary conditions used earlier, namely,

$$
u = v = v' = 0 \qquad \text{when} \quad x = 0
$$

$$
u' = v'' = 0 \qquad \text{when} \quad x = L \qquad\qquad 2.6.11
$$

completely define the problem of the bending of the sandwich beam. In particular, Eq. 2.6.10 gives the rule according to which the shear force W must be distributed at the free end of the cantilever between the faces and the core. The problem set at the beginning of this article is thus reduced to solving two simultaneous differential equations.

Integration of the Differential Equations

With the notation

$$K_1^2 = \frac{Gcb}{EI_1}, \qquad K_2^2 = \frac{Gcb}{2EI_f}, \qquad y = \frac{c+t}{2}v$$

$$I_1 = \tfrac{1}{2}t(c+t)^2b, \qquad 2I_f = \tfrac{1}{6}t^3b \qquad\qquad 2.6.12$$

Equations 2.6.8 to 2.6.10 can be written in the form

$$u'' - K_1^2 u + K_1^2 y' = 0 \qquad\qquad 2.6.13$$

$$y^{iv} - K_2^2 y'' + K_2^2 u' = 0 \qquad\qquad 2.6.14$$

$$\left[y''' - K_2^2 y' + K_2^2 u + \frac{c+t}{4EI_f} W \right]_{x=L} = 0 \qquad\qquad 2.6.15$$

Differentiation of Eq. 2.6.13 and solution for y'' yield

$$y'' = u' - (u'''/K_1^2) \qquad\qquad 2.6.16$$

Substitution of y'' from Eq. 2.6.16 in Eq. 2.6.14 gives

$$u^v - (K_1^2 + K_2^2)u''' = 0$$

With the notation

$$p^2 = K_1^2 + K_2^2 = \frac{Gcb}{EI_1} + \frac{Gcb}{2EI_f} = Gcb\frac{EI_1 + 2EI_f}{EI_1(2EI_f)} \qquad\qquad 2.6.17$$

the differential equation can be written as

$$u^v - p^2 u''' = 0 \qquad\qquad 2.6.18$$

If one introduces the new variable

$$z = u''' \qquad\qquad 2.6.19$$

Eq. 2.6.18 becomes

$$z'' - p^2 z = 0 \qquad\qquad 2.6.20$$

whose solution is well known:

$$z = A \cosh px + B \sinh px = u''' \qquad\qquad 2.6.21$$

where A and B are arbitrary constants of integration. The solution can be verified easily by substitution in Eq. 2.6.20. Two consecutive integrations of $z = u'''$ yield

$$u' = (A/p^2) \cosh px + (B/p^2) \sinh px + Cx + D \qquad\qquad 2.6.22$$

Substitution in Eq. 2.6.16 results in

$$y'' = \frac{K_1^2 - p^2}{K_1^2 p^2} (A \cosh px + B \sinh px) + Cx + D$$

Integration of the two expressions yields

$$u = (1/p^3)(A \sinh px + B \cosh px) + \frac{1}{2} Cx^2 + Dx + F \qquad 2.6.23$$

$$y' = \frac{K_1^2 - p^2}{K_1^2 p^3} (A \sinh px + B \cosh px) + \frac{1}{2} Cx^2 + Dx + H$$

Hence

$$y' - u = \frac{-1}{K_1^2 p} (A \sinh px + B \cosh px) + H - F$$

But according to Eq. 2.6.13

$$y' - u = -u''/K_1^2$$

while from Eq. 2.6.22

$$\frac{u''}{K_1^2} = \frac{1}{K_1^2 p} (A \sinh px + B \cosh px) + \frac{C}{K_1^2}$$

Consequently

$$H - F = -\frac{C}{K_1^2}$$

and thus H can be replaced by $F - (C/K_1^2)$. One more integration of y' yields therefore

$$y = \frac{K_1^2 - p^2}{K_1^2 p^4} (A \cosh px + B \sinh px) + \frac{1}{6} Cx^3 + \frac{1}{2} Dx^2$$
$$+ \left(F - \frac{C}{K_1^2} \right) x + J \qquad 2.6.24$$

Consideration of the boundary conditions 2.6.11 gives:
When $x = 0$:

$$v' = 0 = y' = \frac{K_1^2 - p^2}{K_1^2 p^3} B + F - \frac{C}{K_1^2}$$

$$u = 0 = B/p^3 + F$$

Consequently

$$F = -B/p^3, \qquad C = -B/p \qquad (d)$$

Moreover

$$v = 0 = y = \frac{K_1^2 - p^2}{K_1^2 p^4} A + J$$

and thus

$$-J = \frac{K_1^2 - p^2}{K_1^2 p^4} A \qquad (e)$$

When $x = L$:

$$u' = 0 = \frac{A}{p^2} \cosh pL + \frac{B}{p^2} \sinh pL - \frac{BL}{p} + D \qquad (f)$$

$$v'' = 0 = y'' = \frac{K_1^2 - p^2}{K_1^2 p^2} (A \cosh pL + B \sinh pL) - \frac{BL}{p} + D$$

Consequently

$$v'' - u' = -\frac{1}{K_1^2} (A \cosh pL + B \sinh pL) = 0$$

and

$$B = -A \coth pL \qquad (g)$$

From Eq. f one obtains

$$D = B(L/p) = -A(L/p) \coth pL \qquad (h)$$

Substitution of these values in Eq. 2.6.15 gives

$$A(p/K^2) \coth pL + [(c + t)/4EI_f]W = 0$$

whose solution can be written in the form

$$A = -[(c + t)/2L]pL \tanh pL \, [W/(EI_1 + 2EI_f)] \qquad (i)$$

If the values of the coefficients are substituted in Eqs. 2.6.23 and 2.6.24 and use is made of the relations contained in Eqs. 2.6.12, the following results are obtained after some manipulations:

$$u = \frac{c + t}{2L} \frac{WL^3}{EI_b} \left[\frac{x}{L} - \frac{1}{2} \left(\frac{x}{L}\right)^2 + \frac{\cosh p(L - x) - \cosh pL}{(pL)^2 \cosh pL} \right] \qquad 2.6.25$$

$$v = \frac{WL^3}{3EI_b} \left\{ \frac{3}{2} \left(\frac{x}{L}\right)^2 - \frac{1}{2} \left(\frac{x}{L}\right)^3 \right.$$

$$\left. + 3 \frac{EI_1}{2EI_f} \left[\frac{\sinh p(L - x) - \sinh pL}{(pL)^3 \cosh pL} + \frac{x/L}{(pL^2)} \right] \right\} \qquad 2.6.26$$

where

$$EI_b = EI_1 + 2EI_f \qquad 2.6.27$$

is the bending rigidity of the sandwich beam. The correctness of this solution can be checked by substitution.

End Deflection of Cantilever

Of particular interest is the end deflection of the beam. When $x = L$:

$$v_{x=L} = \frac{WL^3}{3EI_b} \left[1 + \frac{EI_1}{2EI_f} \frac{3}{(pL)^2} \left(1 - \frac{\tanh pL}{pL}\right) \right] \qquad 2.6.28$$

which can also be written in the form

$$v_{x=L} = \psi \frac{WL^3}{3EI_b} \qquad \text{2.6.29}$$

with

$$\psi = 1 + (CEI_1/2EI_f) \qquad \text{2.6.29}a$$

and

$$C = \frac{3}{(pL)^2}\left(1 - \frac{\tanh pL}{pL}\right) \qquad \text{2.6.29}b$$

The factor C is plotted in Fig. 2.6.3. When $\psi = 1$, the deflections of the

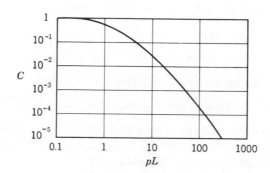

Fig. 2.6.3. Factor C in formula for end deflection.   From author's
paper in *J. Aero. Sciences.*

sandwich beam are in agreement with the values given by the ordinary bending formula.   Hence the second term in the expression for ψ represents the shearing deformations.   The importance of the latter may be judged from a numerical example:

When $L = 10$ in., $c = 1$ in., $t = 0.032$ in., $b = 1$ in., $E = 10.5 \times 10^6$ psi, and $G = 4000$ psi, one obtains

$$EI_1 = 178{,}500 \text{ lb-in.}^2 \qquad 2EI_f = 57 \text{ lb-in.}^2$$
$$Gcb = 4000 \text{ lb} \qquad p^2 = 70 \text{ in.}^{-2}$$
$$pL = 84 \qquad C = 4.2 \times 10^{-4}$$

Consequently

$$\psi = 1 + 1.32 = 2.32$$

and the actual end deflection is 132 per cent higher than that corresponding to a core with perfect shearing rigidity.

Equations 2.6.28 and 2.6.29 can be further simplified for certain values of the non-dimensional parameter pL:

(a) *When $pL < 0.1$.* The hyperbolic tangent can be replaced by the first two terms of its Taylor expansion:

$$\tanh pL = pL - (pL)^3/3 + \cdots$$

One obtains

$$1 - [(\tanh pL)/pL] = (pL)^2/3$$

$$C = 1$$

$$\psi = EI_b/2EI_f$$

and

$$v_{x=L} = \frac{WL^3}{3(2EI_f)} \qquad\qquad 2.6.30$$

The value of pL is small when the shearing rigidity of the core is very small; then the sandwich beam deflects like a beam consisting of only two independent faces. This result could be anticipated, and serves as a check on the fundamental soundness of the assumptions underlying the calculations.

(b) *When $pL > 100$.* In this case $\tanh pL$ is almost equal to unity, and $(\tanh pL)/pL$ is less than 0.01. Hence this term can be safely neglected when it is added to unity. Thus

$$C = 3/(pL)^2$$

and

$$v_{x=L} = \frac{WL^3}{3EI_b} + \frac{WL}{EI_b}\frac{EI_1}{2EI_f}\frac{1}{p^2} = \frac{WL^3}{3EI_b} + \frac{WL}{EI_b}\frac{EI_1}{2EI_f}\frac{EI_1}{Gcb}\frac{2EI_f}{EI_b}$$

$$v_{x=L} = \frac{WL^3}{3EI_b} + \frac{WL}{Gcb}\left(\frac{EI_1}{EI_b}\right)^2 \qquad\qquad 2.6.31$$

Since in most cases EI_b differs little from EI_1, the total end deflection of a sandwich beam with $pL > 100$ can be calculated as the sum of the bending deflection $WL^3/3EI_b$ plus the shearing deflection WL/Gcb. This is also a reasonable result.

(c) *When $pL \to \infty$.* When pL is very large, the second term in the brackets in Eq. 2.6.28 vanishes. Then the deflection is

$$v_{x=L} = WL^3/3EI_b \qquad\qquad 2.6.32$$

that is, the simple bending deflection. Obviously no shearing deformation can occur when the shearing rigidity is indefinitely large.

Discussion of the Advantages of the Minimal Principle

In the last few years several authors have formulated and solved the problems of sandwich structures on the basis of geometric and physical considerations without recourse to the minimal principle. The main difficulty of such an approach lies in deciding what to disregard in setting up the differential equations. If too many factors are neglected, some of the characteristic features of the problem may get lost; on the other hand, the retention of too many details usually leads to complex differential equations that can only be solved for particular boundary conditions with a great deal of effort. The advantage of the strain-energy method lies in the ease with which the essential features can be distinguished from the unessential ones, as was shown in this article. Moreover the minimal principle furnishes all the necessary and sufficient conditions of equilibrium in the form of differential equations as well as boundary conditions. In many problems where the differential equations can be derived with comparative ease from geometry and physical considerations, it is difficult, or even impossible, to stipulate the boundary conditions in any other manner than by resorting to the principle of the minimum of the total potential.

2.7   The Rayleigh-Ritz Method

The Rayleigh Method

It was shown in the preceding articles how the strain-energy method could be used to solve rigorously problems in structural analysis by establishing differential equations and boundary conditions. The strain-energy method has one more field of application, and this is of even greater importance in engineering. It permits the derivation of approximate solutions with comparatively little effort when the establishment and solution of the differential equations present difficulties.

The simplest procedure for obtaining the deflections approximately was devised by Lord Rayleigh. It consists of assuming arbitrarily a reasonable deflected shape involving an undetermined coefficient and equating the strain energy stored in the elastic body to the work done by the external forces. The equation can then be solved for the undetermined constant in the deflection function. The equilibrium conditions of the infinitesimal elements of the structure (which in the preceding sections were represented by the differential equations) are not satisfied in this procedure, but the boundary conditions can be satisfied by choosing suitable deflection functions. The equation from which the undetermined constant is calculated is an expression of the principle of the conservation of energy. In many instances this very simple approach yields results perfectly satisfactory for engineering calculations.

As an example, the deflection of the midpoint of a beam will be calculated which is supported and loaded as shown in Fig. 2.7.1.   In general the most convenient shapes one can assume in the Rayleigh method for the deflections are those represented by trigonometric functions and polynomials. In the present case, for instance, the function

$$y = a \sin (\pi x/L) \qquad\qquad 2.7.1$$

obviously vanishes at $x = 0$ and $x = L$ since $\sin 0 = \sin \pi = 0$.   Its second derivative is

$$d^2y/dx^2 = y'' = -a(\pi/L)^2 \sin (\pi x/L) \qquad\qquad 2.7.2$$

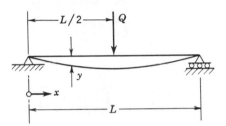

Fig. 2.7.1.   Beam with concentrated load.

and, since the curvature d^2y/dx^2 of a beam is connected with the bending moment by the equation

$$EIy'' = -M \qquad\qquad 2.7.3$$

the bending moment also vanishes at the supports.   The sine function therefore satisfies all the four boundary conditions given in Art. 2.4.   The strain energy stored in the beam during deformations, according to Eq. 2.7.1, is

$$U = \frac{EI}{2} \int_0^L y''^2 \, dx = \frac{EI}{2} a^2 \left(\frac{\pi}{L}\right)^4 \int_0^L \sin^2 \frac{\pi x}{L} \, dx \qquad\qquad 2.7.4$$

Any table of integrals shows that

$$\int \sin^2 z \, dz = \frac{z}{2} - \frac{1}{4} \sin (2z) \qquad\qquad 2.7.5$$

One can write

$$\int_0^L \sin^2 \frac{\pi x}{L} \, dx = \frac{L}{\pi} \int_0^L \sin^2 \frac{\pi x}{L} \, d\frac{\pi x}{L} \qquad\qquad 2.7.6$$

If the substitution

$$z = \pi x/L \qquad\qquad 2.7.7$$

is introduced and it is observed that in accordance with Eq. 2.7.7 the upper limit for z is π when it is L for x, Eq. 2.7.6 can be replaced by

$$\int_0^L \sin^2 \frac{\pi x}{L} \, dx = \frac{L}{\pi} \int_0^\pi \sin^2 z \, dz \qquad 2.7.8$$

which becomes, according to Eq. 2.7.6,

$$\int_0^L \sin^2 \frac{\pi x}{L} \, dx = \frac{L}{\pi} \left[\frac{z}{2} - \frac{1}{4} \sin 2z \right]_0^\pi$$

As the upper limit of the first term is $\pi/2$, that of the second term $\sin 2\pi = 0$, and the two lower limits vanish, one obtains finally

$$\int_0^L \sin^2 \frac{\pi x}{L} \, dx = \left(\frac{L}{\pi} \right) \left(\frac{\pi}{2} \right) = \frac{L}{2} \qquad 2.7.9$$

Consequently Eq. 2.7.4 becomes

$$U = \frac{EI}{2} a^2 \left(\frac{\pi}{L} \right)^4 \frac{L}{2} = \left(\frac{\pi^4}{4} \right) \left(a^2 \frac{EI}{L^3} \right) \qquad 2.7.10$$

The work done during a slow increase of the external load from zero to the value Q is equal to the product of the average value of the load, $Q/2$, and the maximum deflection. (The factor $\frac{1}{2}$ is needed here since the actual work done during the entire loading process is required in the application of the principle of the conservation of energy, and not the potential of the final value of the load as in the principle of virtual displacements.) From Eq. 2.7.1 the deflection at $x = L/2$ is a, and thus the work W is

$$W = aQ/2 \qquad 2.7.11$$

The principle of the conservation of energy requires that

$$U = (\pi^4/4)(a^2 EI/L^3) = W = aQ/2$$

which can be solved for a:

$$a = (2/\pi^4)(QL^3/EI) = 0.02053(QL^3/EI) \qquad 2.7.12$$

Substitution in Eq. 2.7.1 yields

$$y = 0.02053(QL^3/EI) \sin (\pi x/L) \qquad 2.7.13$$

Hence the deflections at the eighth points of the span are:

$x/L = 0$	1/8	1/4	3/8	1/2
$(EI/QL^3)y = 0$	0.008062	0.01452	0.01897	0.02053

The rigorous solution is given in engineering handbooks as

$$y = (WL^3/48EI)[3(x/L) - 4(x/L)^3] \qquad 2.7.14$$

From this equation the exact values of the deflections can be computed. Those given below are correct to four significant figures:

$x/L = 0$	1/8	1/4	3/8	1/2
$(EI/QL^3)y = 0$	0.007650	0.01432	0.01904	0.02083

Comparison with the approximate values shows a remarkably good agreement. This is not necessarily so in every problem, and for this reason it is worth while to find a method by which the first approximation can be improved.

The Rayleigh-Ritz Method

A natural extension of the Rayleigh method, denoted as the Rayleigh-Ritz, or often simply as the Rayleigh method, makes use of a more complex expression for the deflected shape. For instance Eq. 2.7.1 can be replaced by the series

$$y = a_1 \sin (\pi x/L) + a_2 \sin (2\pi x/L) + \cdots + a_n \sin (n\pi x/L) \qquad 2.7.15$$

in which all the n coefficients a_i (where i takes on every integral value from 1 to n) have to be adjusted in such a manner as to approximate best the true deflected shape. The rule from which the best approximation is derived is given by the principle of the minimum of the total potential. It was shown that the true deflected shape differs from all other geometrically possible shapes inasmuch as it corresponds to the minimum value of the total potential. The total potential must therefore be made as small as possible by a suitable choice of the coefficients a_i. This can also be expressed by saying that the total potential must be minimized with respect to the n undetermined coefficients a_i.

The strain energy stored in the beam during deformations corresponding to Eq. 2.7.15 can be written in the form

$$U = \frac{EI}{2} \int_0^L y''^2 \, dx = \left(\frac{EI}{2}\right)\left(\frac{\pi}{L}\right)^4 \int_0^L \left[a_1^2 \sin^2 \frac{\pi x}{L}\right.$$

$$+ \, 2^4 a_2^2 \sin^2 \frac{2\pi x}{L} + \cdots + n^4 a_n^2 \sin^2 \frac{n\pi x}{L}$$

$$+ \, 2 \cdot 2^2 \sin \frac{\pi x}{L} \sin \frac{2\pi x}{L} + 2 \cdot 3^2 \sin \frac{\pi x}{L} \sin \frac{3\pi x}{L} + \cdots$$

$$\left. + \, 2(n-1)^2 n^2 \sin \frac{(n-1)\pi x}{L} \sin \frac{n\pi x}{L}\right] dx \qquad 2.7.16$$

A substitution similar to that presented in Eq. 2.7.7 yields

$$\int_0^L \sin^2 \frac{i\pi x}{L} \, dx = \frac{L}{2} \qquad \qquad 2.7.17$$

for any integral value of i. On the other hand, from a known trigonometric identity

$$\int_0^L \sin \frac{i\pi x}{L} \sin \frac{j\pi x}{L} \, dx = \frac{1}{2} \int_0^L \left[\cos \frac{(i-j)\pi x}{L} - \cos \frac{(i+j)\pi x}{L} \right] dx$$

$$= \frac{1}{2} \left[\frac{L}{(i-j)\pi} \sin \frac{(i-j)\pi x}{L} - \frac{L}{(i+j)\pi} \sin \frac{(i+j)\pi x}{L} \right]_0^L$$

provided i and j are different integers. But in such a case the argument of the sine function is an integral multiple of π at the upper limit, and thus the value of the function there is zero. The function naturally vanishes at the lower limit also with the effect that

$$\int_0^L \sin \frac{i\pi x}{L} \sin \frac{j\pi x}{L} \, dx = 0 \qquad \qquad 2.7.18$$

when i and j are different integers. The properties of the sine functions expressed by Eq. 2.7.17 and 2.7.18 are designated in mathematics as *orthogonality*.

With these definite integrals the strain-energy expression represented by Eq. 2.7.16 can be evaluated:

$$U = (\pi^4/4)(EI/L^3)(a_1^2 + 2^4 a_2^2 + 3^4 a_3^2 + \cdots + n^4 a_n^2) \qquad 2.7.19$$

The potential V of the load Q is by definition $-Qy_m$ with y_m the deflection of the midpoint of the beam where the load is acting. (This is the potential as defined in Eq. 2.2.4, and thus the factor $\frac{1}{2}$ is absent.) For $x = L/2$, one obtains from Eq. 2.7.15

$$V = -Q(a_1 - a_3 + a_5 - a_7 + \cdots + \phi a_{n-1} + \psi a_n) \qquad 2.7.20$$

where

$$\phi = 1 \quad \text{if} \quad (n-2)/2 \text{ is an even number}$$
$$\phi = -1 \quad \text{if} \quad (n-2)/2 \text{ is an odd number} \qquad 2.7.20a$$
$$\phi = 0 \quad \text{if} \quad (n-2)/2 \text{ is not an integer}$$

$$\psi = 1 \quad \text{if} \quad (n-1)/2 \text{ is an even number}$$
$$\psi = -1 \quad \text{if} \quad (n-1)/2 \text{ is an odd number} \qquad 2.7.20b$$
$$\psi = 0 \quad \text{if} \quad (n-1)/2 \text{ is not an integer}$$

The total potential $U + V$ is now represented by a function of the n unknown coefficients a_i.   The analytic expression of the requirement that the total potential have a stationary value is that the partial derivatives of the total potential with respect to the coefficients a_i vanish, as was discussed at the end of Art. 2.2.   Consequently

$$\partial(U + V)/\partial a_1 = (\pi^4/2)(EI/L^3)a_1 - Q \quad = 0$$
$$\partial(U + V)/\partial a_2 = (\pi^4/2)(EI/L^3)2^4 a_2 \quad = 0$$
$$\partial(U + V)/\partial a_3 = (\pi^4/2)(EI/L^3)3^4 a_3 + Q = 0$$
$$\partial(U + V)/\partial a_4 = (\pi^4/2)(EI/L^3)4^4 a_4 \quad = 0 \qquad 2.7.21$$
$$\vdots \qquad\qquad \vdots \qquad\qquad \vdots$$
$$\partial(U + V)/\partial a_n = (\pi^4/2)(EI/L^3)n^4 a_n - \psi Q = 0$$

where ψ has the values given in Eq. 2.7.20b.   It was mentioned at the end of Art. 2.2 that the stationary value of the total potential is a minimum when the equilibrium is stable.   Under such conditions therefore the values of the n coefficients a_1 that satisfy Eqs. 2.7.21 make the total potential as small as possible when the deflected shape is represented by the n trigonometric terms contained in Eq. 2.7.15.   Since the true shape corresponds to the absolute minimum of the total potential, satisfaction of Eqs. 2.7.21 can be expected to yield the best possible approximation compatible with the representation of the deflected shape by the series of Eq. 2.7.15.

Equations 2.7.21 can be readily solved for the unknown coefficients:

$$a_1 = 2QL^3/\pi^4 EI$$
$$a_2 = 0$$
$$a_3 = -(1/3^4)(2QL^3/\pi^4 EI)$$
$$a_4 = 0$$
$$a_5 = (1/5^4)(2QL^3/\pi^4 EI)$$
$$\vdots \qquad \vdots \qquad \vdots$$
$$a_{n-1} = [\phi/(n-1)^4](2QL^3/\pi^4 EI)$$
$$a_n = (\psi/n^4)(2QL^3/\pi^4 EI)$$

The deflected shape can be given therefore after substitution of the values obtained for the coefficients into Eq. 2.7.15, in the form

$$y = (2QL^3/\pi^4 EI)\{\sin(\pi x/L) - (\tfrac{1}{3})^4 \sin(3\pi x/L)$$
$$+ (\tfrac{1}{5})^4 \sin(5\pi x/L) - \cdots + \phi[1/(n-1)^4] \sin[(n-1)\pi x/L]$$
$$+ \psi(1/n)^4 \sin(n\pi x/L)\} \qquad\qquad 2.7.22$$

It can be seen that the first term in this series is identical with the solution given in Eqs. 2.7.12 and 2.7.13. The shape obtained is modified by the $n-1$ additional trigonometric functions. If the first two terms, corresponding to a_1 and a_3, are taken into account, the deflections at the eighth points of the span are:

$x/L = 0$	$\tfrac{1}{8}$	$\tfrac{1}{4}$	$\tfrac{3}{8}$	$\tfrac{1}{2}$
$(EI/QL^3)y = 0$	0.007828	0.01434	0.01907	0.02078

Addition of the third term, namely $(\tfrac{1}{5})^4 \sin(5x/L)$, yields:

$x/L = 0$	$\tfrac{1}{8}$	$\tfrac{1}{4}$	$\tfrac{3}{8}$	$\tfrac{1}{2}$
$(EI/QL^3)y = 0$	0.007858	0.01432	0.01906	0.02081

The numerical values indicate that the approximate solution gradually approaches the exact deflections as the number of terms taken into account in Eq. 2.7.22 is increased. The addition of each term to the series is equivalent to enforcing that the strain energy be a minimum with respect to one more parameter. When the number of terms in the assumption for the deflected shape increases beyond all limits, $n \to \infty$, consideration of the indefinitely large number of coefficients is equivalent to minimizing the total potential with respect to an indefinitely large number of deflection parameters. The important question then arises whether the infinity of deflection functions is capable of describing any arbitrary virtual deflection pattern. If the answer is yes, the total potential is a minimum with respect to all possible virtual displacements. Then all the equilibrium conditions are satisfied, and the solution is exact.

It is shown in texts on mathematics that an infinite trigonometric series of the type given in Eq. 2.7.15, with $n \to \infty$, contains a complete set of functions which means that the series can represent any arbitrary continuous and even many discontinuous functions. Hence the Rayleigh-Ritz method gives an exact solution when an infinite trigonometric series is used in it and all the terms are considered in the calculation.

It can be seen from this discussion that the Rayleigh-Ritz method replaces by a much simpler procedure the often formidable task of minimizing the total potential of a system with the aid of the variational calculus and of solving the differential equations so obtained. In the simpler procedure the solution is assumed in the form of a series, and the total potential is minimized with respect to the coefficients of the series. This minimization makes use only of the methods of the differential calculus and not of those of the variational calculus. Moreover no differential equations need be solved. The solution obtained by the Rayleigh-Ritz method is approximate when a finite number of terms are considered, and often a very small number of terms suffice to obtain a satisfactory solution. The solution is exact when all the terms of an infinite series are taken into account.

In the example given the deflection function was so chosen that every term of it satisfied the geometric boundary conditions (zero deflection at the ends of the beam), as well as the conditions of equilibrium at the boundary (zero moment at the ends of the beam). When the deflection function adopted is complete, only the geometric boundary conditions need be satisfied, because minimization of the total potential enforces the condition of equilibrium that the virtual work vanish with respect to every virtual displacement. Just as in the direct application of the minimal principle, in the Rayleigh-Ritz method the conditions of equilibrium are satisfied by the minimization, and the deflection functions need only take care of the continuity of the deformations. In practical work it is generally preferable to select functions that also satisfy the equilibrium conditions at the boundary, since then, as a rule, a smaller number of terms suffices for an engineering approximation.

2.8   Trigonometric Series

In the solution of problems in structural analysis it is often necessary to represent a function given by an analytical expression or a graph by means of a finite number of terms of a trigonometric series. It is desirable then to choose the coefficients of these terms in such a manner as to obtain the best possible approximation to the given function.

This necessitates naturally the establishment of a suitable criterion for the *best* approximation, and there are many ways in which this can be done. In Fig. 2.8.1, $y = F(x)$, the function to be represented by a finite trigonometric series is shown by a solid line. The approximation obtained by the first N terms of the series

$$y = \sum_{n=1}^{N} a_n \sin \frac{n\pi x}{L} \qquad 2.8.1$$

is schematically represented by the dotted line. One might require that the maximum absolute value of the differences between function and approximation

$$D_1 = [|F(x) - \Sigma a_n \sin (n\pi x/L)|]_{max} \qquad 2.8.2$$

be as small as possible. Alternatively the mean value of these differences in the range $0 \leq x \leq L$

$$D_2 = \frac{1}{L} \int_0^L \left\{ \left| F(x) - \sum a_n \sin \frac{n\pi x}{L} \right| \right\} dx \qquad 2.8.3$$

could be minimized. The difficulty with these definitions lies in the calculation of the absolute values. On the other hand, the omission of the

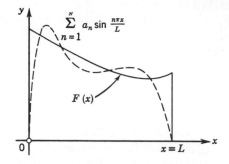

Fig. 2.8.1. Approximation by trigonometric series.

absolute value and the stipulation that the average difference itself

$$D_3 = \frac{1}{L} \int_0^L \left[F(x) - \sum a_n \sin \frac{n\pi x}{L} \right] dx \qquad 2.8.4$$

be made as small as possible leads to an unsuitable criterion. An approximation deviating greatly from the given function in the positive and the negative directions alternately could yield a small or even a vanishing value for D_3, and obviously it would not be good for engineering purposes.

The most generally adopted criterion eliminates the possibility of negative errors canceling positive errors through the use of the square of the difference which is always positive with real functions. It is stipulated that the mean square of the difference between function and approximation

$$D_4 = \frac{1}{L} \int_0^L \left[F(x) - \sum_{n=1}^N a_n \sin \frac{n\pi x}{L} \right]^2 dx \qquad 2.8.5$$

be a minimum with respect to the N coefficients a_n. This can be written in the form of an equation

$$\frac{\partial}{\partial a_i} \left\{ \frac{1}{L} \int_0^L \left[F(x) - \sum_{n=1}^N a_n \sin \frac{n\pi x}{L} \right]^2 dx \right\} = 0 \qquad 2.8.6$$

for all integral values of i between 1 and N. If the differentiation is carried out, one obtains

$$\frac{2}{L} \int_0^L \left[F(x) - \sum_{n=1}^N a_n \sin \frac{n\pi x}{L} \right] \sin \frac{i\pi x}{L} \, dx = 0 \qquad 2.8.7$$

because $F(x)$ and all the terms of the series are independent of a_i except the ith term which is proportional to it. In view of Eqs. 2.7.17 and 2.7.18 the integral of the product of $\sin (i\pi x/L)$ and $a_n \sin (n\pi x/L)$ vanishes for every n that is different from i, and the only integral remaining is

$$-a_1 \int_0^L \sin^2 \frac{i\pi x}{L} \, dx = -a_i \frac{L}{2}$$

Because $\int_0^L F(x) \sin \frac{i\pi x}{L} \, dx$ does not necessarily vanish, Eq. 2.8.7 reduces to

$$\int_0^L F(x) \sin \frac{i\pi x}{L} \, dx - a_i \frac{L}{2} = 0 \qquad 2.8.8$$

This equation can be solved for the coefficient a_i:

$$a_i = \frac{2}{L} \int_0^L F(x) \sin \frac{i\pi x}{L} \, dx \qquad 2.8.9$$

The remarkable fact about this formula is that the ith coefficient in the trigonometric series depends only on the given function and $\sin (i\pi x/L)$, and is independent of all the other coefficients. This means that the coefficients are final, or, in other words, the first N coefficients remain unchanged when the accuracy is increased by addition of M more terms to the series.

Infinite trigonometric series of the type of Eq. 2.8.1 (with $n \to \infty$), whose coefficients are determined from Eq. 2.8.9, are known as Fourier series. In the solution of differential equations it is often desirable to replace a given function, even though it may be quite simple, by its Fourier series expansion. A numerical example will show the details of the procedure. In Fig. 2.8.2 the function is

$$y = \tfrac{1}{2}[1 + (x/L)] \quad \text{when} \quad 0 \le x \le L \qquad 2.8.10$$

It is to be replaced by the infinite series

$$y = a_1 \sin (\pi x/L) + a_2 \sin (2\pi x/L) + \cdots + a_i \sin (i\pi x/L) + \cdots \quad 2.8.11$$

The coefficient of the general term containing a_i can be calculated from Eq. 2.8.9:

$$a_i = \frac{1}{L} \int_0^L \left(1 + \frac{x}{L}\right) \sin\frac{i\pi x}{L}\, dx$$

Since in

$$\int_0^L \sin\frac{i\pi x}{L}\, dx = -\frac{L}{i\pi} \cos\frac{i\pi x}{L}\Big|_0^L$$

the cosine function has as its upper limit -1 when i is odd, and $+1$ when

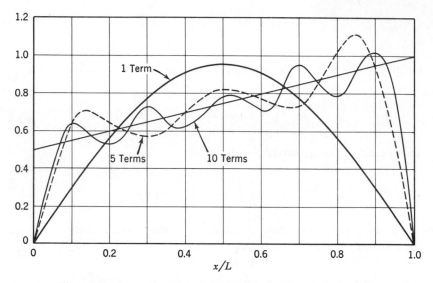

Fig. 2.8.2. Approximation of straight line by trigonometric series.

i is even, and the lower limit is always $+1$, the upper and lower limits add up to -2 when i is odd and cancel out when i is even. In the form of equations

$$\int_0^L \sin\frac{i\pi x}{L}\, dx = \frac{2L}{i\pi} \quad \text{when} \quad i \text{ is odd}$$

$$= 0 \quad \text{when} \quad i \text{ is even} \qquad 2.8.12$$

From tables of integrals

$$\int z \sin z\, dz = \sin z - z \cos z$$

Consequently

$$\int_0^L \frac{x}{L} \sin \frac{i\pi x}{L}\, dx = \frac{L}{i^2\pi^2} \int_0^{i\pi} \frac{i\pi x}{L} \sin \frac{i\pi x}{L}\, d\frac{i\pi x}{L}$$

$$= \frac{L}{i^2\pi^2}\left[\sin \frac{i\pi x}{L} - \frac{i\pi x}{L}\cos \frac{i\pi x}{L}\right]_0^{i\pi}$$

where the limits are the values assumed by $i\pi x/L$ rather than by x.   The sine function vanishes at both limits for all values of i.   The upper limit of the second term in brackets is $-i\pi$ when i is odd and $+i\pi$ when i is even.   The lower limit is always zero since the factor $i\pi x/L$ vanishes.   Thus

$$\int_0^L \frac{x}{L} \sin \frac{i\pi x}{L}\, dx = \frac{L}{i\pi} \qquad \text{when} \quad i \text{ is odd}$$

$$= -\frac{L}{i\pi} \qquad \text{when} \quad i \text{ is even} \qquad 2.8.13$$

Because of Eqs. 2.8.12 and 2.8.13 the coefficient becomes

$$a_i = 3/i\pi \qquad \text{when} \quad i \text{ is odd}$$

$$= -1/i\pi \qquad \text{when} \quad i \text{ is even}$$

The series expression for the function given in Eq. 2.8.10 is therefore

$$y = (1/\pi)[3 \sin (\pi x/L) - \tfrac{1}{2} \sin (2\pi x/L) + \sin (3\pi x/L)$$

$$- \tfrac{1}{4} \sin (4\pi x/L) + \tfrac{3}{5} \sin (5\pi x/L) - \tfrac{1}{6} \sin (6\pi x/L) + \cdots] \qquad 2.8.14$$

Figure 2.8.2 illustrates how the given function is approximated more and more closely as the number of terms considered in the series increases. The difference between the function and the series can be made arbitrarily small if a sufficiently large number of terms are taken.   This statement, however, is not true at $x = 0$ and $x = L$ where the sum of the series always remains zero.   This, as well as other interesting properties of the trigonometric series, is discussed in books on advanced calculus.

The Fourier expansion is useful even when the function to be represented cannot be integrated in closed form, or when it is given as an empirical curve.   The integral indicated in Eq. 2.8.9 can then be evaluated numerically or graphically.   Mechanical and electronic machines have also been constructed for the purpose of determining the Fourier coefficients.   The calculation of the Fourier coefficients is often referred to as a harmonic analysis.

2.9   Galerkin's Method

Derivation of the Method

When the differential equation of a problem is known but is too difficult to solve with the given boundary conditions, an approximate solution can almost always be had with the aid of the Rayleigh-Ritz method.   An alternative approach to the solution of such a problem is available in the Galerkin method.

A differential equation in the theory of structures usually expresses the condition of equilibrium of a force component, or a generalized force, in some direction.   In Art. 2.4, for instance, in the derivation of the beam column equation, the total potential of the system was changed by varying the displacement y along the beam column.   When the beam column is in equilibrium, the work done by all the internal and external forces is zero for any virtual displacement.   Hence setting the change in the total potential, that is, the virtual work, equal to zero for any arbitrary variation δy of the displacements means setting the force resultant in the y direction equal to zero all along the beam column.   Indeed

$$EI(d^4y/dx^4) + P(d^2y/dx^2) = w \qquad 2.9.1$$

that is, Eq. 2.4.6 simply states that the sum of the vertical forces, caused by the elastic resistance of the beam column to bending and by the axial force P in consequence of the change in the shape of the beam column, is in equilibrium with the distributed external load w.

In the Galerkin method, just as in the Rayleigh-Ritz method, an approximate solution is assumed for the unknown quantity which in Eq. 2.9.1 is y.   Substitution of this approximate solution into the differential equation naturally does not satisfy it exactly but leads to an error $e(x)$ which varies along the beam column.   The problem is to make this error as small as possible.

The approximate solution can be assumed as the sum of a number of functions $f(x)$, each satisfying by itself all the boundary conditions of the problem, but none of them, as a rule, satisfying the differential equation. Each function is multiplied by an undetermined coefficient which must be chosen so as to make the error a minimum.   One may write

$$y_{appr} = \sum_{i=1}^{n} a_i f_i(x) \qquad 2.9.2$$

In what follows Q is understood to mean the operator that produces the left-hand member of the differential equation when applied to the unknown function.   For instance, in Eq. 2.9.1

$$Q = EI(d^4/dx^4) + P(d^2/dx^2) \qquad 2.9.3$$

and application of the operator to the approximate solution given in Eq. 2.9.2 results in

$$Q(y_{appr}) = EI(d^4 y_{appr}/dx^4) + P(d^2 y_{appr}/dx^2) \qquad 2.9.3a$$

or

$$Q\left[\sum_{i=1}^{n} a_i f_i(x)\right] = EI \sum_{i=1}^{n} a_i f_i^{iv}(x) + P \sum_{i=1}^{n} a_i f_i''(x) \qquad 2.9.3b$$

where the superscripts iv and '' denote the fourth and second derivatives, respectively. Because of the approximate nature of the solution proposed in Eq. 2.9.2, the equilibrium condition of the vertical forces is not completely satisfied, and the difference between the expression on the right-hand side of Eq. 2.9.3b and w represents the amount by which the vertical force resultant differs from zero. This error $e(x)$ can be calculated from the equation

$$Q(y_{appr}) - w = Q\left[\sum_{i=1}^{n} a_i f_i(x)\right] - w = e(x) \qquad 2.9.4$$

and is simply the vertical force resultant at any point x along the beam.

If $e(x)$ were identically equal to zero, the work done by the vertical forces would be zero for any arbitrary virtual displacement δy. This is not true for an approximate solution, but, since there are n undetermined constants in it, the virtual work can be required to vanish for n displacement patterns. The obvious choice for the displacement patterns is the set of deflection patterns represented by the functions $f_i(x)$. There are exactly n of them, and, as they satisfy the geometric conditions, they all correspond to virtual displacements. The virtual work is then obtained by multiplying the force resultant $e(x)$, as given in Eq. 2.9.4, by one of the displacements represented by $f_k(x)$ (the subscript k is used here rather than i to indicate that any of the n functions can be chosen and not necessarily the ith), and integrating the product along the beam column. The requirement that this integral vanish yields one condition for determining the n unknown coefficients a_i. In the form of an equation one has

$$\int_0^L f_k(x) \, e(x) \, dx = \int_0^L f_k(x) \left\{ Q\left[\sum_{i=1}^{n} a_i f_i(x)\right] - w \right\} dx = 0 \qquad 2.9.5$$

Equation 2.9.5 can be written n times, namely once for each of the n functions $f_k(x)$, and thus n equations are obtained to determine the n unknown coefficients.

Physically the procedure means that the n coefficients in the approximate solution are adjusted in such a manner as to satisfy the equilibrium conditions expressed by the principle of virtual displacements for n virtual displacements. This is all that is required if the system has n degrees of freedom of motion. When the number of degrees of freedom is indefinitely large, as for the bent beam, an exact solution can be obtained by the Galerkin method if an indefinitely large number of terms are included in the assumption, that is, $n \to \infty$ in Eq. 2.9.2, and the condition stated in Eq. 2.9.5 is satisfied for an indefinitely large number of virtual displacements. As long as only a finite number of terms are considered in problems having an indefinitely large number of degrees of freedom of motion, the equilibrium conditions are satisfied only approximately, and the solution obtained is an approximate one. However, the approximation can be quite satisfactory for engineering purposes, as will be shown by an example.

The Stiffness Coefficient of a Beam Column

As the accuracy of the approximate solution can best be checked by comparing it to an exact solution, the stiffness factor of a beam column

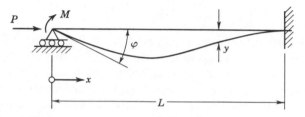

Fig. 2.9.1. Beam column.

will now be calculated by the Galerkin method. Values of the stiffness factor were determined in Art. 2.4 by combining solutions of Eq. 2.4.6 obtained for simply supported beam columns subjected to concentrated end moments. In the present calculations the beam column will be assumed simply supported at its *near* end where the moment M is applied, and rigidly fixed at its *far* end, as shown in Fig. 2.9.1. The differential equation of the problem is Eq. 2.4.6 with the distributed load w set equal to zero:

$$EI(d^4y/dx^4) + P(d^2y/dx^2) = 0 \qquad 2.9.6$$

In addition two end conditions were stipulated in Eq. 2.4.7. The first is in the notation of this problem

$$M = -EI(d^2y/dx^2)_{x=0} \qquad 2.9.6a$$

The second expressed a similar relationship between curvature and applied moment at the far end. This is of no interest with the beam column shown in Fig. 2.9.1, because the far end is rigidly fixed, and thus a reaction moment of any magnitude is available there. As a matter of fact in

Eq. 2.4.4, $d\,\delta y/dx$ is not a virtual displacement at $x = L$ when the far end is fixed. Consequently $(d\,\delta y/dx) = 0$, and all the conditions of equilibrium are satisfied whatever the expression in brackets is at $x = L$.

Equations 2.9.6 and 2.9.6a will now be solved approximately by Galerkin's method with the geometric boundary conditions

$$x = 0, \qquad y = 0, \qquad x = L, \qquad y = 0$$
$$x = L, \qquad dy/dx = 0 \qquad\qquad 2.9.6b$$

A deflected shape satisfying the geometric conditions is

$$y_1 = (L - x)^2 x \qquad\qquad 2.9.7a$$

Similarly the deflected shapes represented by the equations

$$y_2 = (L - x)^2 x^2$$
$$y_3 = (L - x)^2 x^3$$
$$\cdot \qquad\qquad 2.9.7b$$
$$\cdot$$
$$\cdot$$
$$y_n = (L - x)^2 x^n$$

satisfy the conditions stated in Eqs. 2.9.6b. Hence the approximate solution of Eqs. 2.9.6 can be assumed in the form

$$y = (L - x)^2 (a_1 x + a_2 x^2 + a_3 x^3 + \cdots + a_n x^n) \qquad\qquad 2.9.8$$

In this particular problem each term of the assumed deflected shape satisfies the boundary conditions Eqs. 2.9.6b, but the fourth boundary condition, namely Eq. 2.9.6a, represents a condition imposed on the n coefficients a_i. For this reason it will only be possible to satisfy $n - 1$ equilibrium conditions and not n as stated in the preceding discussion. The first derivative of y is

$$y' = -2(L - x)(a_1 x + a_2 x^2 + \cdots + a_n x^n)$$
$$+ (L - x)^2 (a_1 + 2a_2 x + \cdots + na_n x^{n-1}) \qquad 2.9.9$$

The second derivative is

$$y'' = 2(a_1 x + a_2 x^2 + \cdots + a_n x^n) - 4(L - x)(a_1 + 2a_2 x + \cdots + na_n x^{n-1})$$
$$+ (L - x)^2 (2 \cdot 1 \cdot a_2 + 3 \cdot 2 \cdot a_3 x + \cdots + n(n - 1)a_n x^{n-2}) \qquad 2.9.10a$$

This can also be written in the form

$$y'' = 2 \cdot 1(a_2 L^2 - 2a_1 L) + 3 \cdot 2(a_3 L^2 - 2a_2 L + a_1)x$$
$$+ 4 \cdot 3(a_4 L^2 - 2a_3 L + a_2)x^2 + 5 \cdot 4(a_5 L^2 - 2a_4 L + a_3)x^3 + \cdots$$
$$+ n(n - 1)(a_n L^2 - 2a_{n-1}L + a_{n-2})x^{n-2}$$
$$+ (n + 1)n(-2a_n L + a_{n-1})x^{n-1} + (n + 2)(n + 1)a_n x^n \qquad 2.9.10b$$

The fourth derivative is

$$y^{iv} = 4 \cdot 3 \cdot 2 \cdot 1(a_4L^2 - 2a_3L + a_2) + 5 \cdot 4 \cdot 3 \cdot 2(a_5L^2 - 2a_4L + a_3)x$$
$$+ \cdots + n(n-1)(n-2)(n-3)(a_nL^2 - 2a_{n-1}L + a_{n-2})x^{n-4}$$
$$+ (n+1)n(n-1)(n-2)(-2a_nL + a_{n-1})x^{n-3}$$
$$+ (n+2)(n+1)n(n-1)a_nx^{n-2} \qquad 2.9.11$$

Substitution of the expression assumed for the deflections in Eq. 2.9.6 yields

$$e(x) = Q(y) = EI[4 \cdot 3 \cdot 2 \cdot 1(a_4L^2 - 2a_3L + a_2)$$
$$+ 5 \cdot 4 \cdot 3 \cdot 2(a_5L^2 - 2a_4L + a_3)x + \cdots$$
$$+ n(n-1)(n-2)(n-3)(a_nL^2 - 2a_{n-1}L + a_{n-2})x^{n-4}$$
$$+ (n+1)n(n-1)(n-2)(-2a_nL + a_{n-1})x^{n-3}$$
$$+ (n+2)(n+1)n(n-1)a_nx^{n-2}]$$
$$+ P[2 \cdot 1(a_2L^2 - 2a_1L) + 3 \cdot 2(a_3L^2 - 2a_2L + a_1)x$$
$$+ 4 \cdot 3(a_4L^2 - 2a_3L + a_2)x^2 + 5 \cdot 4(a_5L^2 - 2a_4L + a_3)x^3$$
$$+ \cdots + n(n-1)(a_nL^2 - 2a_{n-1}L + a_{n-2})x^{n-2}$$
$$+ (n+1)n(-2a_nL + a_{n-1})x^{n-1}$$
$$+ (n+2)(n+1)a_nx^n] \qquad 2.9.12$$

A first approximation can be obtained by assuming that only a_1 differs from zero. In this case Eq. 2.9.10b becomes, at $x = 0$,

$$y''_{x=0} = -4a_1L$$

and substitution in Eq. 2.9.6a yields

$$M = 4EILa_1$$

Hence

$$a_1 = M/4EIL \qquad 2.9.13a$$

and the slope at the near end ($x = 0$) is, from Eq. 2.9.9,

$$y' = a_1L^2 = ML/4EI \qquad 2.9.13b$$

The stiffness factor S of the Hardy Cross moment-distribution method was defined as the moment that causes unit rotation. Consequently with $y' = 1$ and $M = S$ the last equation can be solved for S:

$$S = 4EI/L \qquad 2.9.14$$

This is the correct solution when the end load P is zero. It represents only a very rough approximation when P is not equal to zero. The inaccuracy of the solution is not surprising if one realizes that the differential equation of the problem, Eq. 2.9.6, was not taken into account at all.

Improvement of the Approximation

A second approximation can be had by considering two terms in the series, namely those multiplied by a_1 and a_2. All the other coefficients are assumed to be zero. Equation 2.9.10b yields now

$$y''_{x=0} = 2a_2L^2 - 4a_1L \qquad\qquad 2.9.15a$$

and substitution in Eq. 2.9.6a results in

$$2a_1 - a_2L = M/2EIL \qquad\qquad 2.9.15b$$

This is a single equation with two unknowns, and thus one more condition is needed to determine the coefficients a_1 and a_2. This second condition is furnished by the principle of virtual displacements if it is required that the virtual work done by the vertical forces represented by $e(x)$ in Eq. 2.9.12 should vanish for one particular set of displacements. The displacements defined by Eq. 2.9.7a are the preferable choice. The virtual work is

$$W = \int_0^L e(x)(L - x)^2x \, dx \qquad\qquad 2.9.16$$

After substitution from Eq. 2.9.12 the following types of integrals are obtained

$$\int_0^L (L - x)^2x^n \, dx = \frac{2L^{n+3}}{(n + 1)(n + 2)(n + 3)} \qquad\qquad 2.9.17$$

In particular

$$\int_0^L (L - x)^2x \, dx = \frac{L^4}{12}$$

$$\int_0^L (L - x)^2x^2 \, dx = \frac{L^5}{30}$$

$$\int_0^L (L - x)^2x^3 \, dx = \frac{L^6}{60}$$

$$\int_0^L (L - x)^2x^4 \, dx = \frac{L^7}{105}$$

$$\int_0^L (L - x)^2x^5 \, dx = \frac{L^8}{168} \qquad\qquad 2.9.18$$

When the only coefficients differing from zero are a_1 and a_2, the condition

$$W = 0$$

yields, after substitutions from Eqs. 2.9.16, 2.9.12, and 2.9.18,

$$4PLa_1 - (60EI - PL^2)a_2 = 0 \qquad\qquad 2.9.19$$

Solution of the two simultaneous equations, Eqs. 2.9.15b and 2.9.19, gives

$$a_1 = \frac{M}{4EIL}\frac{60EI - PL^2}{60EI - 3PL^2} \qquad\qquad 2.9.20a$$

$$a_2 = \frac{MP}{EI}\frac{1}{60EI - 3PL^2} \qquad\qquad 2.9.20b$$

With two non-vanishing coefficients, the slope at the near end, according to Eq. 2.9.9, is again

$$y'_{x=0} = a_1L^2 \qquad\qquad 2.9.21$$

as in Eq. 2.9.13b.   Hence the assumptions

$$y' = 1, \qquad M = S$$

and solution of Eq. 2.9.20a for S yield after some manipulations

$$S = \frac{4EI}{L}\frac{3 - (60/\pi^2)(P_E/P)}{1 - (60/\pi^2)(P_E/P)} \qquad\qquad 2.9.22$$

where P_E stands for Euler's buckling load:

$$P_E = \pi^2EI/L^2 \qquad\qquad 2.9.22a$$

In Fig. 2.9.2 values calculated from Eq. 2.9.22 are compared with the rigorous solution of Art. 2.4.   The agreement is very good as long as P/P_E is smaller than 3.

An even better approximation can be obtained if three terms of the solution are taken into account.   It is easy to see that in this case Eqs. 2.9.15a and b remain unchanged.   On the other hand, the requirement that the virtual work vanish during displacements according to Eq. 2.9.7a yields

$$28a_1 + [7 - (420/\pi^2)(P_E/P)]a_2L + 2a_3L^2 = 0 \qquad\qquad 2.9.23$$

Because there are three undetermined constants in the assumption, the virtual work can be made to vanish for a second set of virtual displacements also. This is chosen as the pattern defined in Eq. 2.9.7b. Hence

$$W = \int_0^L e(x)(L - x)^2 x^2 \, dx = 0 \qquad\qquad 2.9.24$$

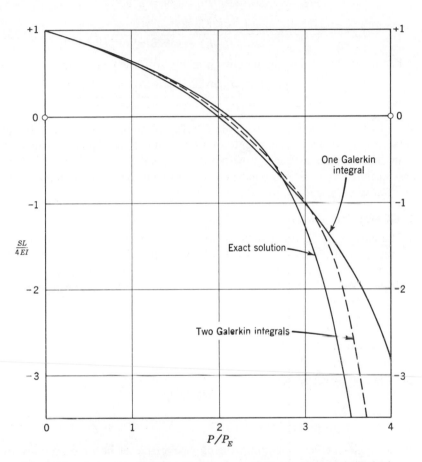

Fig. 2.9.2. Approximate values of stiffness coefficient by Galerkin's method.

Substitution from Eq. 2.9.12 and use of the definite integrals listed under Eqs. 2.9.18 lead after various algebraic manipulations to

$$7a_1 + [4 - (168/\pi^2)(P_E/P)]a_2 L + [2 - (84/\pi^2)(P_E/P)]a_3 L^2 = 0 \qquad 2.9.25$$

Solution of the three simultaneous linear equations, Eqs. 2.9.15b, 2.9.23, and 2.9.25 yields

$$a_1 = \frac{2}{3}\frac{1 - (182/\pi^2)(P_E/P) + (5880/\pi^4)(P_E/P)^2}{3 - (252/\pi^2)(P_E/P) + (3920/\pi^4)(P_E/P)^2}\frac{M}{4EIL} \qquad 2.9.26a$$

$$a_2 L = \frac{14}{3}\frac{1 - (56/\pi^2)(P_E/P)}{3 - (252/\pi^2)(P_E/P) + (3920/\pi^4)(P_E/P)^2}\frac{M}{4EIL} \qquad 2.6.26b$$

$$a_3 L^2 = \frac{21}{3}\frac{1 - (28/\pi^2)(P_E/P)}{3 - (252/\pi^2)(P_E/P) + (3920/\pi^4)(P_E/P)^2}\frac{M}{4EIL} \qquad 2.9.26c$$

From Eq. 2.9.21 and the definition of the stiffness factor one gets

$$S = \frac{3}{2}\frac{3 - (252/\pi^2)(P_E/P) + (3920/\pi^4)(P_E/P)^2}{1 - (182/\pi^2)(P_E/P) + (5880/\pi^4)(P_E/P)^2}\frac{4EI}{L} \qquad 2.9.27$$

Values computed from this formula are also plotted in Fig. 2.9.2. It can be seen that the three-coefficient solution, marked *two Galerkin integrals*, is in much better agreement with the exact solution at values of P/P_E greater that 3 than the two-coefficient solution which is marked in the figure *one Galerkin integral*.

The Carry-over Factor

The carry-over factor is defined in the moment-distribution method as the moment at the fixed end divided by the applied moment. In accordance with the sign convention adopted in Art. 1.17 this ratio must be multiplied by -1. From Eq. 2.9.10a the curvature of the beam column at the far end ($x = L$) is

$$y''_{x=L} = 2(a_1 L + a_2 L^2 + a_3 L^3 + \cdots + a_n L^n) \qquad 2.9.28$$

As the moment is

$$M_{x=L} = -EIy''_{x=L}$$

and the carry-over factor

$$C = -M_{x=L}/M$$

one has

$$C = (2EIL/M)(a_1 + a_2 L + a_3 L^2 + \cdots + a_n L^{n-1}) \qquad 2.9.29$$

In the single-coefficient solution the carry-over factor becomes, in view of Eq. 2.9.13a,

$$C = \tfrac{1}{2} \qquad 2.9.30$$

This is the correct solution in the absence of an end load. When two coefficients differ from zero, Eqs. 2.9.20 and 2.9.29 give

$$C = \frac{1}{2}\frac{(60/\pi^2)(P_E/P) + 3}{(60/\pi^2)(P_E/P) - 3} \qquad 2.9.31$$

Finally with three non-vanishing coefficients, Eqs. 2.9.26 and 2.9.29 yield

$$C = \frac{1}{2} \frac{3 - (56/\pi^2)(P_E/P) + (3920/\pi^4)(P_E/P)^2}{3 - (252/\pi^2)(P_E/P) + (3920/\pi^4)(P_E/P)^2} \qquad \text{2.9.32}$$

Numerical values of the carry-over factor were computed and plotted in Fig. 2.9.3. The solution obtained when only one Galerkin integral was considered is not very accurate in this case, but the two-integral solution

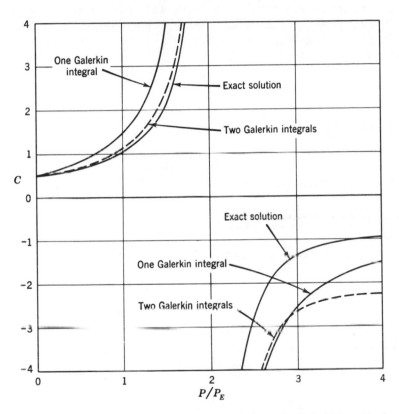

Fig. 2.9.3. Approximate values of carry-over factor by Galerkin's method.

is an excellent approximation when P/P_E is less than 2. It could be expected that the stiffness-factor values obtained by the Galerkin method would be closer to the true values than the values of the carry-over factor. The former were calculated from the slope of the deflection curve, and the latter from its curvature. The slope is the first derivative of the deflections, and the curvature is proportional to the second derivative. Even when the approximate shape of the deflections is very close to the true shape, the first derivative of the approximate solution may differ noticeably, and the second derivative considerably, from the corresponding true values.

Equivalence of Rayleigh-Ritz and Galerkin Methods

It will now be shown that the Galerkin method, as just described, is fully equivalent to the Rayleigh-Ritz method, and that the two necessarily lead to the same results if the same functions are chosen to represent approximately the deflections. When the Rayleigh-Ritz method is used to solve the problem of the stiffness factor of the beam column, the total potential of the system must first be calculated. This consists of the strain energy U stored in the beam column

$$U = \frac{EI}{2} \int_0^L y''^2 \, dx \qquad 2.9.33a$$

the potential V_1 of the axial load P

$$V_1 = -\frac{P}{2} \int_0^L y'^2 \, dx \qquad 2.9.33b$$

and the potential V_2 of the end moment M

$$V_2 = -(My')_{x=0} \qquad 2.9.33c$$

The total potential is therefore represented by

$$U + V = U + V_1 + V_2 = \frac{EI}{2} \int_0^L y''^2 \, dx - \frac{P}{2} \int_0^L y'^2 \, dx - (My')_{x=0}$$

$$2.9.33$$

If the deflected shape is assumed the same as in the calculations by the Galerkin method, and for the sake of brevity the symbol $f_i = f_i(x)$ is introduced as defined by the equation

$$y = \sum_{i=1}^n a_i(L-x)^2 x^i = \sum_{i=1}^n a_i \, f_i(x) = \sum_{i=1}^n a_i f_i \qquad 2.9.34$$

substitution in the expression representing the total potential yields

$$U + V = \frac{EI}{2} \int_0^L \left(\sum_{i=1}^n a_i \, f''_i \right)^2 dx - \frac{P}{2} \int_0^L \left(\sum_{i=1}^n a_i \, f'_i \right)^2 dx - M \left(\sum_{i=1}^n a_i \, f'_i \right)_{x=0}$$

$$2.9.35$$

In the Rayleigh-Ritz method the total potential is made a minimum with respect to all the coefficients a_i in the series by differentiating $U + V$ with respect to each of the coefficients in turn and setting the derivatives equal to zero. Differentiation with respect to the kth coefficient yields

$$\frac{\partial(U+V)}{\partial a_k} = EI \int_0^L \left(\sum_{i=1}^n a_i f''_i \right) f''_k \, dx$$

$$- P \int_0^L \left(\sum_{i=1}^n a_i f'_i \right) f'_k \, dx - (Mf'_k)_{x=0} \qquad 2.9.36$$

because

$$\frac{\partial \left(\sum_{i=1}^n a_i f'_i \right)}{\partial a_k} = \frac{\partial}{\partial a_k} (a_1 f'_1 + a_2 f'_2 + \cdots + a_k f'_k + \cdots + a_n f'_n) = f'_k$$

$$2.9.37$$

and a similar equation holds for the second derivatives also. The derivative can be simplified by integrations by parts:

$$\frac{\partial(U+V)}{\partial a_k} = EI \left[\left(\sum_{i=1}^n a_i f''_i \right) f'_k \right]_0^L - EI \int_0^L \left(\sum_{i=1}^n a_i f'''_i \right) f'_k \, dx$$

$$- P \left[\left(\sum_{i=1}^n a_i f'_i \right) f_k \right]_0^L + P \int_0^L \left(\sum_{i=1}^n a_i f''_i \right) f_k \, dx - (Mf'_k)_{x=0}$$

$$2.9.38$$

The first expression in brackets vanishes at the upper limit because the slope f'_k of the beam column is zero there. It does not vanish at the lower limit since at $x = 0$ both the slope f'_k and the curvature $\sum a_i f''_i$ are arbitrary. The second expression in brackets vanishes at both limits because the displacement f_k is zero for every value of the subscript k at the two supports. If these values are substituted in the equation and one more integration by parts is carried out, the result is

$$\frac{\partial(U+V)}{\partial a_k} = -EI \left[\left(\sum_{i=1}^n a_i f''_i \right) f'_k \right]_{x=0} - EI \left[\left(\sum_{i=1}^n a_i f'''_i \right) f_k \right]_0^L$$

$$+ EI \int_0^L \left(\sum_{i=1}^n a_i f_i^{iv} \right) f_k \, dx + P \int_0^L \left(\sum_{i=1}^n a_i f''_i \right) f_k \, dx - (Mf'_k)_{x=0}$$

$$2.9.39$$

where the second expression in brackets vanishes because the deflection f_k is zero at the two supports. This derivative of the total potential must vanish for any arbitrary choice of the subscript k in agreement with the principle of virtual displacements. After rearranging the terms one has

$$\int_0^L \left[EI \left(\sum_{i=1}^n a_i f_i^{iv} \right) + P \left(\sum_{i=1}^n a_i f''_i \right) \right] f_k \, dx$$

$$- \left[EI \left(\sum_{i=1}^n a_i f''_i \right) + M \right]_{x=0} f'_{k_{x=0}} = 0 \qquad 2.9.40$$

Because of Eqs. 2.9.34 this can also be written as

$$\int_0^L [EIy^{iv} + Py''] f_k \, dx - [EIy'' + M]_{x=0} f'_{k_{x=0}} = 0 \qquad 2.9.41$$

The equation must hold for every value of k from $k = 1$ to $k = n - 1$.

As the integral in Eq. 2.9.41 is a Galerkin integral, the Rayleigh-Ritz method and the Galerkin method as here described are completely equivalent if the deflection functions selected satisfy not only the geometric boundary conditions, but also the conditions of equilibrium at the boundary.

PART THREE

The Calculation of
Buckling Loads

3.1 Stability

Stability of a Rigid Ball

In the preceding two parts of this book the equilibrium of various systems was investigated. In this article the properties of the state of equilibrium will be examined. It will be found that some states are stable, and thus the designer can count upon them with confidence; others are unstable and are upset by the slightest disturbance.

It is common knowledge that the pendulum of a grandfather's clock is in a stable state of equilibrium after it has stopped. When the weights reach the bottom of the clock and no more driving power is available, the pendulum settles in a vertical position. If it is moved from this equilibrium position, it returns to it again after some oscillations. On the other hand, the ball on the top of a rod supported on the nose of a juggler is in an unstable state of equilibrium which can be maintained only through the unusual skill of the juggler; without his continuous effort the rod would topple. Nevertheless an examination of the two systems, for instance by means of the minimal principle, leads to the same result, namely that both are in equilibrium. Since no designer wants to incorporate members in his structure whose equilibrium is as precarious as that of the juggler's rod, it is of great importance to him that criteria be established that can distinguish between the two types of equilibrium.

The concept of stability can be explained with the aid of the example of a ball lying on uneven ground as shown in Fig. 3.1.1a. If both the ball and the contour are assumed to be completely rigid, and frictional forces are disregarded, the ball is in equilibrium in positions A, B, C, and D, and is not in equilibrium in position E. The force equilibrium is illustrated in Fig. 3.1.1b which shows that the weight W of the ball and the normal reaction are equal and opposed, and have a common line of action.

This is true for all the positions marked except E where in the absence of friction the reaction R must be perpendicular to the contour and cannot be in equilibrium with W (see Fig. 3.1.1c).

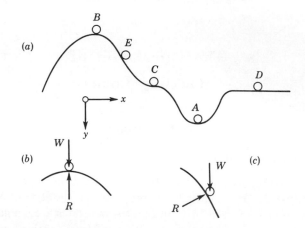

Fig. 3.1.1.   Ball on uneven ground.

If the minimal principle is applied to the problem, the total potential is the potential of the weight W of the ball.   The potential of R is zero since a reaction at a rigid support does no work when virtual displacements alone are considered; moreover no strain energy can be stored in a perfectly rigid system.   Hence

$$U + V = -Wy \qquad\qquad 3.1.1$$

and the minimal principle requires that for equilibrium

$$\delta(U + V) = -W\,\delta y = 0 \qquad\qquad 3.1.2$$

The virtual displacement δy must satisfy the condition that the ball must not leave the contour; it is not permitted to bounce nor to penetrate the rigid contour.   If the equation defining the contour is

$$y = f(x) \qquad\qquad 3.1.3$$

then a virtual displacement from a position characterized by the abscissa x has a vertical downward component

$$\delta y = y(x + \delta x) - y(x)$$
$$= y'(x)\delta x + (1/2)y''(x)(\delta x)^2 + (1/3!)y'''(x)(\delta x)^3 + \cdots \qquad 3.1.4$$

Substitution of δy from Eq. 3.1.4 in Eq. 3.1.2 yields

$$\delta(U + V) = -W[y'(x)\delta x + (1/2)y''(x)(\delta x)^2 + (1/3!)y'''(x)(\delta x)^3 + \cdots] = 0$$
3.1.5

It was shown in Part 2 that only the term containing the first power of the variation is to be considered when the equilibrium of a system is investigated by means of the minimal principle. (In the present example the terms containing the higher powers of δx define the shape of the contour away from the point under consideration and thus cannot influence the direction of the virtual displacement.) Omission of the terms containing powers of δx higher than the first results in

$$\delta(U + V) = -Wy'(x)\,\delta x = 0$$
3.1.6

Since W is not zero and the magnitude of δx is arbitrary, the identity can be satisfied only if

$$y'(x) = 0$$
3.1.7

Consequently the requirement of equilibrium is that the contour have a horizontal tangent. This requirement is fulfilled at points A, B, C, and D.

The next task is to examine the properties of the equilibrium in positions A, B, C, and D. Position E need not be considered, since at E the ball is not in equilibrium, and stability is a property of equilibrium; the definition of stability makes no sense when the system is not in equilibrium.

In agreement with everyday usage, stability will be defined in the following manner: Let the equilibrium of the ball be disturbed by moving the ball slightly out of its original position. If the ball is left alone in its new position, it may or may not return to its original position. If it does return, the equilibrium is stable; if, on the other hand, a slight displacement causes larger ones to follow, and the ball rolls farther and farther away from the original position of equilibrium, the original equilibrium is unstable. Thousands of years of practical experience have taught mankind that the ball returns into its position A after a disturbance, while it rolls down the slope, never to return, if it is pushed slightly to the side from position B.

The connection between the stability of the system and the properties of the expression for the total potential will now be investigated. Consideration of Eq. 3.1.7 changes Eq. 3.1.5 into

$$\delta(U+V) = -W[(1/2)y''(x)(\delta x)^2+(1/3!)y'''(x)(\delta x)^3+(1/4!)y^{iv}(x)(\delta x)^4+\cdots]$$
3.1.8

According to Eq. 3.1.8 the change in the total potential is small of the second order. This expression means that a small displacement of the order δx causes a change in the total potential that is of the order $(\delta x)^2$. In accordance with the minimal principle this must of course be true whenever the system investigated is in equilibrium.

When the virtual displacement δx is sufficiently small, the first non-vanishing term in the brackets in Eq. 3.1.8 determines the sign of the change in the total potential If the first non-vanishing term contains a derivative of even order, the total potential must have either a maximum or a minimum. This can be seen from Eq. 3.1.8 in which the derivatives of even order are multiplied by even powers of δx. Consequently, when $y''(x)$ is positive, the total potential decreases whether the ball is displaced in the positive or negative direction (through a distance δx or $-\delta x$). This might be the case at point B. To the left of B the slope $y'(x)$ is negative; at B it is zero; and to the right of B it is positive. Hence $y'(x)$ increases with x in the neighborhood of B, and $y''(x)$, which is the rate of change of y' with x, is positive unless both $y''(x)$ and $y'''(x)$ vanish at B and $y^{iv}(x)$ is positive; or more generally the first non-vanishing derivative is of even order and is positive. Because of the negative sign of W, the change in the total potential is negative in the neighborhood of B; that is, the total potential has a maximum at B. Thus a maximum of the total potential is associated with unstable equilibrium.

The situation is different at A where the slope $y'(x)$ decreases from positive values through zero to negative values. Here $y''(x)$ is negative (unless both $y''(x)$ and $y'''(x)$ vanish and $y^{iv}(x)$ is negative at A, or more generally the first non-vanishing derivative is of even order and is negative), $\delta(U + V)$ is positive, and the total potential is a minimum. The minimum of the total potential is therefore associated with stable equilibrium.

When the first non-vanishing derivative is of an odd order, the sign of $\delta(U + V)$ depends on the sign of δx because odd derivatives of $y(x)$ are multiplied by odd powers of δx in Eq. 3.1.8. Such a situation exists at point C. The total potential decreases when δx is positive, and increases when it is negative. This state of affairs is just as undesirable for the designer as a maximum of the total potential, since any oscillations of the ball will bring it eventually into a position to the right of point C from which it will roll down into the valley. Because the inflection point at C with a horizontal tangent appears to be a minimum from one direction and a maximum from the other, the curve is said to have a minimax at C. The ball situated at a minimax is an unstable configuration according to age-old experience.

The three positions A, B, and C corresponding to minimum, maximum, and minimax do not exhaust all the possible categories of a stationary value of the total potential characterized by $y'(x) = 0$. A fourth possibility is shown at D where the contour is a horizontal straight line. Here all the derivatives of $y = f(x)$ vanish, and the ball will remain in any position in which it is displaced. This kind of equilibrium is known as neutral equilibrium.

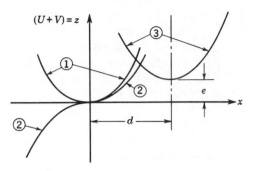

Fig. 3.1.2.   Examples of total potential.

A few examples should clarify further the properties of the total potential. In Fig. 3.1.2, line 1 has the equation

$$U + V = z = ax^2, \qquad a > 0 \qquad\qquad (a)$$

At $x = 0$ the first derivative of the function, $z' = 2ax$, vanishes. Since z represents the total potential of the system, the system is in equilibrium at $x = 0$ where z has a stationary value. Moreover, the equilibrium is stable, since the second derivative, $z'' = 2a$, is positive, indicating that the stationary value is a minimum.

The equation of line 2 is

$$U + V = z = bx^3, \qquad b > 0 \qquad\qquad (b)$$

When $x = 0$ the first non-vanishing derivative is the third:

$$z''' = 6b \qquad\qquad (c)$$

Correspondingly the system whose total potential is represented by z is in an unstable equilibrium at $x = 0$ where the total potential has a minimax.

Finally the equation of line 3 is

$$U + V = z = c(x - d)^4 + e, \qquad c > 0, \quad d > 0, \quad e > 0 \qquad (d)$$

The derivatives of z are

$$z' = 4c(x - d)^3$$
$$z'' = 12c(x - d)^2$$
$$z''' = 24c(x - d)$$
$$z^{\mathrm{iv}} = 24c \qquad\qquad (e)$$

The system whose total potential is represented by z is in equilibrium at $x = d$ since there the z curve has a stationary value ($z' = 0$). This equilibrium is stable because the first non-vanishing derivative, namely the fourth, is of even order and is positive.

Here it might be mentioned that in most problems encountered in engineering the total potential is a quadratic expression in the displacement quantities. Consequently, its second derivative is a constant whose sign determines the stability of the system, and derivatives of a higher order vanish identically. Such a system can have only a single stationary value and not several, as shown in Fig. 3.1.1.

Stability of Elastic Systems

The investigation of the stability of an elastic system is more involved than that of a rigid system because elastic stability does not depend on the geometry alone, but also on the magnitude and direction of the forces. For instance, a straight bar is known to be in a stable state of equilibrium when it is subjected to two equal and opposite tensile forces applied to its ends. The string of a violin is such a bar; its stability can be examined by plucking the middle of the string. Of course, such a disturbance of the originally straight-line state of equilibrium causes rapid oscillations of the string about the equilibrium position. After a while the oscillations subside, and the string resumes its original position. Hence the violin string is in a stable state of equilibrium.

When the magnitude of the tensile end loads is first decreased to zero, and then increasing compressive end loads are applied to the ends of the bar, the stability of the system changes over to instability. The value of the load at which the change takes place depends on the geometric and mechanical properties of the bar. For instance, a thin bamboo walking stick doubles up and collapses when a heavy man leans on it, while a thick solid hickory axe shaft can safely carry much greater loads. Similarly every column placed in a compression testing machine is stable as long as the compressive end load applied to it is sufficiently small. This can be ascertained by deflecting the center of the column and then releasing it; after a few oscillations the column resumes its original position. On the other hand, a small transverse displacement imposed on a heavily loaded column results in rapidly increasing deflections; eventually the stress in the column exceeds the yield stress of the material, and the column fails.

The load at which the stability of a structural element changes over to instability is known as the critical or the buckling load. When the compressive end load of a column is smaller than the critical value, a disturbance of the equilibrium is generally followed by oscillations of decreasing amplitude, and eventually the column resumes its original straight-line shape.   When the end load is greater than the critical value, the slightest disturbance of the straight-line state of equilibrium leads to increasing deflections and collapse.   Hence in the limiting case of the critical load itself small deflections imposed on the column can neither increase nor decrease; at the critical load a small disturbance must carry over the column into a neighboring new state of equilibrium.   For this reason the column is in a neutral state of equilibrium under the action of the critical load; this state corresponds to position D of the ball in Fig. 3.1.1.

Because the critical load is characterized by the existence of a neutral state of equilibrium, its magnitude can be calculated by means of the minimal principle.   The value of the load must be found under which the structure can be in equilibrium not only in the original, but also in an adjacent position.

The novel problem of examining the stability of an equilibrium configuration is thus reduced to the simpler problem of investigating the equilibrium in a neighboring configuration; this second problem can be solved by using the minimal principle as discussed in detail in Part 2.

Elastic Column

In Fig. 3.1.3 the lower end of the column, $x = 0$, is attached to a pivot fixed to the ground, while the upper end, $x = L$, is guided vertically in such a manner that it can rotate freely but cannot deflect horizontally. The problem is to find the conditions under which the column subjected to end loads P can be in equilibrium in a deflected position $y = y(x)$ adjacent to the original straight-line equilibrium position.

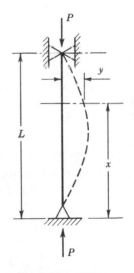

From the minimal principle the deflected position $y = y(x)$ is an equilibrium position if a variation δy of the deflections does not change the value of the total potential of the system.   Hence one has to calculate the total potential in a deflected state $y = y(x)$, determine its change corresponding to the variation δy, and set the change equal to zero. In the calculation of the total potential the axis of the column can be considered inextensible, as was shown in Art. 2.4.

The strain energy of bending can be calculated from the formula

$$U = \frac{1}{2}\int_0^L EIy''^2 \, dx$$

Fig. 3.1.3.   Column.

The potential of the external load P is, in agreement with Eq. 2.4.2,

$$V = -\frac{1}{2} P \int_0^L y'^2 \, dx$$

Hence the total potential is

$$(U + V) = \frac{1}{2} \int_0^L EIy''^2 \, dx - \frac{1}{2} P \int_0^L y'^2 \, dx \qquad 3.1.9a$$

The change in the total potential can be determined in the same way as in Art. 2.4:

$$\delta(U + V) = \int_0^L EIy'' \, \delta y'' \, dx - P \int_0^L y' \, \delta y' \, dx \qquad 3.1.9b$$

The derivatives of the variation of the deflections can be eliminated by integrations by parts:

$$\int_0^L y' \, \delta y' \, dx = y' \, \delta y \Big|_0^L - \int_0^L y'' \, \delta y \, dx \qquad (f)$$

The first term in the right-hand member of Eq. f vanishes since δy is zero when $x = 0, L$; a transverse displacement of the end points would infringe upon the geometric constraints. Similarly

$$\int_0^L y'' \, \delta y'' \, dx = y'' \, \delta y' \Big|_0^L - \int_0^L y''' \, \delta y' \, dx \qquad (g)$$

where the first term in the right-hand member vanishes because y'' is zero when $x = 0, L$. This must be so, since y'' is the curvature of a flat arc, and the curvature of an originally straight bar is proportional to the bending moment. But the two pivots cannot transmit any bending moments, and thus y'' must be zero at the two ends of the column. A repetition of the integration yields

$$-\int_0^L y''' \, \delta y' \, dx = -y''' \, \delta y \Big|_0^L + \int_0^L y^{iv} \, \delta y \, dx \qquad (h)$$

Here again the first term in the right-hand member is zero since δy is zero at the end points of the column.

Consequently the change in the total potential can be written in the form

$$\delta(U+V) = \int_0^L EIy^{iv} \, \delta y \, dx + \int_0^L Py'' \, \delta y \, dx = \int_0^L (EIy^{iv} + Py'')\delta y \, dx \quad 3.1.10$$

If the deflected shape $y = y(x)$ is a configuration of equilibrium, the change in the total potential must vanish identically for any arbitrary choice of δy.

A reasoning similar to that used in Art. 2.4 shows that this is possible only if

$$EI y^{\mathrm{iv}} + P y'' = 0 \qquad\qquad 3.1.11$$

This differential equation, together with the end conditions

$$y = y'' = 0 \quad \text{when} \quad x = 0, L \qquad\qquad 3.1.12$$

is the condition of equilibrium in the slightly deflected configuration. The next step in the solution is therefore the integration of the differential equation.

Derivation of the Euler Load

If the notation is introduced

$$z = y'' \qquad\qquad 3.1.13$$

Eq. 3.1.11 assumes the form

$$EI z'' + P z = 0 \qquad\qquad 3.1.14$$

This is the well-known equation of harmonic vibrations discussed in Art. 2.4 (see Eq. n).   Its solution is

$$z = A' \cos kx + B' \sin kx \qquad\qquad 3.1.15$$

where A' and B' are arbitrary constants of integration and

$$k^2 = P/EI \qquad\qquad 3.1.16$$

Two consecutive integrations of $y'' = z$ give

$$y = A \cos kx + B \sin kx + Cx + D \qquad\qquad 3.1.17$$

where A, B, C, and D are arbitrary constants of integration.   Substitution of this solution in Eq. 3.1.11 yields an identity.

Two differentiations of y result in

$$y'' = -k^2(A \cos kx + B \sin kx)$$

Since y'' must vanish when $x = 0$, one obtains

$$A = 0 \qquad\qquad 3.1.18$$

With $A = 0$ the deflection when $x = 0$ is

$$y_{x=0} = D$$

But the end deflections must vanish, and thus

$$D = 0$$ 3.1.19

At $x = L$ one gets

$$y_{x=L} = B \sin kL + CL$$

and, since this deflection also vanishes,

$$C = -(B/L) \sin kL$$ 3.1.20

Consequently

$$y = B[\sin kx - (x/L) \sin kL]$$ 3.1.21

and

$$y'' = -Bk^2 \sin kx$$

The fourth boundary condition requires that y'' vanish at $x = L$:

$$-Bk^2 \sin kL = 0$$ 3.1.22

Since k^2 is not zero by assumption, this equation has only two kinds of solutions. One is

$$B = 0$$ 3.1.23

which is known as the trivial solution. When $B = 0$ is substituted in Eq. 3.1.21, the deflection y becomes identically equal to zero. This means that the original straight-line shape is an equilibrium configuration, which of course was known. In addition to the straight-line shape, there are deflected shapes which also correspond to equilibrium when the solution of Eq. 3.1.22 is

$$\sin kL = 0$$ 3.1.24

Then Eq. 3.1.21 becomes

$$y = B \sin kx$$ 3.1.25

where B may have any arbitrary value. Equation 3.1.24 is satisfied when

$$kL = 0, \pi, 2\pi, 3\pi, \cdots, n\pi, \cdots$$ 3.1.26

where n is any integer. The value $kL = 0$ must be ruled out because it would mean $P = 0$ according to Eq. 3.1.16. The physical meaning of Eq. 3.1.26 can be understood if one writes

$$k^2 L^2 = n^2 \pi^2$$

and substitutes the value of k^2 from Eq. 3.1.16:

$$PL^2/EI = n^2 \pi^2$$ 3.1.27

Equation 3.1.27 represents the connection between the physical and mechanical properties of the column and the compressive end load that must be satisfied when deflected shapes according to Eq. 3.1.25 are equilibrium configurations. The equation can be solved for the load

$$P = n^2\pi^2 EI/L^2 \qquad\qquad 3.1.28$$

under which the column is in neutral equilibrium. There are an indefinitely large number of such loads, and the one of greatest practical interest is the smallest of them. This corresponds to $n = 1$ and is called the Euler load

$$P_E = \pi^2 EI/L^2 \qquad\qquad 3.1.29$$

in honor of the mathematician who first calculated the value of the buckling load.

Since it is known that under sufficiently small loads the column is in stable equilibrium, and since the stability cannot change over to instability without passing through a neutral state of equilibrium, the Euler load is the smallest value of the end load under which the stability of the column vanishes. Hence under compressive loads smaller than the Euler load the column must be in a stable state of equilibrium. The Euler load is the danger signal; the designer must be careful to keep the loading of his columns below the Euler load.

If both members of Eq. 3.1.29 are divided by the cross-sectional area A and the symbol ρ is introduced to designate the radius of gyration of the section defined as

$$\rho^2 = I/A \qquad\qquad 3.1.30$$

the Euler formula assumes the alternate form

$$\sigma_E = \pi^2 E/(L/\rho)^2 \qquad\qquad 3.1.31$$

where σ_E is the Euler stress.

3.2   The Behavior of Columns

Consequences of Idealization

It was found in Art. 3.1 that a straight column remains straight under compressive end loads until their value reaches the Euler load. Under the Euler load the straight shape still corresponds to equilibrium, but there are other, slightly curved, equilibrium configurations also in the neighborhood of the original straight one. Since at the Euler load $n = 1$ and from Eq. 3.1.26

$$kL = \pi$$

the deflected shape is from Eq. 3.1.25 with $k = \pi/L$

$$y = B \sin (\pi x/L) \qquad\qquad 3.2.1$$

As long as the end load is less than the Euler load, the column remains stable in its straight-line configuration, but at the Euler load the stability becomes neutral since half sinc-wave deflections according to Eq. 3.2.1 with an arbitrary amplitude B also become possible states of equilibrium. The only restriction on B is that it must remain small. This is so, because in the calculation of the strain energy stored in the bent beam the curvature was assumed to be equal to y'', which is true only for small deflections; also, the difference between the length of the arc and that of its projection upon the x axis was calculated from a formula that does not hold for large deflections.

This theoretical behavior of the column is unexpected since perfect straightness up to a particular load and a sudden change to sinusoidal deformations of arbitrary amplitude at that load do not correspond to practical experience. But the theory was derived for idealized conditions; in reality perfectly straight and perfectly centrally loaded columns do not exist in engineering structures. Better agreement with reality can be had if either of these idealizations is discarded. To demonstrate this, the behavior of a column under compressive end loads will be investigated when the initial center line of the column deviates from a straight line because of the unavoidable inaccuracies of the manufacturing process.

Initially Crooked Column

In Fig. 3.2.1 the dashed line represents the centroidal axis of the initially crooked column. The initial deflections y_0 are shown to an exaggerated scale; they can be represented conveniently by a Fourier series (see Art. 2.8). Naturally the initial deflections increase when the end load is applied; the additional deflections are designated by y. The problem on hand is a simple equilibrium problem, and not a problem of stability: What additional deflections y develop in consequence of the end load? The conditions of equilibrium under any load P can be written as the known differential equation of the bent beam, or they can be stipulated by means of the minimal principle. If the latter procedure is followed, the calculations of the preceding article can be utilized. The only difference between the columns of Fig. 3.1.3 and Fig. 3.2.1 is the initial deflection y_0. The effect of y_0 on the total potential can be determined in the following manner:

It is a fact well established in strength of materials that the deflections and the curvature of a slightly curved beam can be calculated from the formulas valid for straight beams. Hence the strain energy stored in a slightly curved beam is

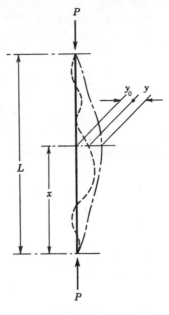

$$U = \frac{1}{2} \int_0^L EI y''^2 \, dx$$

where y'' is the curvature corresponding to the additional deflections. For this reason the first term in the right-hand member of Eq. 3.1.9a can be taken over unchanged when the expression is written for the total potential in the present problem. The only difference is that in Eq. 3.1.9a y represented the deflections while now it designates the additional deflections.

The potential of the external load P is the negative of the product of the load by the difference in the length of the projection upon the x axis of the curved centroidal axis of the column before and after the application of the load P. It follows from Eq. 2.4.2 that the difference

Fig. 3.2.1.
Variation of deflected shape.

between the length of the centroidal axis of the column and the length of its projection upon the x axis after application of the load P is

$$\Delta L_1 = \frac{1}{2} \int_0^L (y_0 + y)'^2 \, dx$$

The difference between the arc length and its projection before the load P is applied is

$$\Delta L_2 = \frac{1}{2} \int_0^L y'_0{}^2 \, dx$$

Hence the difference sought is

$$\Delta L = \Delta L_1 - \Delta L_2 = \frac{1}{2} \int_0^L [(y'_0 + y')^2 - y'_0{}^2] \, dx$$

$$= \frac{1}{2} \int_0^L (2 y'_0 y' + y'^2) \, dx$$

Consequently the total potential of Eq. 3.1.9a is replaced by the following expression:

$$U + V = \frac{1}{2} \int_0^L EI y''^2 \, dx - \frac{1}{2} P \int_0^L (2 y'_0 y' + y'^2) \, dx \qquad 3.2.2$$

The change in the total potential corresponding to a variation δy in the additional deflections is (the initial deflections y_0 cannot change since they are caused by inaccurate workmanship and must be considered as a given property of the column):

$$\delta(U + V) = \int_0^L EIy'' \, \delta y'' \, dx - P \int_0^L (y'_0 \, \delta y' + y' \, \delta y') \, dx$$

$$= \int_0^L EIy'' \, \delta y'' \, dx - P \int_0^L (y'_0 + y') \, \delta y' \, dx \qquad 3.2.3$$

Integration by parts of the second term of the last member of Eq. 3.2.3 yields

$$\int_0^L (y'_0 + y') \, \delta y' \, dx = (y'_0 + y') \, \delta y \Big|_0^L - \int_0^L (y''_0 + y'') \, \delta y \, dx$$

Here again the first term in the right-hand member vanishes since δy is zero for $x = 0$ and $x = L$. The strain-energy term is simplified in the same manner as in Art. 3.1. Consequently the differential equation replacing Eq. 3.1.11 is

$$EIy^{iv} + P(y'' + y''_0) = 0 \qquad 3.2.4$$

and the end conditions remain unchanged

$$y = y'' = 0 \qquad \text{when} \qquad x = 0, L \qquad 3.2.5$$

Since y_0 is a given function of x, it is convenient to write Eq. 3.2.4 in the form

$$EIy^{iv} + Py'' = -Py''_0 \qquad 3.2.6$$

If the right-hand member were zero, the equation would be homogeneous. As a matter of fact it would be identical with Eq. 3.1.11 whose general solution was found to be (see Eq. 3.1.17)

$$y_c = A \cos kx + B \sin kx + Cx + D \qquad 3.2.7$$

where

$$k^2 = P/EI \qquad 3.2.7a$$

The subscript c is used to indicate that Eq. 3.2.7 contains the complementary solution of Eq. 3.2.6. The complete solution is obtained if a particular solution y_p of the complete inhomogeneous equation is added to the complementary solution. Before a particular solution can be found, the initial deflection y_0 must be expressed in a suitable form.

Representation of Deflections by Fourier Series

A particularly convenient representation of any arbitrary initial deflected shape can be had by means of a Fourier series. It was shown in Art. 2.8

that any function having the properties of the dashed line in Fig. 3.2.1 can be expanded in a Fourier series

$$y_0 = a_1 \sin \frac{\pi x}{L} + a_2 \sin \frac{2\pi x}{L} + a_3 \sin \frac{3\pi x}{L} + \cdots = \sum_{n=1}^{\infty} a_n \sin \frac{n\pi x}{L} \qquad 3.2.8$$

For the time being the analysis will be continued for an initial deflected shape corresponding to the typical term of the Fourier series rather than to the entire series:

$$y_0 = a_n \sin (n\pi x/L) \qquad (a)$$

If Eq. a represented the complete initial deflection, it would be easy to find a particular solution of Eq. 3.2.6; one has to observe only that the left-hand member contains terms with y'' and y^{iv} only, and that the second and fourth derivatives of a sine function are again sine functions. Consequently the assumption

$$y_p = F \sin (n\pi x/L) \qquad (b)$$

with F an unknown constant is certain to be successful. The second derivative of y_p is

$$y_p'' = -F(n\pi/L)^2 \sin (n\pi x/L) \qquad (c)$$

and its fourth derivative is

$$y_p^{iv} = F(n\pi/L)^4 \sin (n\pi x/L) \qquad (d)$$

while the second derivative of y_0 from Eq. a is

$$y''_0 = -a_n(n\pi/L)^2 \sin (n\pi x/L) \qquad (e)$$

Substitutions result in

$$EIF(n\pi/L)^4 \sin (n\pi x/L) - PF(n\pi/L)^2 \sin (n\pi x/L)$$
$$= Pa_n(n\pi/L)^2 \sin (n\pi x/L) \qquad (f)$$

Since y_p in Eq. b is a solution of Eq. 3.2.6 only if Eq. f is fulfilled identically for every value of x, both sides of the equation can be divided by $\sin (n\pi x/L)$. After a further division by $(n\pi/L)^2$ the resulting equation can be solved for F:

$$F = \frac{Pa_n}{EI(n\pi/L)^2 - P}$$

Division of both numerator and denominator by P gives

$$F = \frac{a_n}{(n^2 P_E/P) - 1}$$

where

$$P_E = \pi^2 EI/L^2$$

Consequently the particular solution sought is

$$y_p = \frac{a_n \sin (n\pi x/L)}{(n^2 P_E/P) - 1} \qquad (g)$$

Equation g is, of course, valid for any value of n. For this reason the particular solution corresponding to the entire Fourier series in Eq. 3.2.8 is

$$y_p = \sum_{n=1}^{\infty} \frac{a_n \sin (n\pi x/L)}{(n^2 P_E/P) - 1} \qquad 3.2.9$$

This is a consequence of the linearity of the differential equation Eq.3.2.6 which permits the superposition of the solutions corresponding to the individual terms of the Fourier series. The correctness of the solution can be checked by substitution of y_p from Eq. 3.2.9 in Eq. 3.2.6.

The complete solution of the differential equation Eq. 3.2.6 is the sum of the complementary solution (Eq. 3.2.7) and of the particular solution (Eq. 3.2.9):

$$y = A \cos kx + B \sin kx + Cx + D + \sum_{n=1}^{\infty} \frac{a_n \sin (n\pi x/L)}{(n^2 P_E/P) - 1} \qquad 3.2.10$$

The values of the four constants of integration A, B, C, and D must next be determined from the boundary conditions Eqs. 3.2.5.

The second derivative of the solution is

$$y'' = -Ak^2 \cos kx - Bk^2 \sin kx - \sum_{n=1}^{\infty} \frac{a_n (n\pi x/L)^2 \sin (n\pi x/L)}{(n^2 P_E/P) - 1} \qquad (h)$$

When $x = 0$, then y'' must vanish. Substitution of $x = 0$ in Eq. h yields

$$y''_{x=0} = -Ak^2 = 0$$

hence

$$A = 0$$

Also, when $x = 0$, then y must vanish. Substitution of $x = 0$ in Eq. 3.2.10 leads to

$$y_{x=0} = D = 0$$

Moreover $y'' = 0$ when $x = L$. Thus from Eq. h

$$y''_{x=L} = -Bk^2 \sin kL = 0 \qquad (i)$$

since $\sin (n\pi L/L) = \sin n\pi$ is zero. For particular values of k the function $\sin kL$ may vanish. This was investigated in Art. 3.1. For all other values of k corresponding to non-vanishing values of P (see Eq. 3.2.7a) the only solution of Eq. i is

$$B = 0$$

With $A = B = D = 0$ the deflection y at $x = L$ becomes from Eq. 3.2.10

$$y_{x=L} = CL$$

since the expression under the summation sign again vanishes because $\sin(n\pi) = 0$. Consequently all the four constants of integration are zero, and the solution of the differential equation is simply

$$y = \sum_{n=1}^{\infty} \frac{a_n \sin(n\pi x/L)}{(n^2 P_E/P) - 1}$$

$$= \frac{a_1 \sin(\pi x/L)}{(P_E/P) - 1} + \frac{a_2 \sin(2\pi x/L)}{(4P_E/P) - 1} + \frac{a_3 \sin(3\pi x/L)}{(9P_E/P) - 1} + \cdots \qquad 3.2.11$$

Magnification of Initial Deflections

Comparison of the initial deflected shape as given in Eq. 3.2.8 with the additional deflections of Eq. 3.2.11 reveals that, as P increases, each component of the initial deflections is increased in a different ratio. For example, when the compressive load P is one-half the Euler load P_E, the multiplier of the first Fourier term is $1/[(P_E/P) - 1] = 1$. At the same time the multiplier of the second term is $1/(8 - 1) = 1/7$, and that of the third term $1/(18 - 1) = 1/17$. Hence at one-half the Euler load the additional deflections can be obtained by adding to the initial first component one seventh of the initial second component, one seventeenth of the initial third component, and so on.

When the end load is nine tenths of the Euler load, the multiplying factors are 9, 9/31, 1/9, and so on. At 99/100 of the Euler load the factors are 99, 99/301, 99/801, and so on. It is seen that, however small the first component $a_1 \sin(\pi x/L)$ of the initial deflections may have been, it is exaggerated more and more as the load approaches the Euler load, until the deflected shape resembles the half-sine wave corresponding to $y = \sin(\pi x/L)$. Moreover the deflections increase beyond all limits as P approaches P_E.

The behavior of an initially crooked column differs therefore considerably from that of the perfectly straight column. The initial deflections of the column are increased as soon as the compressive end load is applied, but with increasing load the deflections are magnified very slowly at first, and more and more rapidly as the load approaches the Euler load. At the Euler load the denominator of the first term in the right-hand member of Eq. 3.2.11 becomes zero, and thus the equation has no meaning. As a matter of fact the validity of the equation is lost earlier, as soon as the deflections become so large that the assumptions underlying the derivation of the differential equation are not fulfilled. However, the large deflections give rise to large bending stresses which together with the compressive stress, may cause the column to yield before Eq. 3.2.11 ceases to be valid. This will be shown by a numerical example.

Failure of Initially Crooked Column

For the sake of simplicity let it be assumed that the initial deviations from straightness follow exactly the half-sine-wave pattern:

$$y_0 = a_1 \sin (\pi x/L) \qquad\qquad 3.2.12$$

The additional deflections are, from Eq. 3.2.11,

$$y = \frac{a_1 \sin (\pi x/L)}{(P_E/P) - 1} \qquad\qquad 3.2.13$$

If the cross section of the column is square with a side length $a = \frac{1}{2}$in., the moment of inertia of the section is

$$I = a^4/12 = 1/192 = 0.00521 \text{ in.}^4$$

and its section modulus

$$Z = 1/48 = 0.0208 \text{ in.}^3$$

If the column length is $L = 25$ in. and the material is SAE-1025 carbon steel, the Euler load is

$$P_E = \pi^2 \times 29 \times 10^6 \times 0.00521/25^2 = 2380 \text{ lb}$$

since the modulus of the material of the column is 29×10^6 psi. The yield stress is 36,000 psi.

The maximum deflection occurs at the middle of the column. In the calculation of the bending moment the total deviation y_{tot} from the straight axis is needed. This is the sum of the initial deviation y_0 and the elastic deflection y. Hence at $x = L/2$

$$y_{tot\,max} = (y_0 + y)_{x=L/2} = a_1 + \frac{a_1}{(P_E/P) - 1}$$

$$= \frac{P_E/P}{(P_E/P) - 1} a_1 \qquad\qquad 3.2.14$$

Consequently the maximum bending moment M_{max} is reached at the middle of the column with the value

$$M_{max} = \frac{P_E/P}{(P_E/P) - 1} a_1 P \qquad\qquad 3.2.15$$

The maximum bending stress is

$$\sigma_b = \pm \frac{M_{max}}{Z} = \pm 48 a_1 P \frac{P_E/P}{(P_E/P) - 1}$$

and the uniform compressive stress is

$$\sigma_c = -P/A = -4P$$

where A is the cross-sectional area of the column. The maximum combined stress is consequently

$$\sigma_{max} = \sigma_b + \sigma_c = -48 a_1 \frac{P_E}{(P_E/P) - 1} - 4P_E(P/P_E)$$

$$= -114{,}200 \frac{a_1}{(P_E/P) - 1} - 9520(P/P_E) \qquad\qquad 3.2.16$$

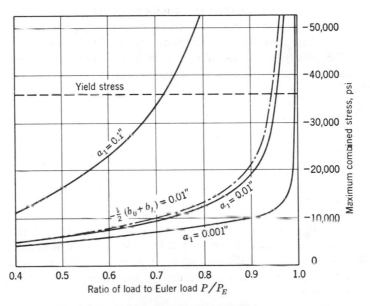

Fig. 3.2.2. Stresses in imperfect columns.

If the maximum initial eccentricity is assumed to be 0.001 in., 0.01 in., and 0.1 in. in turn, and the bending stress is calculated from Eq. 3.2.16 for several values of the ratio P_E/P, the results can be plotted in a diagram as shown in Fig. 3.2.2. The three values chosen correspond to a column manufactured and fitted with an unusually high accuracy, to a more or less average column, and to one that is visibly crooked. The maximum combined stress increases very slowly in the first column until the Euler load is closely approached, while in the visibly crooked column the stresses attain considerable values well below the Euler load. The yield stress is reached in the latter at 72 per cent of the Euler load, at 96 per cent in the average column, and at over 99 per cent in the exceptionally accurate column. Consequently average columns can be expected to become deformed permanently when the compressive end load reaches values between 90 per cent and 100 per cent of the Euler load. Very carefully manufactured test specimens can support loads deviating from the Euler load by less than 1 per cent. On the other hand, grossly inaccurate columns may fail at loads considerably smaller than the Euler load.

This investigation explains the practical significance of the Euler load of a column. It is the limiting value of the load that the column can support if it is manufactured and fitted with extreme care. This limit can be approached within 5 or 10 per cent if ordinary care is exercised in the construction. Of course the numerical values shown in Fig. 3.2.2 are valid only for the particular cross section, length, and material, but qualitatively they represent the behavior of all columns.

The deflections of the midpoint of the column corresponding to the yield stress condition can be calculated from Eq. 3.2.13:

$$y_{max} = \frac{a_1}{(P_E/P) - 1}$$

Hence the deflection of the crooked column is

$$y_{max_{1/10}} = \frac{\frac{1}{10}}{(1/0.72) - 1} = 0.257 \text{ in.}$$

For the other columns one obtains similar values. Since these values are approximately equal to one half of the side length of the square section or to one hundredth of the length of the column, the deflections can be considered small, and thus the validity of the theory for these deflections is established.

Eccentric Load Application

The behavior of a real column, as distinct from an idealized theoretical column, can be investigated from another standpoint by assuming its center line to be perfectly straight but considering the load to be applied eccentrically as shown in Fig. 3.2.3. The load P at $x = 0$ can be replaced by a centrally applied load P plus an end moment $M_0 = -Pb_0$, and at $x = L$ one has a centrally applied load P plus an end moment $M_L = -Pb_L$.

The difference $M_L - M_0$ is balanced by a couple produced by horizontal reactions at the two ends of the column. These reactions are not shown in the figure.

The deflections of a column subjected to two end moments as well as to a compressive end load are a special case of the deflections of the beam column given in Eq. 2.4.10. Substitution of $w = 0$ yields

$$y = (1/k^2EI)[M_0(\cos kx - 1) + (M_L - M_0 \cos kL)$$
$$\times (\sin kx/\sin kL) + (M_0 - M_L)(x/L)] \qquad 3.2.17$$

Because of the validity of the principle of superposition for constant values of the end load P, the effects of the two end moments can be treated separately. When $M_0 = 0$, one obtains

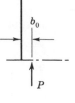

Fig. 3.2.3.
Eccentrically
loaded column.

$$y = (1/k^2EI)[M_L(\sin kx/\sin kL) - M_L(x/L)] \qquad 3.2.18$$

At the middle of the column the deflection is

$$y_{\text{middle}} = \frac{M_L}{k^2EI}\left[\frac{\sin (kL/2)}{\sin kL} - \frac{1}{2}\right]$$

Since $k^2 = P/EI$ and $M_L = -Pb_L$, one has

$$M_L/k^2EI = -b_L$$

and because

$$\sin kL = 2 \sin \frac{kL}{2} \cos \frac{kL}{2}$$

the deflection becomes

$$y_{\text{middle}} = -\frac{b_L}{2}\left[\frac{1}{\cos (kL/2)} - 1\right]$$

or

$$y_{\text{middle}} = -\frac{b_L}{2}\left(\sec \frac{kL}{2} - 1\right) \qquad 3.2.19$$

When $M_L = 0$ and M_0 is different from zero, considerations of symmetry or substitution in Eq. 3.2.17 yield

$$y_{\text{middle}} = -\frac{b_0}{2}\left(\sec \frac{kL}{2} - 1\right) \qquad 3.2.20$$

Consequently the total deflection at the middle of the column is

$$y_{\text{tot}} = -\frac{1}{2}(b_0 + b_L)\left(\sec \frac{kL}{2} - 1\right) \qquad 3.2.21$$

Since the bending moment at the middle of the column in the non-deflected state shown in Fig. 3.2.3 is

$$-\tfrac{1}{2}(b_0 + b_L)P$$

addition of the bending moment Py_{tot} caused by the elastic deflections results in a total bending moment

$$M_{tot} = -\tfrac{1}{2}(b_0 + b_L)P \sec (kL/2) \qquad\qquad 3.2.22$$

The maximum bending stress is

$$\sigma_b = \pm M_{tot}/Z \qquad\qquad 3.2.23$$

Because

$$kL/a = (L/2)\sqrt{P/EI} = (\pi/2)\sqrt{PL^2/\pi^2 EI} = (\pi/2)\sqrt{P/P_E}$$

Eq. 3.2.23 can be written with the aid of Eq. 3.2.22 as

$$\sigma_b = \mp \frac{1}{2}(b_0 + b_L) \left(\frac{P_E}{Z}\right) \left(\frac{P}{P_E}\right) \sec \left(\frac{\pi}{2}\sqrt{\frac{P}{P_E}}\right) \qquad\qquad 3.2.24$$

The maximum combined stress is obtained by adding the maximum compressive stress

$$\sigma_c = -P/A = -(P_E/A)(P/P_E) \qquad\qquad 3.2.25$$

to the maximum bending stress.   If the values of the numerical example are substituted, the result is

$$\sigma_{max} = -571(b_0 + b_L)\frac{P}{P_E}\left(\sec \frac{\pi}{2}\sqrt{\frac{P}{P_E}}\right) - 9520\frac{P}{P_E} \qquad\qquad 3.2.26$$

The maximum stress calculated from Eq. 3.2.26 with $\tfrac{1}{2}(b_0 + b_L) = 0.01$ in. was plotted in Fig. 3.2.2.   It can be seen that the average eccentricity of the load application has about the same effect on the maximum stress as an equal amount of initial crookedness.   Consequently permanent deformations will occur well below the Euler load if the end loads are applied off the center of the column.   If the eccentricity is reduced to very small values, the permanent deformations can be delayed until the value of the end load approaches the Euler load within 2 or 3 per cent.

It is of interest to note that the deflections caused by eccentric load application also approach the half-sine-wave form as the load approaches the Euler load.   This can be seen from Eq. 3.2.18.   The second term in the brackets increases in proportion to the load P since $M_L = Pb_L$.   The first term contains the same multiplying factor, but it also has sin kL in the denominator.   Since $kL = \pi\sqrt{P/P_E}$, ovbiously kL approaches π as P

approaches P_E. But $\sin \pi = 0$, and thus the denominator of the first term becomes very small when the end load attains values close to the Euler load. Consequently the first term becomes very large and governs the deflected shape at high loads. On the other hand,

$$kx = kL(x/L) = (\pi x/L)\sqrt{P/P_E}$$

approaches $\pi x/L$ as P approaches P_E. This means that in the neighborhood of the Euler load the deflected shape is substantially proportional to $\sin (\pi x/L)$. It is therefore a half-sine wave.

Because the deflections caused by eccentric load application are very much like those caused by initial curvature, a simple technique can be used to improve the accuracy of column tests. If the deflection of the midpoint of the column is large and positive at about one-half the Euler load, the point of application of the load must be shifted at one end of the column twice that distance in the positive direction. Alternatively one end of the column may be shifted in the negative direction relative to the point of application of the load. A few adjustments of the column in the testing machine suffice to eliminate the major part of the deflections, with the result that the bending stresses remain small until the end load closely approaches the Euler load.

The Southwell Plot

The preponderance of the first Fourier component in the deflection pattern and the similarity of the eccentricity curve in Fig. 3.2.2 to the initial crookedness curve permit a graphic construction which is useful in finding the correct value of the Euler load in experiments. Assuming that the elastic transverse displacement of the midpoint of the column is represented by the equation

$$y_m = \frac{e}{(P_E/P) - 1} \qquad\qquad 3.2.27$$

where e is the difference between the amplitude a_1 of the first Fourier component of the initial deviations from the straight axis and the average eccentricity $\frac{1}{2}(b_0 + b_L)$ of the load application, one can multiply both sides of the equation by the denominator and rearrange terms to obtain

$$y_m = P_E(y_m/P) - e \qquad\qquad 3.2.28$$

If the mid-point deflections y_m measured in a column test are plotted in a diagram against the ratio y_m/P where P is the measured load corresponding to the deflection y_m, a straight line is obtained in accordance with Eq. 3.2.28.

The straight-line diagram is shown in Fig. 3.2.4.   The difference between the ordinates of points 2 and 1 is

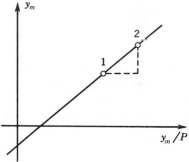

Fig. 3.2.4.   Southwell plot.

$$y_{m2} - y_{m1}$$

while the difference of their abscissas is

$$(y_m/P)_2 - (y_m/P)_1$$

If Eq. 3.2.28 is written first for point 2 and then for point 1 and the latter equation is subtracted from the former, the result is

$$y_{m2} - y_{m1} = P_E[(y_m/P)_2 - (y_m/P)_1]$$

Consequently the Euler load P_E is the slope of the straight line:

$$P_E = \frac{y_{m2} - y_{m1}}{(y_m/P)_2 - (y_m/P)_1} \qquad 3.2.29$$

This procedure was devised by Southwell.   It permits a satisfactory experimental determination of the Euler load when a low yield point, a high buckling stress, and a large eccentricity or initial crookedness combine to cause yielding of the material well before the Euler load is reached.

Distributed Transverse Loads

The last point to be discussed in this article is the behavior of a beam subjected simultaneously to compressive end loads and a uniformly distributed transverse load w.   This problem is again a special case of the beam column investigated in Art. 2.4.   If the beam is supported on pivots at its ends and the end loads are applied centrally, the equation for the deflections Eq. 2.4.10 reduces to

$$y = (1/k^2 EI)[(w/k^2)(\cos kx - 1) + (w/k^2)(1 - \cos kL)(\sin kx/\sin kL)$$
$$- (wL/2)(L - x)(x/L)]$$
$$= (w/P)\{(1/k^2)[\cos kx - 1 + (1 - \cos kL)(\sin kx/\sin kL)]$$
$$- (L/2)(L - x)(x/L)\}$$

Manipulations result in an equation valid when $x = L/2$:

$$y_{L/2} = \frac{w}{P}\left[\frac{1}{k^2}\left(\sec\frac{kL}{2} - 1\right) - \frac{L^2}{8}\right] \qquad 3.2.30$$

In the absence of an end load the bending moment at midspan is

$$M_w = wL^2/8$$

The additional bending moment is the end load P multiplied by the deflection $y_{L/2}$:

$$M_P = \frac{w}{k^2} \left(\sec \frac{kL}{2} - 1 \right) - \frac{wL^2}{8}$$

Consequently the total bending moment is

$$M_{tot} = \frac{w}{k^2} \left(\sec \frac{kL}{2} - 1 \right) \qquad \text{3.2.31}$$

Because

$$kL/2 = (\pi/2)\sqrt{P/P_E}$$

Eq. 3.2.31 can also be written in the form

$$M_{tot} = \frac{wL^2}{8} \frac{2}{(kL/2)^2} \left(\sec \frac{kL}{2} - 1 \right)$$

or

$$M_{tot} = C(wL^2/8) \qquad \text{3.2.32}$$

with

$$C = \frac{8}{\pi^2} \left(\frac{P_E}{P} \right) \left[\sec \left(\frac{\pi}{2} \sqrt{\frac{P}{P_E}} \right) - 1 \right] \qquad \text{3.2.32a}$$

Since $wL^2/8$ is the maximum bending moment when there is no end load, the factor C represents the effect of the end load on the maximum bending moment. It is plotted in Fig. 3.2.5. It can be seen from the figure that C, and consequently the bending moment, increases rapidly when the end load approaches the Euler load P_E. Depending on the values of the moment of inertia, the section modulus, the cross-sectional area of the beam, and the magnitude of the transverse load w, the yield stress in the beam will be reached at different values of the ratio P/P_E. When w is very small and the stress corresponding to the Euler load is considerably smaller than the yield stress, the maximum end load that the beam can support will be close to the Euler load.

Again the deflected shape approaches the half-sine-wave form when P approaches P_E. This can be seen from the equation preceding Eq. 3.2.30 in which the dominant term at high values of P is the one whose denominator contains $\sin kL$. The proof can be given in the same manner as was done when the deflected shape caused by eccentric loads was investigated. At any prescribed value of the end load P the dominant term is a constant times $\sin kx$, and, as in the earlier derivation, $\sin kx$ approaches $\sin (\pi x/L)$ when P approaches P_E. Consequently in the limit the deflected shape is a half-sine wave.

Conclusions

In conclusion it can be said that the theoretical buckling load is independent of the initial crookedness of the column, of the eccentricity of the load application, and of the magnitude of the transverse loading. It represents the maximum compressive load that the column can support when crookedness, eccentricity, and transverse load are small, and the

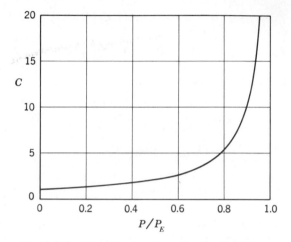

Fig. 3.2.5.   Magnification factor for bending moment.

yield stress of the material is high.   The column becomes permanently distorted under loads smaller than the buckling load when any of the three quantities mentioned is large.   This is not surprising, since obviously the yield stress can be reached in a beam even without an end load if the transverse load w is sufficiently large.

Finally it should be noted that all the equations presented in this article were derived on the assumption that Hooke's law is valid.   Strictly speaking, this restricts the validity of the discussion to stresses smaller than the proportional limit of the material.   But inclusion of stresses up to the yield point is permissible when only approximate results are sought and when the stress-strain curve of the material shows little permanent deformation below the yield point.   This situation exists with mild steel.   A further restriction is that the compressive stress corresponding to the Euler load must be smaller than the proportional limit of the material. Problems of buckling involving critical stresses higher than the proportional limit will be investigated in Art. 3.9.

3.3   The Buckling Load of a Sandwich Column

Derivation of the Differential Equations

The bending deformations of a sandwich beam were investigated in some detail in Art. 2.6.   When the load W in Fig. 2.6.1 is replaced by a compressive end load P, uniformly distributed over the end surfaces of the two faces, the sandwich column is likely to buckle, provided the end compression P is large enough.   In the fully compressed state prevailing just before buckling takes place the column is still straight.   As the total cross-sectional area of the faces is $2bt$, the uniformly distributed compressive stress is $\sigma = P/2bt$.

The strain energy stored in the column can be expressed by the same equations as in Art. 2.6.   In the expression for the total potential (Eq. 2.6.6) the potential of the transverse load W must be replaced by the potential of the end load P.   The latter quantity is the negative of the product of force P and displacement of the end section $x = L$ in consequence of the deflections v.   The longitudinal displacements u do not contribute a term to the potential of P becuase they are of opposite direction in the two faces as shown in Fig. 2.6.1.   No displacement coordinate is provided to represent the uniform compression corresponding to the state before buckling, and the strain energy in that state is not calculated.   The change in this part of the strain energy during the variation of the deflections is not needed in view of the analysis presented in the last section of Art. 2.4. It would simply furnish an equation requiring that the average compressive strain in the sandwich column remain unchanged during buckling.

The shortening in the distance between the two end sections of the column can be calculated from Eq. 2.4.2 if y is replaced by v:

$$\Delta L = \frac{1}{2} \int_0^{T_i} v'^2 \, dx$$

where the prime indicates differentiation with respect to x.   The potential of the compressive end load is therefore

$$V = -\sigma bt \int_0^L v'^2 \, dx \qquad\qquad 3.3.1$$

The column is assumed simply supported at its ends.   By this is meant that no bending stresses can be present at the ends in the column as a whole, and thus

$$u' = 0 \qquad\qquad 3.3.2a$$

and no bending can occur in the individual faces; that is

$$v'' = 0 \qquad\qquad 3.3.2b$$

The conditions of equilibrium are obtained from the minimal principle. When the displacements u and v are varied, the change in the total potential is again represented by Eq. 2.6.7 except that the term $-W\,\delta v_{x=L}$ is replaced by

$$\delta V = -2\sigma bt \int_0^L v'\,\delta v'\,dx \qquad\qquad 3.3.3$$

The derivatives of δu and δv are again eliminated through integration by parts, and the same integrals are obtained as in Art. 2.6. However, with the new end conditions Eqs. 3.3.2 and the geometric constraint imposed on the variations at the ends of the column

$$\delta v = 0 \qquad\qquad 3.3.4$$

all the terms referring to $x = L$ in Art. 2.6 vanish. Integration by parts of the right-hand member of Eq. 3.3.3 yields

$$\delta V = -2\sigma bt v'\,\delta v \Big|_0^L + 2\sigma bt \int_0^L v''\,\delta v\,dx \qquad\qquad 3.3.5$$

The first term in the right-hand member of this equation vanishes because of Eq. 3.3.4. Hence the total change in the total potential is

$$\delta(U + V) = \int_0^L \left[-2Ebtu'' + \frac{2Gbc}{c+t}\left(\frac{2u}{c+t} - v' \right) \right] \delta u\,dx$$
$$+ \int_0^L \left[E\frac{bt^3}{6} v^{\mathrm{iv}} + 2\sigma btv'' + Gbc\left(\frac{2u'}{c+t} - v'' \right) \right] \delta v\,dx \qquad 3.3.6$$

As this change must vanish for any arbitrary variation in u and v, the expressions in brackets must vanish individually. If the notation is re-introduced

$$K_1^2 = Gbc/EI_1, \qquad K_2^2 = Gbc/2EI_f, \qquad y = \tfrac{1}{2}(c+t)v$$
$$I_1 = \tfrac{1}{2}t(c+t)^2 b, \quad 2I_f = \tfrac{1}{6}t^3 b, \qquad\qquad 3.3.7$$

the conditions of equilibrium can be written in the form of the two differential equations

$$u'' - K_1^2 u + K_1^2 y' = 0 \qquad\qquad 3.3.8a$$

$$y^{\mathrm{iv}} + (\alpha - K_2^2)y'' + K_2^2 u' = 0 \qquad\qquad 3.3.8b$$

where

$$\alpha = 2\sigma bt/2EI_f \qquad\qquad 3.3.8c$$

These equations must be solved together with the boundary conditions

$$v = v'' = u' = 0 \qquad \text{when} \quad x = 0, L \qquad\qquad 3.3.8d$$

Solution of the Differential Equations

As the integration of the differential equations is a rather lengthy but routine process, details of it will not be given here. The solution can be written in the form

$$u = A_1 \cosh p_1 x + A_2 \sinh p_1 x + A_3 \cos p_2 x + A_4 \sin p_2 x + A_5 \quad 3.3.9$$

$$y = (1/p_1)[1 - (p_1^2/K_1^2)](A_1 \sinh p_1 x + A_2 \cosh p_1 x)$$
$$+ (1/p_2)[1 + (p_2^2/K_1^2)](A_3 \sin p_2 x - A_4 \cos p_2 x) + A_5 x + A_6$$

$$3.3.10$$

where

$$p_1 = (\mu + \nu)^{1/2} \qquad\qquad p_2 = (\mu - \nu)^{1/2}$$

$$\mu = \tfrac{1}{2}[(K_1^2 + K_2^2 - \alpha)^2 + 4\alpha K_1^2]^{1/2}$$

$$\nu = \tfrac{1}{2}(K_1^2 + K_2^2 - \alpha) \qquad\qquad 3.3.11$$

and the square root must be taken with the positive sign. The correctness of the solution can be checked by substitution.

A detailed examination of the end conditions shows that two solutions are possible. In one, all the integration constants vanish; this trivial solution means simply that the column is in equilibrium in the straight-line configuration. In the second solution, all the integration constants but A_3 vanish. In this case, however, $\sin p_2 L$ must also vanish in order to satisfy the requirement $v - v'' = 0$ at $x = L$. This happens when

$$p_2 L = n\pi \qquad\qquad 3.3.12$$

where n is any integer. Because $p_2 = (\mu - \nu)^{1/2}$, substitutions, manipulations, and solution of the equation for the critical load $P_{cr} = 2\sigma bt$ yield

$$P_{cr} = n^2 P_1 \frac{n^2 P_2 + Gbc[1 + (P_2/P_1)]}{n^2 P_1 + Gbc} \qquad\qquad 3.3.13$$

The lowest critical load corresponds to $n = 1$:

$$P_{cr} = P_1 \frac{P_2 + Gbc[1 + (P_2/P_1)]}{P_1 + Gbc} \qquad\qquad 3.3.14$$

where

$$P_1 = \pi^2 (EI)_1/L^2 \qquad\qquad 3.3.14a$$

$$P_2 = \pi^2 (2EI)_f/L^2 \qquad\qquad 3.3.14b$$

P_1 is the buckling load of a sandwich beam whose normal stress-carrying material is concentrated into two vanishingly thin faces and whose core is perfectly rigid in shear. P_2 is the buckling load of a sandwich beam consisting of two independent faces.   Finally Gbc is the buckling load of a sandwich beam that buckles in a pure shear mode without causing any extensions in the faces.

The following particular cases are of some interest:

1. When the bending rigidity of the two individual faces $(2EI)_f$ is small, Eq. 3.3.14 reduces to

$$P_{cr} = P_1 Gbc/(P_1 + Gbc) \qquad 3.3.15$$

The the buckling load is one-half the harmonic mean of P_1 and Gbc.

2. When the shearing rigidity Gbc is large, Eq. 3.3.14 becomes

$$P_{cr} = \pi^2 [EI_1 + 2(EI)_f]/L^2 \qquad 3.3.16$$

This is the ordinary Euler formula for the sandwich column which neglects the capacity of the core to carry normal stresses and disregards the shearing deformations.

3. When the shearing rigidity Gbc is negligibly small, the buckling load is

$$P_{cr} = \pi^2 (2EI)_f/L^2 = P_2 \qquad 3.3.17$$

With a vanishing shearing rigidity, naturally the two faces have to act as independent columns.

3.4   The Rayleigh Method

Criticism of Stability Analysis

Two shortcomings of the procedure developed in Art. 3.1 for the calculation of the buckling load should be noted.   The critical load was found as the force under which the state of equilibrium of the system is neutral.   Although it is reasonable to assume that a physical system must pass through neutral equilibrium before it can reach an unstable state of equilibrium, the converse of the statement in the sense that a state of neutral equilibrium is always a limiting case between stability and in-stability does not necessarily follow from the considerations presented.   It may be possible for the change in the total potential corresponding to a virtual displacement to decrease with increasing load until it vanishes and the system reaches a neutral state of equilibrium, and yet upon a further increase in the load the system may again become stable.   Then the critical load calculated according to the procedure of Art. 3.1 is not the limiting value of the load above which the system is unstable.   On the other hand, below this critical load the system is undoubtedly stable in every case.   If the designer wishes to avoid a neutral state of equilibrium, which permits large deformations to develop, just as he wants to rule out unstable states of equilibrium, then the critical load calculated according to Art. 3.1 can serve as a danger signal, as an upper limit of the loading, which should never be reached in structures and machinery.

Another objection to the method of Art. 3.1 is that it establishes the limit of stability for only very small disturbances. The properties of the equilibrium were judged from the first non-vanishing term in the brackets in Eq. 3.1.8, which is the correct procedure only when the displacement δx from the equilibrium position is so small that its higher powers can be neglected compared to its lower powers. However, a system can be stable when the disturbances are small, and unstable when they are large.

As an example, a ball might be cited that is resting at the bottom of a small, shallow hole at the top of a steep hill. As long as the disturbances move the ball around inside the hollow, the equilibrium is stable; but, when the disturbance results in pushing the ball over the rim of the hole, the ball will start rolling down the hillside and will never again return to its original position of equilibrium. As a matter of fact, position A in Fig. 3.1.1 corresponds to a stable equilibrium only if the disturbances are smaller than the distance from A to B. If displacements from A to or beyond B are considered possible disturbances of the state of equilibrium, the equilibrium at A is unstable.

In practical problems it is of course difficult to estimate the magnitude of the disturbance. In testing a column in the testing machine one would have to know, for instance, the amplitude of the vibrations transmitted to the column from the engine actuating the testing machine, and from the foundation that might shake when a streetcar passes by the building. Fortunately in most problems encountered by engineers the situation is much simpler than that shown in Fig. 3.1.1. When the forces and deflections are connected by a linear law, the expression for the total potential of the system is a quadratic expression in the displacement quantities. Consequently it can have a single extremum only, and not a maximum, a minimum, and an inflection point with a horizontal tangent like the hill in Fig. 3.1.1. Then stability for small disturbances involves stability for large disturbances also. On the other hand, it must not be forgotten that in many practical systems the linear relationship between force and displacement changes to a non-linear one when the displacements become large.

In a complete and rigorous analysis the stability of a system should be investigated in accordance with the definition of stability given in Art. 3.1. A displacement from the equilibrium configuration should be assumed and the ensuing vibrations determined by means of the dynamic equations of motion. If these show that the structure will eventually again reach its original position of equilibrium, then stability is established. Of course such an investigation is practically impossible except for the very simplest systems. Even the rigorous establishment of the neutral state of equilibrium by means of the variational calculus given in Arts. 3.1 and 3.3 is far too complex in many cases of practical importance. Then a much simpler and very practical method due to Lord Rayleigh can be used to advantage.

The Rayleigh Method

When the Rayleigh method is applied to the column, a simple deflected shape is assumed, and the change in the total potential is calculated during the transition from the initial state of equilibrium into the deflected configuration. If the initial state is a state of equilibrium, the first-order terms in the expression for the change of the total potential must be zero for any virtual displacement, and thus also for the displacement pattern chosen. This is just a restatement of the minimal principle. The original state of equilibrium is stable, if the higher-order terms are positive, since they signify a minimum of the total potential. The system can be displaced from a configuration corresponding to a minimum of the total potential only if some additional force not belonging to the system does work on the system. After the force is removed, the system returns to its original configuration of equilibrium. The original state is unstable if the higher-order terms are negative since they correspond to a maximum of the total potential. From a configuration corresponding to a maximum of the total potential, the system can be shifted without work being done by extraneous forces; on the contrary, energy is freed in the process. Between stability and instability is the limiting case of neutral equilibrium characterized by the vanishing of the higher-order terms in the expression for the change in the total potential. During a displacement from a neutral state of equilibrium energy is neither absorbed nor freed by the system.

The procedure of calculating the buckling load from the condition that the total potential remain unchanged even when higher-order terms are taken into consideration would be an exact method if the column had only one degree of freedom of motion permitting it to assume the deflected shape chosen. In reality the column has an indefinitely large number of degrees of freedom of motion, and stability is established only if the total potential is a minimum with respect to all of them. If the equilibrium position corresponding to a horizontal tangent plane of the total potential surface is situated on a saddle-shaped portion of the total potential surface, and the value of the total potential is a maximum even with respect to one single displacement coordinate, while it is a minimum with respect to all the others, the equilibrium is unstable as was indicated earlier. This is the reason why stability as well as equilibrium can be established rigorously only through an investigation of all the possible deflected configurations, as was done when the calculus of variations was used.   When an analysis by the calculus of variations is too difficult, or when it leads to differential equations that are inconvenient or impossible to solve, the stress analyst is compelled to turn to an approximate method.   Fortunately Lord Rayleigh showed that in many problems, in- cluding all linear column problems, the stability limit, calculated from the changes in the value of the total potential, is rather insensitive to the particular choice of the deflected shape.   For this reason good approximate values of the buckling load can be obtained by the Rayleigh method if a reasonable deflected shape is chosen.

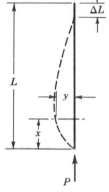

Fig. 3.4.1.   Defor- mations of column.

An example will clarify the procedure.   The pin-ended straight column in Fig. 3.4.1 is sub- jected to the axial compressive loads P.   If it is assumed that the buckled shape can be represented by the equation

$$y = a \sin (\pi x/L) \qquad 3.4.1$$

then the strain energy stored in the column due to bending is

$$\delta U - \frac{1}{2} EI \int_0^L y''^2 \, dx = \frac{1}{2} EI \int_0^L a^2 \left(\frac{\pi}{L}\right)^4 \sin^2 \frac{\pi x}{L} \, dx$$

With

$$\int_0^L \sin^2 \frac{\pi x}{L} \, dx = \frac{L}{2}$$

one obtains

$$\delta U = \tfrac{1}{4}\pi^4 a^2 (EI/L^3) \qquad 3.4.2$$

This is the total change in the strain energy during buckling, since the strain energy of compression does not change as the column deflects.

The decrement in the potential of the load P is equal to the load times the distance ΔL through which P is lowered.   Since

$$\Delta L = \frac{1}{2} \int_0^L y'^2 \, dx = \frac{1}{2} \int_0^L a^2 \left(\frac{\pi}{L}\right)^2 \cos^2 \frac{\pi x}{L} \, dx$$

and thus with

$$\int_0^L \cos^2 \frac{\pi x}{L} \, dx = \frac{L}{2}$$

one has

$$\Delta L = \tfrac{1}{4}\pi^2 (a^2/L) \qquad 3.4.3$$

the change in the potential of the external load is

$$\delta V = -P \, \Delta L = -\tfrac{1}{4}\pi^2 a^2 (P/L) \qquad\qquad 3.4.4$$

The change in the total potential during the deflection of the originally straight column is therefore

$$\delta(U + V) = \tfrac{1}{4}\pi^4 a^2 (EI/L^3) - \tfrac{1}{4}\pi^2 a^2 (P/L) \qquad\qquad 3.4.5$$

The change in the total potential contains only terms with a^2. Hence displacements of the order a cause a change in the total potential of the order a^2. The absence of the first-order terms is a proof of the equilibrium in the original straight configuration. (Of course this proof is not needed since the equilibrium of the straight bar can be established by much simpler methods.) In Eq. 3.4.5 the change $\delta(U + V)$ is obviously positive when P is small. Consequently the total potential is a minimum, and the column is stable when subjected to small compressive loads. But $\delta(U + V)$ can certainly be made negative by choosing P sufficiently large. Thus the total potential is a maximum, and the column is unstable under large compressive loads. The stability changes into instability when $\delta(U + V)$ vanishes. The equation

$$\tfrac{1}{4}\pi^4 a^2 (EI/L^3) - \tfrac{1}{4}\pi^2 a^2 (P/L) = 0 \qquad\qquad 3.4.6$$

can be solved for P. One obtains the critical value of the end load as

$$P_{\text{cr}} = \pi^2 EI/L^2 \qquad\qquad 3.4.7$$

Two remarks are in order in connection with these calculations. First, it is not surprising that an approximate calculation yielded the exact value of the buckling load because the deflected shape chosen for the demonstration of the Rayleigh method was the exact shape of the buckled column. Through the use of the exact shape of deformations at buckling, the stability limit was established in an exact manner. Second, it is of interest to note that the same conclusions are reached if the different steps in the derivation are reinterpreted in the following way:

Instead of considering the right-hand member of Eq. 3.4.5 as the change $\delta(U + V)$ in the total potential during the transition from the straight-line shape to the slightly curved shape, one can take it to represent the total potential $(U + V)$ of the column in the deflected configuration. If this is done, only the strain energy of uniform compression is neglected, and in Art. 2.4 this was found permissible. If this slightly deflected shape is also a configuration of equilibrium, then the change in the total potential

$$U + V = \tfrac{1}{4}\pi^2 a^2 [(\pi^2 EI/L^3) - (P/L)]$$

corresponding to the variation δy, that is,

$$\delta(U + V) = \tfrac{1}{2}\pi^2 a\ \delta a[(\pi^2 EI/L^3) - (P/L)]$$

must vanish. If a is zero, the trivial solution is obtained; δa is arbitrary by assumption; consequently the expression in brackets must vanish if a non-trivial solution is to be obtained. This last requirement yields the Euler load through the same manipulations as those required to obtain Eq. 3.4.7 from Eq. 3.4.6.

In this second approach, equilibrium in the deflected configuration, involving neutral equilibrium in the original straight configuration, was established incompletely through an application of the minimal principle. The application was incomplete because only the magnitude of the deflections was varied but not the shape. Keeping the chosen shape unchanged is equivalent to adding imaginary geometric constraints to the system. Hence the system investigated by the Rayleigh method is stiffer than the actual one, and the actual unrestrained system may be unstable when the Rayleigh analysis finds stability. On the other hand, the structural element is certainly unstable when it is found to be so in a calculation by the Rayleigh method. Thus the Rayleigh method always yields a too high critical load unless by chance the investigation is carried out for the accurate deflected shape at buckling in which case the stability limit obtained by the Rayleigh method is exact.

Use of the Law of Conservation of Energy

A third interpretation of the calculations can be given on the basis of the law of the conservation of energy. When a column which carries a load P is deflected through a small but finite distance according to Eq. 3.4.1, the strain energy stored in the column is

$$U = \tfrac{1}{4}\pi^4 a^2(EI/L^3)$$

At the same time the work done by the load P is

$$W = \tfrac{1}{4}\pi^2 a^2(P/L)$$

For a given value of the amplitude a, the strain energy is a constant in the first equation while in the second the work is proportional to P. Hence for small values of P the strain energy stored during a disturbance of the straight-line equilibrium configuration is greater than the work necessary to raise the load P into its original position. The column will therefore straighten out, and the excess energy will be dissipated by friction during the ensuing vibrations.

On the other hand, the strain energy U is less than the work W when the load P is large. Then a disturbance resulting in deflections according to

Eq. 3.4.1 stores an elastic-energy quantity insufficient to do the work required to raise P into its original position.   Consequently the deflections will increase further, and in the process the excess work will increase proportionately to a^2.   The accelerated motion of the column will presumably stop only when the load reaches the ground.

Between the small values of P corresponding to stability and the large values connected with instability there is a limiting value P_{cr} characterized by the equality of work and strain energy.   When the critical load is acting on the column, it is possible to deflect it according to the half-sine pattern with an arbitrary though small amplitude without any absorption or release of energy.   Solution for P of the equation expressing the equality of work and strain energy yields the critical load.

If the deflected shape assumed happens to be the correct shape of deformations at the moment of buckling, the buckling load calculated is the correct critical load.   When the assumed shape differs from the actual shape of deflections, the calculations yield an exaggerated value for the buckling load in agreement with the considerations given earlier. Fortunately the difference is small in many problems of practical interest.

Deflections Represented by Polynomial

The effect of the shape assumed for the deflections can be investigated by recalculating the buckling load of the straight column for a different deflected shape.   A rather convenient assumption is a polynomial because of the simplicity of the integrations involved.   The deflected shape should preferably satisfy not only the geometric boundary conditions (deflections vanish at the end points), but also the conditions of equilibrium at the boundary (the bending moments vanish at the end points). The four conditions can be satisfied if the expression contains four constants, and a fifth constant is needed in order to have an indeterminate parameter corresponding to a in Eq. 3.4.1.   Hence the new assumption is

$$y = ax^4 + bx^3 + cx^2 + dx + e \qquad\qquad 3.4.8$$

The second derivative of the function is

$$y'' = 12ax^2 + 6bx + 2c$$

The requirement

$$y'' = 0 \qquad \text{when} \quad x = 0$$

yields

$$c = 0 \qquad\qquad (a)$$

Similarly it follows from

$$y'' = 0, \qquad \text{when} \quad x = L$$

that

$$b = -2aL \qquad (b)$$

Consequently Eq. 3.4.8 becomes

$$y'' = ax^4 - 2aLx^3 + dx + e$$

The condition

$$y = 0 \qquad \text{when} \quad x = 0$$

yields

$$e = 0 \qquad (c)$$

and from

$$y = 0 \qquad \text{when} \quad x = L$$

it follows that

$$d = aL^3 \qquad (d)$$

Therefore the assumption contained in Eq. 3.4.8 satisfies all the end conditions if the values of the constants are substituted from Eqs. a, b, c, and d:

$$y = a(x^4 - 2Lx^3 + L^3x) \qquad 3.4.9$$

Since

$$y' = a(4x^3 - 6Lx^2 + L^3)$$

squaring and integration give

$$\Delta L = \frac{1}{2} \int_0^L y'^2 \, dx = \frac{17}{70} a^2 L^7$$

The work done by the external load is

$$\delta W = P \, \Delta L = \tfrac{17}{70} a^2 L^7 P$$

From

$$y'' = 12a(x^2 - Lx)$$

it follows that

$$\int_0^L y''^2 \, dx = \frac{72}{15} a^2 L^5$$

Thus the change in the strain energy is

$$\delta U = \frac{1}{2} EI \int_0^L y''^2 \, dx = \frac{36}{15} a^2 L^5 EI$$

The critical load is characterized by the equality of work done and strain energy stored. Hence

$$\tfrac{17}{70} a^2 L^7 P = \tfrac{36}{15} a^2 L^5 EI$$

Solution for P gives the critical value of the force

$$P_{\text{cr}} = (\tfrac{36}{15})(\tfrac{14}{71})(EI/L^2) = 9.8823(EI/L^2) \qquad 3.4.10$$

Since the exact value of the Euler load is (see Eq. 3.1.29)

$$P_{cr} = \pi^2(EI/L^2) = 9.8696(EI/L^2)$$

the difference between the exact and the approximate results is $0.0127(EI/L^2)$ which is only 0.129 per cent of the exact value.

Comparison of the Law of the Conservation of Energy with the Minimal Principle in Buckling-Load Calculations

At the end of Art. 2.2 attention was called to the important difference between the minimal principle and the law of the conservation of energy. Yet in the present article use of the two different approaches to the solution of the problem not only led to the same results, but involved even the identical integrals and algebraic manipulations. The reason for this merging of the two principles is the constancy of the end load P during deflections in a state of neutral equilibrium. All external loads are considered constant when the minimal principle is applied. Hence the virtual work done by P is $P \Delta L$ when the column assumes its bent shape. On the other hand, in the application of the law of the conservation of energy the actual, and not the virtual, work must be calculated. Under ordinary conditions deflections increase only when the external loads increase, and thus the displacement times the final value of the load is not the work done by the load. As a matter of fact, the strain-energy formulas in Art. 2.1, which were calculated from the work done by the external loads, contain the factor $\frac{1}{2}$. This factor becomes unity when the actual work done by P is calculated in the buckling problem because $P = P_{cr}$ remains constant during the deformations that take place in the buckling process. For this reason calculations by the law of the conservation of energy become identical with those based on the minimal principle.

3.5 The Buckling Load of a Column of Variable Cross Section

Calculation by the Rayleigh Method

The Rayleigh method permits the calculation of the buckling load of columns of variable cross section with comparatively little effort. Details of the procedure will be given for the column in Fig. 3.5.1. It is assumed that the moment of inertia of the column varies linearly from its initial value I_0 at $x = 0$ to its maximum value I_m which is reached when $x = L/3$. It is constant from $L/3$ to $2L/3$, and decreases again linearly to I_0 at $x = L$. Hence

$$
\begin{aligned}
I &= I_0 + (I_m - I_0)(3x/L) & \text{when} & & 0 \leq x \leq (L/3) \\
I &= I_m & \text{when} & & (L/3) \leq x \leq (2L/3) \\
I &= I_0 + (I_m - I_0)(L - x)(3/L) & \text{when} & & (2L/3) \leq x \leq L & \quad 3.5.1
\end{aligned}
$$

The first step in the calculation is the assumption of a deflected shape. A sine function seems to be appropriate since in the limiting case, when $I_0 = I_m$, the column is known to deflect according to the sine law when it buckles. One can write

$$y = a \sin (\pi x / L) \qquad 3.5.2$$

where a, an indeterminate constant, represents the unknown deflection of the midpoint of the column. In the evaluation of the strain-energy integral the second derivative of y is needed:

$$y'' = -a(\pi / L)^2 \sin (\pi x / L) \qquad 3.5.3$$

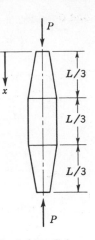

Fig. 3.5.1. Column with variable bending rigidity.

Because the moment of inertia is expressed in three equations, the integral representing the strain energy stored in the column during the deformations must be broken up into three parts:

$$U = \frac{1}{2} \int_0^{L/3} EI y''^2 \, dx + \frac{1}{2} \int_{L/3}^{2L/3} EI y''^2 \, dx + \frac{1}{2} \int_{2L/3}^{L} EI y''^2 \, dx \qquad 3.5.4$$

The expression may be assumed to represent the total amount of the strain energy stored in the column since it was shown in the last section of Art. 2.4 that the strain energy of uniform compression can be disregarded. Substitution of the value of I from Eq. 3.5.1 and that of y'' from Eq. 3.5.3 changes the first integral of the right-hand member into

$$\frac{1}{2} \int_0^{L/3} EI y''^2 \, dx = \frac{EI_0}{2} a^2 \left(\frac{\pi}{L} \right)^4 \int_0^{L/3} \sin^2 \frac{\pi x}{L} \, dx$$

$$+ \frac{3}{2} E(I_m - I_0) a^2 \left(\frac{\pi}{L} \right)^4 \int_0^{L/3} \frac{x}{L} \sin^2 \frac{\pi x}{L} \, dx \qquad 3.5.5$$

The first definite integral is

$$\int_0^{L/3} \sin^2 \frac{\pi x}{L} \, dx = \frac{L}{\pi} \int_0^{L/3} \sin^2 \frac{\pi x}{L} \, d \frac{\pi x}{L}$$

$$= \left[\frac{x}{2} - \frac{L}{4\pi} \sin \frac{2\pi x}{L} \right]_0^{L/3}$$

$$= \frac{L}{6} - \frac{L}{4\pi} \sin \frac{2\pi}{3}$$

Since

$$\sin (2\pi/3) = \sin 120° = \sin 60° = \sqrt{3}/2$$

one obtains

$$\int_0^{L/3} \sin^2{(\pi x/L)} \, dx = (L/6) - (\sqrt{3}L/8\pi) = 0.0977L \qquad (a)$$

The second definite integral can be written as

$$\frac{L}{\pi^2} \int_0^{L/3} \frac{\pi x}{L} \sin^2{\frac{\pi x}{L}} \, d\frac{\pi x}{L} = \left[\frac{x^2}{4L} - \frac{x}{4\pi} \sin{\frac{2\pi x}{L}} - \frac{L}{8\pi^2} \cos{\frac{2\pi x}{L}} \right]_0^{L/3}$$

$$= \frac{L}{36} - \frac{L}{12\pi} \sin{\frac{2\pi}{3}} - \frac{L}{8\pi^2} \cos{\frac{2\pi}{3}} + \frac{L}{8\pi^2}$$

Because

$$\cos{(2\pi/3)} = \cos{120°} = -\cos{60°} = -\tfrac{1}{2}$$

the integral becomes

$$\int_0^{L/3} \frac{x}{L} \sin^2{\frac{\pi x}{L}} dx = \frac{L}{36} - \frac{\sqrt{3}}{24\pi} L + \frac{L}{16\pi^2} + \frac{L}{8\pi^2} \qquad (b)$$

$$= 0.0238L$$

Hence Eq. 3.5.5 can be written in the form

$$\frac{1}{2} \int_0^{L/3} EI y''^2 \, dx = \frac{EI_0}{2} a^2 \left(\frac{\pi}{L}\right)^4 (0.0977L)$$

$$+ \frac{3}{2} E(I_m - I_0)a^2 \left(\frac{\pi}{L}\right)^4 (0.0238L)$$

$$= a^2 EI_0 \frac{\pi^4}{L^3} (0.0128 + 0.0357k) \qquad (c)$$

where

$$k = I_m/I_0$$

The second integral in Eq. 3.5.4 is

$$\frac{1}{2} \int_{L/3}^{2L/3} EI_m a^2 \left(\frac{\pi}{L}\right)^4 \sin^2{\frac{\pi x}{L}} \, dx$$

$$= \frac{1}{2} EI_m a^2 \left(\frac{\pi}{L}\right)^4 \left[\frac{x}{2} - \frac{L}{4\pi} \sin{\frac{2\pi x}{L}} \right]_{L/3}^{2L/3}$$

$$= \frac{1}{2} EI_m a^2 \left(\frac{\pi}{L}\right)^4 \left[\frac{L}{6} - \frac{L}{4\pi} (\sin{240°} - \sin{120°}) \right]$$

$$= a^2 EI_0 \frac{\pi^4}{L^3} (0.1522)k \qquad 3.5.6$$

Evaluation of the third integral is unnecessary since it is equal to the first one owing to the symmetry of both the column and the deflection function about the line $x = L/2$. Consequently the strain energy stored in the column during the displacement y is

$$U = [0.0256 + 0.224k]a^2(\pi^4/L^3)EI_0 \qquad 3.5.7$$

The shortening of the distance between the end points of the column depends only on the deflection function chosen and not on the variation of the moment of inertia. In the present problem, since the deflections were assumed to be the same as in the first example in Art. 3.4, the shortening ΔL, and consequently the potential of the load P, must be identical with those found earlier. Making use of Eq. 3.4.4, one can write

$$V = -\tfrac{1}{4}\pi^2 a^2(P/L) \qquad 3.5.8$$

The total potential is

$$U + V = (0.0256 + 0.224k)a^2(\pi^4/L^3)EI_0 - \tfrac{1}{4}\pi^2 a^2(P/L) \qquad 3.5.9$$

The change in the total potential corresponding to a variation δa of a vanishes when

$$(0.0256 + 0.224k)(\pi/L)^2 EI_0 = P/4$$

Satisfaction of this equation is the condition of equilibrium in the deflected

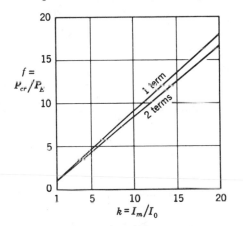

Fig. 3.5.2.   Buckling load coefficient.

position and thus of a state of neutral equilibrium in the original straight-line position. Hence solution for P yields the critical load of the column:

$$P_{cr} = (0.104 + 0.896k)\pi^2 EI_0/L^2 = f\pi^2 EI_0/L^2 = fP_E \qquad 3.5.10$$

where

$$k = I_m/I_0 \qquad 3.5.10a$$

The variation of the buckling load with the ratio I_m/I_0 is plotted in Fig. 3.5.2.

In agreement with the statements made in Art. 3.4 the buckling load obtained must be equal to or greater than the actual buckling load. The load calculated by the Rayleigh method is the correct theoretical buckling load when the deflection function assumed represents the actual deflected shape at buckling. In the present problem this is obviously true when $I_m = I_0$, and in this case Eq. 3.5.10 does yield the Euler load. However, it can be expected that an increase in the bending rigidity of the middle third of the column will flatten out the middle portion of the deflection curve. For this reason Eq. 3.5.10 is not likely to be very accurate when the ratio I_m/I_0 is large; it may then be desirable to obtain a better approximation.

The Rayleigh-Timoshenko Method

In Fig. 3.5.3 the dashed line is the sum of a half-sine wave of large amplitude and a three-half-sine wave of small amplitude. It can be represented by the equation

$$y = a \sin (\pi x/L) + b \sin (3\pi x/L) \qquad 3.5.11$$

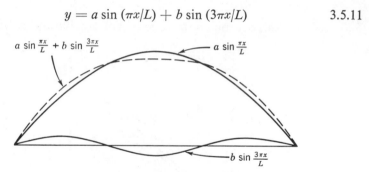

Fig. 3.5.3. Superposition of deflected shapes.

This deflected shape is probably closer to the actual shape at buckling when I_m/I_0 is larger than the shape represented by Eq. 3.5.2; for this reason it should lead to a better approximation to the critical load. The analysis is now repeated on the basis of Eq. 3.5.11 as the deflected shape at buckling. For the time being, nothing is stipulated regarding the magnitude of the coefficient b, it will be determined later from the minimal principle.

The second derivative of y is

$$y'' = -a(\pi/L)^2 \sin (\pi x/L) - b (3\pi/L)^2 \sin (3\pi x/L) \qquad 3.5.12$$

The square of y'' comprises three terms

$$y''^2 = a^2(\pi/L)^4 \sin^2 (\pi x/L) + 2ab(\pi/L)^2(3\pi/L)^2 \sin (\pi x/L) \sin (3\pi x/L)$$

$$+ b^2(3\pi/L)^4 \sin^2 (3\pi x/L) \qquad (d)$$

Consequently Eq. 3.5.5 has to be replaced by

$$\frac{1}{2}\int_0^{L/3} EIy''^2\,dx = \frac{EI_0}{2}\int_0^{L/3}\left[a^2\left(\frac{\pi}{L}\right)^4\sin^2\frac{\pi x}{L}\right.$$

$$\left. + 2ab\left(\frac{\pi}{L}\right)^2\left(\frac{3\pi}{L}\right)^2\sin\frac{\pi x}{L}\sin\frac{3\pi x}{L} + b^2\left(\frac{3\pi}{L}\right)^4\sin^2\frac{3\pi x}{L}\right]dx$$

$$+ \frac{3}{2}E(I_m - I_0)\int_0^{L/3}\left[a^2\left(\frac{\pi}{L}\right)^4\frac{x}{L}\sin^2\frac{\pi x}{L}\right.$$

$$+ 2ab\left(\frac{\pi}{L}\right)^2\left(\frac{3\pi}{L}\right)^2\frac{x}{L}\sin\frac{\pi x}{L}\sin\frac{3\pi x}{L}$$

$$\left. + b^2\left(\frac{3\pi}{L}\right)^4\frac{x}{L}\sin^2\frac{3\pi x}{L}\right]dx \qquad\qquad 3.5.13$$

In addition to Eqs. a and b the following definite integrals are needed for evaluating Eq. 3.5.13:

$$\int_0^{L/3}\sin\frac{\pi x}{L}\sin\frac{3\pi x}{L}\,dx = \frac{1}{2}\int_0^{L/3}\left(\cos\frac{2\pi x}{L} - \cos\frac{4\pi x}{L}\right)dx$$

$$= \frac{1}{2}\left[\frac{L}{2\pi}\sin\frac{2\pi x}{L} - \frac{L}{4\pi}\sin\frac{4\pi x}{L}\right]_0^{L/3} = 0.1033L \qquad (e)$$

$$\int_0^{L/3}\sin^2\frac{3\pi x}{L}\,dx = \frac{L}{3\pi}\int_0^{L/3}\sin^2\frac{3\pi x}{L}\,d\frac{3\pi x}{L}$$

$$= \frac{L}{3\pi}\left[\frac{3\pi x}{2L} - \frac{1}{4}\sin\frac{6\pi x}{L}\right]_0^{L/3} = 0.166L \qquad (f)$$

$$\int_0^{L/3}\frac{x}{L}\sin\frac{\pi x}{L}\sin\frac{3\pi x}{L}\,dx = \frac{1}{2}\int_0^{L/3}\frac{x}{L}\cos\frac{2\pi x}{L}\,dx$$

$$- \int_0^{L/3}\frac{x}{L}\cos\frac{4\pi x}{L}\,dx = \frac{L}{8\pi^2}\left[\cos\frac{2\pi x}{L}\right.$$

$$\left. + \frac{2\pi x}{L}\sin\frac{2\pi x}{L}\right]_0^{L/3} = 0.0202L \qquad (g)$$

$$\int_0^{L/3}\frac{x}{L}\sin^2\frac{3\pi x}{L}\,dx = \frac{L}{9\pi^2}\int_0^{L/3}\frac{3\pi x}{L}\sin^2\frac{3\pi x}{L}\,d\frac{3\pi x}{L}$$

$$= \frac{L}{9\pi^2}\left[\frac{9\pi^2 x^2}{4L^2} - \frac{3\pi x}{4L}\sin\frac{6\pi x}{L} - \frac{1}{8}\cos\frac{6\pi x}{L}\right]_0^{L/3} = 0.0278L \quad (h)$$

Substitution of the definite integrals in Eq. 3.5.13 yields

$$\frac{1}{2}\int_0^{L/3} EIy''^2\, dx = \frac{\pi^4}{2}\frac{E}{L^3}[I_0(0.0265a^2 + 0.768ab + 6.70b^2)$$

$$+ I_m(0.0714a^2 + 1.092ab + 6.75b^2)] \qquad (i)$$

Equation 3.5.6 must be replaced by

$$\frac{1}{2}\int_{L/3}^{2L/3} EI_m y''^2\, dx = \frac{1}{2}\int_{L/3}^{2L/3} EI_m \left[a^2\left(\frac{\pi}{L}\right)^4 \sin^2\frac{\pi x}{L}\right.$$

$$+ 2ab\left(\frac{\pi}{L}\right)^2\left(\frac{3\pi}{L}\right)^2 \sin\frac{\pi x}{L}\sin\frac{3\pi x}{L}$$

$$\left.+ b^2\left(\frac{3\pi}{L}\right)^4 \sin^2\frac{3\pi x}{L}\right] dx \qquad 3.5.14$$

The integrals determined earlier permit the evaluation of this definite integral. One obtains

$$\frac{1}{2}\int_{L/3}^{2L/3} EI_m y''^2\, dx = \pi^4\frac{EI_m}{L^3}(0.152a^2 - 1.86ab + 6.75b^2) \qquad (j)$$

The total strain energy stored in the column is twice the definite integral of Eq. *i* plus once the definite integral of Eq. *j*:

$$U = \pi^4(E/L^3)[(0.0265a^2 + 0.768ab + 6.70b^2)I_0$$

$$+ (0.223a^2 - 0.768ab + 13.5b^2)I_m] \qquad 3.5.15$$

The potential of the external loads

$$V = -P\frac{1}{2}\int_0^L y'^2\, dx \qquad (k)$$

can be calculated without breaking up the integral into three parts because *P* is constant and y'^2 is expressed by a single equation over the entire length of the column. With

$$y' = a(\pi/L)\cos(\pi x/L) + b(3\pi/L)\cos(3\pi x/L)$$

and

$$y'^2 = a^2(\pi/L)^2\cos^2(\pi x/L) + 2ab(\pi/L)(3\pi/L)\cos(\pi x/L)\cos(3\pi x/L)$$

$$+ b^2(3\pi/L)^2\cos^2(3\pi x/L)$$

the integration indicated in Eq. k can be carried out if it is observed that

$$\int_0^L \cos^2 \frac{\pi x}{L}\, dx = \frac{L}{\pi}\int_0^L \cos^2 \frac{\pi x}{L}\, d\frac{\pi x}{L} = \frac{L}{\pi}\left[\frac{\pi x}{2L} + \frac{1}{4}\sin \frac{2\pi x}{L}\right]_0^L = \frac{L}{2} \quad (l)$$

$$\int_0^L \cos \frac{\pi x}{L} \cos \frac{3\pi x}{L}\, dx = \frac{1}{2}\int_0^L \left[\cos \frac{4\pi x}{L} + \cos \frac{2\pi x}{L}\right] dx$$

$$= \frac{1}{2}\left[\frac{L}{4\pi}\sin \frac{4\pi x}{L} + \frac{L}{2\pi}\sin \frac{2\pi x}{L}\right]_0^L = 0 \quad (m)$$

$$\int_0^L \cos^2 \frac{3\pi x}{L}\, dx = \frac{L}{3\pi}\int_0^L \cos^2 \frac{3\pi x}{L}\, d\frac{3\pi x}{L} = \frac{L}{3\pi}\left[\frac{3\pi x}{2L}\right.$$

$$\left. + \frac{1}{4}\sin \frac{6\pi x}{L}\right]_0^L = \frac{L}{2} \quad (n)$$

Consequently

$$V = -(P/2)[a^2(\pi/L)^2(L/2) + b^2(3\pi/L)^2(L/2) = -(\pi^2/4)(P/L)(a^2 + 9b^2)$$

$$3.5.16$$

The total potential is

$$U + V = \pi^4(E/L^3)[(0.0265a^2 + 0.768ab + 6.70b^2)I_0$$

$$| \ (0.223a^2 - 0.768ab + 13.5b^2)I_m]$$

$$- (\pi^2/4)(P/L)(a^2 + 9b^2) \qquad\qquad 3.5.17$$

The deflected shape given in Eq. 3.5.11 is an equilibrium configuration, and thus the original straight-line equilibrium is neutral, if the change in the total potential is zero for any arbitrary virtual displacement. Since by the procedure followed in these calculations, the freedom of motion of the column was restricted to the single and the three-half-wave patterns represented by Eq. 3.5.11, only the amplitudes a and b can be varied. The change in the total potential corresponding to a variation of a is

$$\delta(U + V) = \{\pi^4(E/L^3)[(0.053a + 0.768b)I_0 + (0.446a - 0.768b)I_m]$$

$$+ (\pi^2/2)(P/L)a\}\delta a = 0$$

When b is varied, one obtains

$$\delta(U + V) = \{\pi^4(E/L^3)[(0.768a + 13.4b)I_0 + (-0.768a + 27b)I_m]$$

$$- 9(\pi^2/2)(P/L)b\}\delta b = 0$$

Since δa and δb are arbitrary, the expressions in braces must vanish. After manipulations the two equations can be written in the form

$$[0.053 + 0.446k - \tfrac{1}{2}(P/P_E)]a + (0.768 - 0.768k)b = 0$$

$$[0.768 - 0.768k]a + [13.4 + 27k - \tfrac{9}{2}(P/P_E)]b = 0$$

3.5.18

where

$$P_E = \pi^2 EI_0/L^2 \quad \text{and} \quad k = I_m/I_0 \qquad 3.5.19$$

Solution by Cramer's Rule

The column is in neutral equilibrium if Eqs. 3.5.18 are satisfied. The solution of a set of simultaneous linear equations

$$p_1 x + q_1 y = r_1$$

$$p_2 x + q_2 y = r_2$$

3.5.20

can be written in the form of quotients of determinants:

$$x = \frac{\begin{vmatrix} r_1 & q_1 \\ r_2 & q_2 \end{vmatrix}}{\begin{vmatrix} p_1 & q_1 \\ p_2 & q_2 \end{vmatrix}} \qquad 3.5.21$$

$$y = \frac{\begin{vmatrix} p_1 & r_1 \\ p_2 & r_2 \end{vmatrix}}{\begin{vmatrix} p_1 & q_1 \\ p_2 & q_2 \end{vmatrix}} \qquad 3.5.22$$

If all the right-hand members, namely r_1 and r_2, are simultaneously zero, the determinants in the numerators vanish. Then the unknowns must vanish,

$$x = y = 0$$

unless at the same time the denominator determinant also vanishes. When this happens, the unknowns are obtained as ratios of zero over zero, and they may have finite values. Hence a set of linear equations with vanishing right-hand members can have a non-trivial solution, that is, a solution in which not all the unknowns are zero, only if its denominator determinant vanishes. When the denominator determinant is finite, only the trivial solution exists in which all the unknowns are zero. These statements are known as Cramer's rule.

In Eqs. 3.5.18 the two amplitudes a and b are the unknown quantities. Since the right-hand members of the equations are zero, one possible solution is

$$a = b = 0$$

Substitution of these values in Eq. 3.5.11 results in no deformations whatsoever. Thus the trivial solution means that the original straight configuration corresponds to equilibrium; this was of course known.

A non-trivial solution is possible only if the denominator determinant vanishes:

$$\begin{vmatrix} [0.053 + 0.446k - \frac{1}{2}(P/P_E)] & 0.768(1-k) \\ 0.768(1-k) & [13.4 + 27k - \frac{9}{2}(P/P_E)] \end{vmatrix} = 0 \qquad 3.5.23$$

This equation can be written explicitly as

$$(P/P_E)^2 - (3.08 + 6.9k)(P/P_E) + 5.07k^2 + 3.81k + 0.05 = 0$$

$$3.5.24$$

The equation was solved for various values of k, and the results were plotted in Fig. 3.5.2. It can be seen that the buckling loads calculated for the deflected shape of Eq. 3.5.11 are slightly lower than those corresponding to the original single half-wave assumption. This is to be expected since the new deflected shape with two degrees of freedom permits a better approximation to the actual shape of buckling than the single-degree-of-freedom pattern of Eq. 3.5.2, and because any restriction of the freedom of motion is bound to lead to a higher buckling load. However, the difference between the two buckling loads is surprisingly small even when $k = I_m/I_0$ is a large quantity. For instance, at $k = 20$ the first approximation based on Eq. 3.5.2 is only 7 per cent higher than the improved value corresponding to Eq. 3.5.11.

The close agreement between the original and the improved buckling loads leads one to believe that even the first approximation is sufficiently accurate for engineering purposes in this particular example. It also seems to indicate that the deflected pattern contains little of the three-sine wave. This latter conjecture can be easily checked by substituting the value of the buckling load calculated by the improved method, namely,

$$P = 16.84 P_E \qquad \text{when} \quad k = 20$$

in one of Eqs. 3.5.18. If the second equation is chosen, one obtains

$$-14.6a + 477.6b = 0$$

The solution is

$$b/a = 0.0306$$

which is indeed very small.

Figure 3.5.2 reveals that the buckling load increases almost in proportion to the moment of inertia of the middle portion. For instance, the buckling load is 16.84 times the Euler load of a constant section column of moment of inertia I_0 when the moment of inertia of the middle section is $20I_0$. Consequently tapering the column is a good means of saving weight without decreasing substantially the load-carrying capacity.

3.6   Buckling of a Column with Elastic End Fixation

The End Restraint

The ends of bars in actual, not ideal, frameworks are welded or riveted together rather than connected by pin joints. When a compression member of a framework buckles, its ends are consequently not free to rotate,

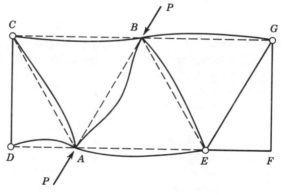

Fig. 3.6.1.   Framework.

but are restrained by the other members. Naturally this restraint is not absolutely rigid, there is some give in the system, and the joints to which the compression member is attached rotate slightly and elastically because of moments exerted on them by the member that buckles. It is possible to calculate the proportionality factor between the moment exerted on a joint and the angle of rotation if simplifying assumptions are made.

In Fig. 3.6.1 bar AB is the one whose buckling load is to be investigated. Its end point A is restrained from rotation by members AC, AD, and AE. Let it be assumed for the time being that these three bars are not subjected to axial loads and that their end points C, D, and E are attached by pin joints to the rest of the framework. A moment M_{AC} applied to bar AC at A would cause a rotation

$$\alpha_{AC} = \tfrac{1}{3}(ML/EI)_{AC} \qquad\qquad 3.6.1$$

in agreement with item 3 in Fig. 1.12.8.   This is true provided point A is not displaced from its original position during the application of M_{AC}.   It was shown in Art. 1.18 that joint displacements can be disregarded in the calculation of the deformations of frameworks when the loading consists of moments applied to the bars at the joints.

Similarly the rotations of the other two bars are

$$\alpha_{AD} = \tfrac{1}{3}(ML/EI)_{AD}$$

$$\alpha_{AE} = \tfrac{1}{3}(ML/EI)_{AE}$$

if they are subjected to end moments M_{AD} and M_{AE} at joint A.   Because of the rigid connection between the bars joined at A the condition

$$\alpha_{AC} = \alpha_{AD} = \alpha_{AE}$$

must be satisfied.   Consequently

$$M_{AD}/M_{AC} = (EI/L)_{AD}/(EI/L)_{AC}$$

$$M_{AE}/M_{AC} = (EI/L)_{AE}/(EI/L)_{AC}$$

A moment M applied to the three bars AC, AD, and AE at their juncture is therefore distributed according to the equations

$$M_{AC} = (EI/L)_{AC}/\Sigma(EI/L)$$

$$M_{AD} = (EI/L)_{AD}/\Sigma(EI/L)$$

$$M_{AE} = (EI/L)_{AE}/\Sigma(EI/L)$$

where

$$\Sigma(EI/L) = (EI/L)_{AC} + (EI/L)_{AD} + (EI/L)_{AE}$$

The angle of rotation is

$$\alpha = \alpha_{AC} = \alpha_{AD} = \alpha_{AE} = M/\Sigma(3EI/L) \qquad\qquad 3.6.2$$

In reality, of course, the *far ends* C, D, and E of the bars are not pinned, but riveted or welded to the rest of the framework.   Because of the elasticity of the rest of the framework, the restraint from rotation is not complete.   It is somewhere between no restraint corresponding to pin attachment and perfect restraint corresponding to rigid end fixation.   In the latter case Eq. 3.6.1 must be replaced by

$$\alpha_{AC} = \tfrac{1}{4}(ML/EI)_{AC} \qquad\qquad 3.6.3$$

as was shown in Art. 1.15 (see Eq. 1.15.4).   A derivation paralleling that given above would lead to

$$\alpha = \alpha_{AC} = \alpha_{AD} = \alpha_{AE} = M/\Sigma(4EI/L) \qquad\qquad 3.6.4$$

Equations 3.6.2 and 3.6.4 can be replaced by

$$\alpha = M/\Sigma S \qquad\qquad 3.6.5$$

where S is the stiffness of a bar as defined in the Hardy Cross method (see Art. 1.17). It is $4EI/L$ when the far end of the bar is fixed, as is usually assumed in the Hardy Cross method, and $3EI/L$ when the far end is pinned. The difference between the two extremes is small, and for this reason a guess as to the actual value of the numerical coefficient cannot be too far off the mark.

When the bars AC, AD, and AE are also subjected to axial loads, the values of S are altered. For compressive loads and the fixed-end condition Eq. 2.4.14 and Fig. 2.4.6 give values for the stiffness coefficient. When the far end of a compression bar is pinned, Eq. 2.4.19 and Fig. 2.4.8 should be used. The effect of gusset plates on the stiffness coefficient has also been evaluated; the results are shown in Figs. 2.4.9 and 2.4.11.

This means that an estimate can always be made of the stiffness of each bar, and the resistance of any assembly of bars, supporting the compressive member whose buckling load has to be calculated, can be determined approximately. If the notation

$$C = \Sigma S \qquad 3.6.6$$

is introduced, solution of Eq. 3.6.5 for the moment M corresponding to an angle of twist α yields

$$M = C\alpha \qquad 3.6.7$$

It can be seen therefore that $C = \Sigma S$ is the spring constant of the assembly of bars restraining the principal compression member. The spring, of course, resists rotation rather than displacement.

Analysis by the Rayleigh-Timoshenko Method

Figure 3.6.2 shows schematically the compression member at the moment of buckling. Its end points 1 and 2 cannot be displaced laterally. The action of the other members of the framework is represented by the two coiled springs which are chosen so as to exert exactly the same reaction moments upon the compression member as would the omitted portion of the framework when the ends of the compression member undergo rotations. In the general case the spring constant C has different values at 1 and 2. Equation 3.6.7 should therefore be replaced by

$$M_1 = C_1\alpha_1 \qquad 3.6.8a$$

at the lower end, and by

$$M_2 = C_2\alpha_2 \qquad 3.6.8b$$

at the upper end of the bar.

The following considerations can be offered regarding the buckled shape: Without restraint at the ends the buckled shape would be a half-sine wave as shown in Fig. 3.6.3a. With the notation of the figure the deflected shape can be represented by the equation

$$y = a \sin (\pi x/L)$$

If both ends are rigidly fixed, buckling occurs according to the pattern shown in Fig. 3.6.3b. The equation of the curve in the figure is

$$y = -(b/2)[\cos (2\pi x/L) - 1]$$

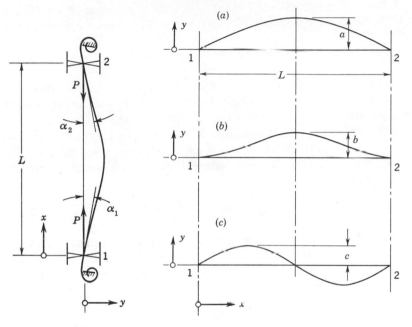

Fig. 3.6.2.   Elastically
restrained column.

Fig. 3.6.3.   Components of buckled shape.

The buckling pattern of any elastically restrained column can be obtained by combining in the correct ratio patterns *a* and *b*, provided the restraint is equal at ends 1 and 2. If the end fixation is more rigid at one end than at the other, there will be an antisymmetric component in the deflected shape in addition to components *a* and *b*. The antisymmetric component can be represented by the equation

$$y = c \sin (2\pi x/L)$$

It is shown in Fig. 3.6.3c.

The most general deflected shape can thus be described by the equation

$$y = a \sin (\pi x/L) - (b/2)[\cos (2\pi x/L) - 1] + c \sin (2\pi x/L) \qquad 3.6.9.$$

The buckling load of the elastically restrained column can now be calculated by the following procedure: The originally straight column is assumed to undergo a virtual displacement defined in Eq. 3.6.9 in which the constant coefficients a, b, and c are considered as small quantities. The strain energy δU_b stored in the bar in consequence of the flexure and the strain energy δU_{spr} stored in the springs because of the rotations of the ends are then determined, and the virtual work δW done by the compressive loads P during the shortening of the distance between end points 1 and 2 is calculated. If the change of the total potential of the system $\delta(U + V)$ from the original straight form to the assumed deflected form

$$\delta(U + V) = \delta U_b + \delta U_{spr} - \delta W \qquad 3.6.10$$

does not contain terms that are first-order small, that is, proportional to the coefficients a, b, or c, the original straight form of the bar corresponds to a possible pattern of equilibrium according to the minimal principle. If the second-order small terms, that is, those proportional to the squares or the products of the coefficients a, b, and c, add up to a positive quantity, the equilibrium is stable, but it is unstable if their sum is negative. Neutral equilibrium prevails when

$$\delta U_b + \delta U_{spr} - \delta W = 0 \qquad 3.6.11$$

It will be seen that the sign of the quantity on the left-hand member of Eq. 3.6.11 is a function of the load P. The load can therefore assume a value that satisfies Eq. 3.6.11. This value is the critical, or buckling, load P_{cr}.

Strain Energy

The calculations can now be carried out without difficulty. The strain energy of bending is obtained from the formula

$$\delta U_b = \frac{EI}{2} \int_0^L \left(\frac{d^2y}{dx^2}\right)^2 dx$$

The derivatives of the deflections follow from Eq. 3.6.9:

$$dy/dx = y' = (\pi/L)[a \cos(\pi x/L) + b \sin(2\pi x/L) + 2c \cos(2\pi x/L)]$$

$$3.6.12$$

$$d^2y/dx^2 = y'' = (\pi/L)^2[-a \sin(\pi x/L) + 2b \cos(2\pi x/L) - 4c \sin(2\pi x/L)]$$

$$3.6.13$$

The following integrals are needed in the calculation of the change in the potential:

$$\int \cos^2 x \, dx = \frac{x}{2} + \frac{1}{4} \sin 2x$$

$$\int \sin^2 x \, dx = \frac{x}{2} - \frac{1}{4} \sin 2x$$

$$\int \cos mx \sin mx \, dx = -\frac{1}{4m} \cos 2mx$$

$$\int \cos mx \cos nx \, dx - \frac{\sin (m - n)x}{2(m - n)} + \frac{\sin (m + n)x}{2(m + n)}$$

$$\int \sin mx \sin nx \, dx = \frac{\sin (m - n)x}{2(m - n)} - \frac{\sin (m + n)x}{2(m + n)}$$

$$\int \sin mx \cos nx \, dx = -\frac{\cos (m - n)x}{2(m - n)} - \frac{\cos (m + n)x}{2(m + n)}$$

provided that $\qquad\qquad m \neq n \qquad\qquad$ 3.6.14

Hence the following definite integrals can be written

$$\int_0^L \cos^2 \frac{m\pi x}{L} \, dx = \frac{L}{2}$$

$$\int_0^L \sin^2 \frac{m\pi x}{L} \, dx = \frac{L}{2}$$

$$\int_0^L \cos \frac{m\pi x}{L} \sin \frac{m\pi x}{L} \, dx = 0$$

$$\int_0^L \cos \frac{m\pi x}{L} \cos \frac{n\pi x}{L} \, dx = 0$$

$$\int_0^L \sin \frac{m\pi x}{L} \sin \frac{n\pi x}{L} \, dx = 0$$

$$\int_0^L \cos \frac{\pi x}{L} \sin \frac{2\pi x}{L} \, dx = \frac{4}{3\pi} L$$

$$\int_0^L \cos \frac{2\pi x}{L} \sin \frac{\pi x}{L} \, dx = -\frac{2}{3\pi} L$$

3.6.15

The integral of the square of the expression in the right-hand member of Eq. 3.6.13 can now be calculated between the limits 0 and L. One obtains

$$\delta U_b = \frac{EIL}{4} (\pi/L)^4 \left[a^2 + 4b^2 + 16c^2 + \frac{16}{3\pi} ab \right] \qquad 3.6.16$$

In the calculation of the strain energy stored in the springs, the angles α_1 and α_2 are needed. Of these α_1 is the slope of the curve when $x = 0$. Hence from Eq. 3.6.12

$$\alpha_1 = y'_1 = (\pi/L)(a + 2 c) \qquad 3.6.17$$

The angle α_2 has the same magnitude as the slope at $x = L$, but its sign is negative when y'_2 is positive:

$$\alpha_2 = -y'_2 = - (\pi/L)(-a + 2c) \qquad 3.6.18$$

The strain energy stored in a spring is equal to the work done by the moment acting on the spring during the rotation. Thus the strain energy stored in the two springs is

$$\delta U_{\text{spr}} = \tfrac{1}{2}(M_1\alpha_1 + M_2\alpha_2) \qquad 3.6.19$$

If Eqs. 3.6.8a and 3.6.8b are used to express M_1 and M_2 in terms of α_1 and α_2, and Eqs. 3.6.17 and 3.6.18 are duly taken into account, the following expression is obtained

$$\delta U_{\text{spr}} = \tfrac{1}{2}(C_1 y_1'^2 + C_2 y_2'^2) = \tfrac{1}{2}(\pi/L)^2[(a^2 + 4c^2)(C_1 + C_2) + 4ac(C_1 - C_2)]$$
$$3.6.20$$

Potential of the End Loads

Finally the virtual work done by the loads P must be calculated. In the application of the minimal principle, as well as in that of the principle of virtual displacements, the external loads are always considered as constants. Hence the work done by P is $P \Delta L$, where ΔL is the difference between the length of the curve defined by Eq. 3.6.9 and the vertical distance between the end points of the curve. Since

$$\Delta L = \frac{1}{2} \int_0^L (y')^2 \, dx$$

substitution of the right-hand member of Eq. 3.6.12 and consideration of the definite integrals listed earlier yield

$$\Delta L = \left(\frac{L}{4}\right) \left(\frac{\pi}{L}\right)^2 [a^2 + b^2 + 4c^2 + (16/3\pi)ab] \qquad 3.6.21$$

Consequently

$$\delta W = (PL/4)(\pi/L)^2[a^2 + b^2 + 4c^2 + (16/3\pi)ab] \qquad 3.6.22$$

The Buckling Load

It may be noticed that δU_b, δU_{spr}, and δW, as given in Eqs. 3.6.16, 3.6.20, and 3.6.22, contain only terms that are second-order small in the coefficients a, b, and c. Therefore the original straight-line shape of the loaded column is a configuration of equilibrium, as was, of course, known. The stability of the equilibrium depends on the sign of the second-order terms. Their sum is positive as long as P is small. When P is sufficiently large, δW is the prevailing term in the left-hand member of Eq. 3.6.11, and the sum of the second-order terms is negative. The critical load P_{cr} is defined as the value of P corresponding to the change from positive to negative values, that is, from stability to instability. This change takes place when Eq. 3.6.11 is satisfied.

If the strain-energy and work quantities just calculated are substituted in Eq. 3.6.11, P_{cr} is written in place of P, and the equation so obtained is solved for P_{cr}, one arrives at the formula

$$P_{cr} = P_E\, N/[a^2 + b^2 + 4c^2 + (16/3\pi)ab] \qquad 3.6.23$$

where P_E denotes the Euler load

$$P_E = \pi^2 EI/L^2 \qquad 3.6.23a$$

and

$$N = a^2 + 4b^2 + 16c^2 + (16/3\pi)ab + (2/\pi^2)(L/EI)(C_1 + C_2)(a^2 + 4c^2)$$
$$+ (2/\pi^2)(L/EI)(C_1 - C_2)4ac \qquad 3.6.23b$$

The critical load of the restrained column given in Eq. 3.6.23 is equal to the Euler load times a multiplying factor. The factor depends on the non-dimensional quantities EI/LC_1 and EI/LC_2 which are ratios of the bending rigidity of the column to the stiffnesses of the end restraints. These ratios will be represented by the symbols Ω according to the defining equations

$$\Omega_1 = EI/LC_1, \qquad \Omega_2 = EI/LC_2 \qquad 3.6.24$$

and will be referred to as relative stiffness factors. The buckling load depends not only on the relative stiffness factors but also on the parameters a, b, and c. It is readily ascertained that the buckling load is given correctly by Eq. 3.6.23 in the limiting cases of pin-jointed ends and rigid fixation. In the former case C_1, C_2, b, and c are zero, and the equation reduces to

$$P_{cr} = P_E$$

In the latter case a and c are zero, and Eq. 3.6.23 becomes

$$P_{cr} = 4P_E$$

Minimization of the Buckling Load

In the intermediate cases of elastic end fixation different choices for a, b, and c result in different buckling loads P_{cr}. The column will actually buckle at the lowest possible buckling load according to the buckling pattern that corresponds to it. Hence the smallest possible value of the multiplier of P_E in the right-hand member of Eq. 3.6.23 must be calculated. The determination of this value is a typical minimum problem of the calculus. Before actually minimizing P_{cr} one may note that both numerator and denominator can be divided by a^2. If the symbols u and v are introduced according to the definitions

$$u = b/a, \qquad v = c/a \qquad\qquad 3.6.25$$

and the relative stiffness factors of Eqs. 3.6.24 are used, Eq. 3.6.23 becomes

$$P_{cr} = P_E N^*/[1+u^2+4v^2+(16/3\pi)u] \qquad\qquad 3.6.26$$

where

$$N^* = 1 + 4u^2 + 16v^2 + (16/3\pi)u + (2/\pi^2)[(1/\Omega_1) + (1/\Omega_2)](1 + 4v^2)$$
$$+ (2/\pi^2)[(1/\Omega_1) - (1/\Omega_2)]4v \qquad 3.6.26a$$

According to Eq. 3.6.26 the critical load is a function of the given quantities Ω_1 and Ω_2 and the unknown ratios u and v. The two latter characterize the deflected shape, and show in what proportion the three components a, b, and c of Fig. 3.6.3 are mixed in the final buckled pattern. The value of a is just a scale factor which is of no importance as far as the buckling load is concerned, since the actual magnitude of the deflections is immaterial in the state of neutral equilibrium.

According to the calculus, the necessary condition that P_{cr} be a minimum is the vanishing of the derivative of P_{cr} with respect to each parameter. If Eq. 3.6.26 is written in the form

$$P_{cr} = P_E f(u, v)$$

where $f(u, v)$ is the symbolic representation of the factor of P_E which is a function of both u and v, the minimum conditions are

$$\partial P_{cr}/\partial u = P_E[\partial f(u, v)/\partial u] = 0$$

$$\partial P_{cr}/\partial v = P_E[\partial f(u, v)/\partial v] = 0$$

These minimum conditions can be brought into a slightly more convenient form. If the notation is used

$$P_{cr} = P_E f(u, v) = P_E(p/q) \qquad\qquad 3.6.27$$

where p and q represent the numerator and the denominator, respectively of the factor in the right-hand member of Eq. 3.6.27, and both p and q are functions of u and v, the first minimum condition gives

$$\frac{\partial P_{cr}}{\partial u} = P_E \frac{q(\partial p/\partial u) - p(\partial q/\partial u)}{q^2} = 0$$

The equation can be satisfied only if

$$q(\partial p/\partial u) - p(\partial q/\partial u) = 0$$

which means that

$$\frac{p}{q} = \frac{\partial p/\partial u}{\partial q/\partial u} \tag{3.6.28}$$

In a similar manner the second minimum condition yields

$$\frac{p}{q} = \frac{\partial p/\partial v}{\partial q/\partial v} \tag{3.6.29}$$

Consequently, the buckling load is a minimum when the ratio p/q has the same value as the ratio of the derivative of p with respect to a parameter to the derivative of q with respect to the same parameter.

Differentiation with respect to u yields

$$P_{cr} = P_E \frac{8u + (16/3\pi)}{2u + (16/3\pi)} \tag{3.6.30}$$

Similarly, differentiation with respect to v gives

$$P_{cr} = P_E \frac{4v + (2/\pi^2)[(1/\Omega_1) + (1/\Omega_2)]v + (1/\pi^2)[(1/\Omega_1) - (1/\Omega_2)]}{v} \tag{3.6.31}$$

Equations 3.6.26, 3.6.30, and 3.6.31 contain three unknowns, namely P_{cr}, u, and v. The equations can be solved for the unknowns when the values of Ω_1 and Ω_2 are given. The solution, however, entails a great deal of work, since u and v are not linear in the expressions. It is preferable to assume values for u and v, solve for Ω_1 and Ω_2 and prepare charts for everyday use. The procedure is now shown by means of two numerical examples.

Preparation of Buckling Charts

In the first example a symmetric buckling configuration is considered which necessarily prevails when $\Omega_1 = \Omega_2$. In this case c, and thus v, must vanish. If u is assumed to have the value 0.2, Eq. 3.6.30 yields $P_{cr} = 1.57P_E$. Substitution of 0.2 for u in Eq. 3.6.26 results in

$$P_{cr} = [1.087 + (0.294/\Omega)]P_E$$

where Ω is the common value of the two relative stiffness factors.   Hence

$$1.57 = [1.087 + (0.294/\Omega)]$$

from which

$$\Omega = 0.609$$

follows immediately.   In a similar manner, through other assumptions for u, a number of corresponding values of Ω and the multiplying factor of P_E can be calculated.   If these values are plotted in a diagram, the multiplying factor can be taken from the curve for any given value of Ω.   Instead of the multiplying factor itself, the square root of its reciprocal was used as

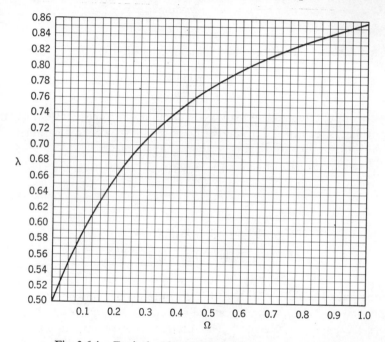

Fig. 3.6.4.   Equivalent-length diagram.   Symmetric case.
From author's paper in *J. Roy. Aero. Soc.*

the ordinate in Fig. 3.6.4.   This quantity is denoted by the Greek letter λ. Its physical meaning is the ratio of the distance between inflection points at buckling to the total length of the column as can be easily shown.   If the letter K is used to designate the multiplying factor of P_E,

$$P_{cr} = KP_E \qquad (a)$$

and the known expression for the Euler load is substituted for P_E, one obtains

$$P_{cr} = K\pi^2 EI/L^2 \qquad (b)$$

This can also be written in the form

$$P_{cr} = \pi^2 EI/(L/\sqrt{K})^2 \qquad (c)$$

This means that the buckling load of the restrained column, as given by Eqs. *a* and *b*, is the same as that of an equivalent pin-ended column whose length is $L/\sqrt{K}$. But the inflection points are the points at which the curvature of the column, and consequently the bending moment acting on the column, is zero. The portion of the column extending from one inflection-point to the other is therefore in equilibrium under the action of the end load P_{cr} in the absence of any end moments. Moreover, the deflected shape is a half-sine wave. Consequently P_{cr} is the buckling load of the pin-ended column whose length is equal to the distance between the inflection points and

$$\lambda = 1/\sqrt{K} \qquad (d)$$

The curve can be used in practical computations in the following manner: Let the column be elastically restrained at both ends, and let the relative stiffness factor Ω be 0.4. From Fig. 3.6.4 the equivalent length coefficient

$$\lambda = 0.746$$

Consequently the buckling load P_{cr} is

$$P_{cr} = \pi^2 EI/(0.746L)^2$$

When the relative stiffness factors differ at the two ends, the computations are a little lengthier. For instance, when $u = 0.4$ and $v = 0.2$, Eqs. 3.6.26, 3.6.30, and 3.6.31 become

$$P_{cr} = \{1.48 + 0.1175[(1/\Omega_1) + (1/\Omega_2)] + 0.081[(1/\Omega_1) - (1/\Omega_2)]\}P_E \qquad (e)$$

$$P_{cr} = 1.96P_E \qquad (f)$$

$$P_{cr} = \{4 + 0.2025[(1/\Omega_1) + (1/\Omega_2)] + 0.505[(1/\Omega_1) - (1/\Omega_2)]\}P_E \qquad (g)$$

If the multiplying factors of P_E in Eqs. *e* and *g* are equated to 1.96, the following two equations are obtained

$$0.2025[(1/\Omega_1) + (1/\Omega_2)] + 0.505[(1/\Omega_1) - (1/\Omega_2)] = -2.04$$

$$0.1175[(1/\Omega_1) + (1/\Omega_2)] + 0.081[(1/\Omega_1) - (1/\Omega_2)] = 0.48$$

The solution for the sum and the difference is

$$(1/\Omega_1) + (1/\Omega_2) = 9.5$$
$$(1/\Omega_1) - (1/\Omega_2) = -7.86$$

Consequently

$$\Omega_1 = 1.22, \qquad \Omega_2 = 0.115$$

The equivalent length coefficient is

$$\lambda = 1/\sqrt{1.96} = 0.713$$

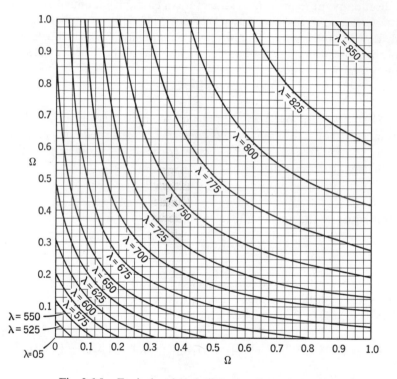

Fig. 3.6.5.   Equivalent-length diagram.   General case.
From author's paper in *J. Roy. Aero. Soc.*

In a similar manner many other corresponding values of the relative stiffness factor and the equivalent length coefficient can be computed. Their values are given as the faired-in curves shown in Fig. 3.6.5. For the convenience of the designer two more graphs are presented in Figs. 3.6.6 and 3.6.7. They contain the equivalent length coefficient plotted against the relative stiffness factor of one end of the column when the other end is pin-jointed or rigidly fixed.

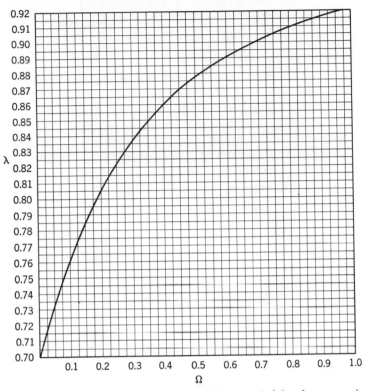

Fig. 3.6.6.   Equivalent-length diagram.   Column pin-jointed at one end.
From author's paper in *J. Roy. Aero. Soc.*

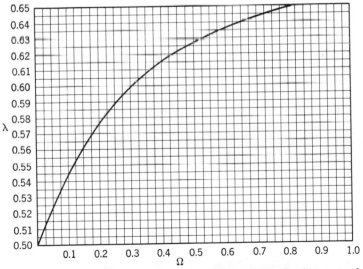

Fig. 3.6.7.   Equivalent-length diagram.   Column rigidly fixed at one end.
From author's paper in *J. Roy. Aero. Soc.*

3.7   Torsional-Flexural Buckling of Thin-Walled Open-Section Columns

Physical Explanation of the Phenomenon

It has been observed that thin-walled open-section columns often buckle by twisting, or some combination of twisting and flexure, rather than according to the pure flexural pattern of the Euler type so well known to structural engineers. At first thought it might seem odd that a column would twist under the action of an axial compressive force, but further examination indicates that this phenomenon is just as logical as flexural buckling. If the cross section of the column is doubly symmetric, either type of buckling destroys the symmetry of the system consisting of the column and the centrally applied compressive force. For this reason buckling must be initiated by some imperfection in the symmetry of the system, such as an inhomogeneity of the material or inaccuracy of workmanship or load application, or by a non-symmetric disturbing force.

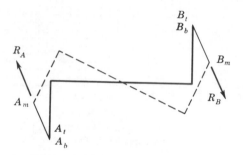

Fig. 3.7.1. Projection on horizontal plane of three cross sections of originally straight vertical column. From author's discussion in *J. Applied Mechanics.*

Figure 3.7.1 should be helpful to an understanding of how an axial thrust can maintain a column in a twisted state. It shows, in plan view, three cross sections of an originally straight vertical column after buckling. The top and bottom sections cannot twist because they are attached to the testing machine; they are shown in the figure by full lines in their original positions. The middle section, indicated by the dashed lines, is twisted. After buckling, the edge fiber connecting points A_t of the top, A_m of the middle, and A_b of the bottom section is curved. If for the sake of simplicity the curved line is imagined to be replaced by the two straight lines drawn from A_t to A_m and from A_m to A_b, the axial components of the compressive forces in the edge fiber balance each other at point A_m, but the transverse components add up to the resultant R_A. Similarly a resultant R_B is acting at point B_m of the middle section of the column because of the change in direction of the edge fiber $B_tB_mB_b$. The two resultants form a couple which tries to increase the angle of twist. This tendency is resisted by the torsional rigidity of the column. Since the rigidity is independent of the magnitude of the compressive load while the resultants R_A and R_B are proportional to it, it can be readily understood that torsional buckling can take place if the compressive load is sufficiently large.

Of course in reality the edge fibers do not consist of two straight lines joined at the middle section, but this argument is not essential to the proof. When line $A_tA_mA_b$ is curved, transverse forces in the direction of R_A exist all along the fiber because of the continuous changes in the slope of the curve. Torsional buckling is likely to occur if the torsional rigidity of the section is low; for this reason it is often observed with thin-walled open sections but not with solid sections or thin-walled closed sections. Fortunately the theory of non-uniform torsion of thin-walled open sections has been developed in sufficient detail to permit an analysis of the torsional buckling of columns having such sections.

The argument offered in connection with Fig. 3.7.1 is satisfactory for the purpose of giving a qualitative explanation of the phenomenon of torsional instability, but it is not suitable for an exact derivation of the buckling loads. That task will be accomplished with the aid of the Rayleigh-Timoshenko method; the deflections of the column will be represented by infinite series, and the condition of neutral stability will be determined from considerations of the changes in the total potential of the system.

Analysis by the Rayleigh-Timoshenko Method

The most general deformed shape of the column can be reached by two translations and one rotation of each of its sections as long as the cross-sectional shape remains unchanged. In the analysis it is advantageous to use the shear center† of the section as the origin of the coordinates and to let the x and y axes coincide with the principal axes of inertia of the section; with this choice the strain-energy expressions become particularly simple. Figure 3.7.2 shows an arbitrary thin-walled open section in its initial position as well as in the position reached after a translation u in the x direction, a translation v in the y direction, and a rotation β about an axis perpendicular to the xy plane and passing through the shear center of the section. The direction of the perpendicular is the z direction.

† The shear center is the point of the cross section through which a shear force must pass to cause bending and shearing, but no twisting. The theory of the shear center is discussed in the references listed in the appendix.

The deflections are represented by the expressions

$$u = \sum_{n=1}^{\infty} a_n \sin \frac{n\pi z}{L} \qquad\qquad 3.7.1a$$

$$v = \sum_{n=1}^{\infty} b_n \sin \frac{n\pi z}{L} \qquad\qquad 3.7.1b$$

$$\beta = \sum_{n=1}^{\infty} c_n \sin \frac{n\pi z}{L} \qquad\qquad 3.7.1c$$

Each term of these series satisfies boundary conditions resembling the pin-end conditions of Euler's flexural buckling theory. The displacements u and v vanish at the two end points where $z = 0$ and $z = L$, respectively. The bending moment also vanishes at these points since the

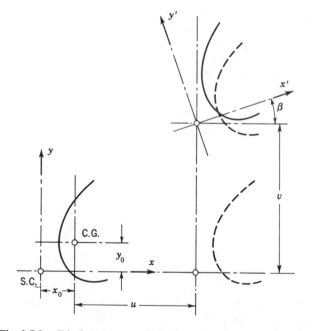

Fig. 3.7.2.   Displacements and rotation of section.   From author's paper in Q. *Applied Mathematics.*

curvature is zero in both the xz and yz planes. This is true as long as the deformations are small enough to permit the use of the quantities d^2u/dz^2 and d^2v/dz^2 in place of the exact curvature expressions. The rotation β also vanishes at the ends of the column, and so does $d^2\beta/dx^2$ which is proportional to the normal stress caused by non-uniform warping. The load is assumed to be distributed uniformly over the section, and the load per square inch of the cross section is designated by σ.

The increment δU in strain energy corresponding to the change from the initial straight configuration to the deformed shape defined by Eqs. 3.7.1 is

$$\delta U = \frac{1}{2} EI_x \int_0^L \left(\frac{d^2u}{dz^2}\right)^2 dz + \frac{1}{2} EI_y \int_0^L \left(\frac{d^2v}{dz^2}\right)^2 dz$$

$$+ \frac{1}{2} GC \int_0^L \left(\frac{d\beta}{dz}\right)^2 dz + \frac{1}{2} E\Gamma \int_0^L \left(\frac{d^2\beta}{dz^2}\right)^2 dz \qquad 3.7.2$$

The first two terms of the right-hand member represent flexural strain energies, the third term torsional strain energy in accordance with the St.-Venant theory of uniform torsion, and the last term is the strain energy of extension and compression accompanying non-uniform torsion. EI_x is the bending rigidity of the column for flexure in the xz plane, EI_y is the bending rigidity for flexure in the yz plane, GC is the torsional rigidity according to the St.-Venant theory of uniform torsion, and $E\Gamma$ is the warping rigidity (see Eq. 2.1.22).

The expressions appearing under the integral signs are squares of trigonometric series, since both the first and the second derivatives of the expressions in Eqs. 3.7.1 are again trigonometric series. For instance,

$$(d^2u/dz^2)^2 = a_1^2(\pi/L)^4 \sin^2(\pi z/L) + a_2^2(2\pi/L)^4 \sin^2(2\pi z/L) + \cdots$$

$$+ 2a_1a_2(\pi/L)(2\pi/L) \sin(\pi z/L) \sin(2\pi z/L)$$

$$+ 2a_1a_3(\pi/L)(3\pi/L) \sin(\pi z/L) \sin(3\pi z/L) + \cdots$$

It was shown in Art. 2.7 that the integral between the limits zero and L of the products in this expression is zero (see Eq. 2.7.18); and the integral of the squares of the sine functions is $L/2$ (see Eq. 2.7.17). Consequently

$$\int_0^L \left(\frac{d^2u}{dz^2}\right)^2 dz = \frac{L}{2} \left(\frac{\pi}{L}\right)^4 (a_1^2 + 2^4a_2^2 + 3^4a_3^2 + \cdots)$$

$$= \frac{L}{2} \left(\frac{\pi}{L}\right)^4 \sum_{n=1}^{\infty} n^4 a_n^2 \qquad 3.7.3$$

Integrations similar to those performed in Art. 2.7 yield the following results:

$$\int_0^L \cos^2 \frac{i\pi x}{L} dx = \frac{L}{2} \qquad 3.7.4$$

$$\int_0^L \cos \frac{i\pi x}{L} \cos \frac{j\pi x}{L} dx = 0 \qquad 3.7.5$$

provided
$$j \neq i$$
The integrals indicated in Eq. 3.7.2 can now be evaluated:

$$\delta U = EI_x \frac{L}{4} \left(\frac{\pi}{L}\right)^4 \sum_{n=1}^{\infty} n^4 a_n^2 + EI_y \frac{L}{4} \left(\frac{\pi}{L}\right)^4 \sum_{n=1}^{\infty} n^4 b_n^2$$

$$+ GC \frac{L}{4} \left(\frac{\pi}{L}\right)^2 \sum_{n=1}^{\infty} n^2 c_n^2 + E\Gamma \frac{L}{4} \left(\frac{\pi}{L}\right)^4 \sum_{n=1}^{\infty} n^4 c_n^2 \qquad 3.7.6$$

The Potential of the Applied Load

The next quantity to be calculated is the change in the potential of the external load. The load acting on a fiber whose cross-sectional area is dA is $\sigma\, dA$, and the change in the distance between the end points of the fiber can be denoted by ΔL. Then the increase in the potential of the total load acting on the cross section is

$$\delta V = - \int_A \sigma\, \Delta L\, dA \qquad 3.7.7$$

where the letter A under the integral sign indicates that the integration must be extended over the entire cross-sectional area. The value of ΔL was calculated in Art. 2.4 for the case when displacements occurred only in a single plane. Now displacements must be taken into account in both the x and y directions, and the derivations of Art. 2.4 must be correspondingly modified. An element of a fiber of length dz is shown in Fig. 3.7.3. The x displacement of the lower end of the element is designated as u^*, and its y displacement as v^*. At the upper end of the element the corresponding quantities are $u^* + du^*$ and $v^* + dv^*$. From the Pythagorean theorem the length ds of the element is then

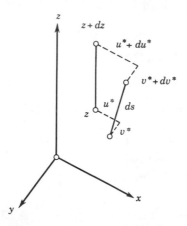

Fig. 3.7.3. Line element in space.

$$ds^2 = dz^2 + du^{*2} + dv^{*2}$$

or

$$ds = dz[1 + (du^*/dz)^2 + (dv^*/dz)^2]^{1/2}$$

The square root of the expression in brackets can be given in a simpler form if the squares of the slopes du^*/dz and dv^*/dz are small compared to unity. An argument paralleling the one given in Art. 2.4 would show that in good approximation

$$ds = dz[1 + \tfrac{1}{2}(du^*/dz)^2 + \tfrac{1}{2}(dv^*/dz)^2]$$

The length of the curved fiber is

$$L + \Delta L = \int_0^L ds = L + \frac{1}{2} \int_0^L \left(\frac{du^*}{dz}\right)^2 dz + \frac{1}{2} \int_0^L \left(\frac{dv^*}{dz}\right)^2 dz$$

and thus the difference ΔL between the length of the curved fiber and its projection upon the vertical is

$$\Delta L = \frac{1}{2} \int_0^L \left[\left(\frac{du^*}{dz}\right)^2 + \left(\frac{dv^*}{dz}\right)^2\right] dz \qquad 3.7.8$$

The difference can be considered as the displacement of the point of application of the load $\sigma\, dA$ during the deformations that take place at buckling.

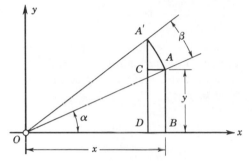

Fig. 3.7.4.   Displacements caused by rotation.

Naturally the displacements u^* and v^* vary along the median line of the thin-walled section because of the rotation of the section.   The connection between these quantities and the displacements u, v of the shear center and the rotation β of the section can be easily derived.   In Fig. 3.7.4 point A is moved to A' through a rotation β.   The y component of the displacement of point A is $A'C = A'D - CD = A'D - AB$.   If L is used to denote the distance $OA = OA'$ one has

$$A'C = L \sin(\alpha + \beta) - L \sin\alpha = L(\sin\alpha\cos\beta + \cos\alpha\sin\beta - \sin\alpha)$$

The x component of the displacement of point A is $-AC = -(BO - DO)$. This can be written in the form

$$-AC = -L\cos\alpha + L\cos(\alpha + \beta)$$
$$= L(-\cos\alpha + \cos\alpha\cos\beta - \sin\alpha\sin\beta)$$

As all deformations during buckling are considered small, the angle β is a small quantity, and thus one can set

$$\cos\beta = 1, \qquad \sin\beta = \beta$$

Substitution of these values in the expressions obtained for the displacements yields

$$A'C = \beta L \cos \alpha$$

$$-AC = -\beta L \sin \alpha$$

As L is the distance AO, obviously

$$L \cos \alpha = y, \qquad L \sin \alpha = x$$

and thus

$$A'C = x\beta$$

$$-AC = -y\beta$$

These two quantities are the displacements of a point x, y on the median line of the section as caused by the rotation β. Since the point undergoes the translations u and v of the shear center at the same time, the total displacements of point x, y are

$$u^* = u - y\beta \qquad\qquad 3.7.9a$$

$$v^* = v + x\beta \qquad\qquad 3.7.9b$$

Substitution in Eq. 3.7.8 yields

$$\Delta L = \frac{1}{2} \int_0^L \left\{ \left[\frac{du}{dz} - y \frac{d\beta}{dz} \right]^2 + \left[\frac{dv}{dz} + x \frac{d\beta}{dz} \right]^2 \right\} dz$$

If the infinite series assumed for u, v, and β in Eqs. 3.7.1 are substituted and Eqs. 3.7.4 and 3.7.5 are taken into account, one obtains after some manipulations

$$\Delta L = \frac{\pi^2}{4L} \left(\sum_{n=1}^{\infty} n^2 a_n^2 + \sum_{n=1}^{\infty} n^2 b_n^2 - 2y \sum_{n=1}^{\infty} n^2 a_n c_n \right.$$

$$\left. + 2x \sum_{n=1}^{\infty} n^2 b_n c_n + x^2 \sum_{n=1}^{\infty} n^2 c_n^2 + y^2 \sum_{n=1}^{\infty} n^2 c_n^2 \right) \qquad 3.7.10$$

The value thus obtained for ΔL must now be substituted in Eq. 3.7.7, and the integration over the cross section must be carried out. During the integration the expressions under the summation signs remain unchanged, and the only variables are x and y. If the constants are disregarded for the moment, the following integrals have to be evaluated:

$$\int_A dA = A$$ This is the total cross-sectional area of the column.

$$\int_A y \, dA = y_0 A$$ This is the static moment of the cross-sectional area with respect to the x axis passing through the shear center.

$$\int_A x \, dA = x_0 A$$ This is the static moment of the cross-sectional area with respect to the y axis passing through the shear center.

$$\int_A (x^2 + y^2) \, dA = \rho^2 A$$ This is the polar moment of inertia of the cross-sectional area with respect to the shear center, and ρ is the polar radius of gyration based on the shear center.

It can be seen that each integral contains A. Since $\sigma A = P$, and in Eq. 3.7.7 the constant compressive stress σ is one of the factors in the expression under the integral sign, the total compressive force acting on the column is a common multiplier of all the terms obtained through integration of Eq. 3.7.7. This integral can now be written in the form

$$\delta V = -\frac{\pi^2 P}{4L} \left(\sum_{n=1}^{\infty} n^2 a_n^2 + \sum_{n=1}^{\infty} n^2 b_n^2 - 2y_0 \sum_{n=1}^{\infty} n^2 a_n c_n \right.$$

$$\left. | \; 2x_0 \sum_{n=1}^{\infty} n^2 b_n c_n + \rho^2 \sum_{n=1}^{\infty} n^2 c_n^2 \right) \qquad 3.7.11$$

The Buckling Condition

The following notation is now introduced:

$$\pi^2 E I_x / L^2 = N \qquad\qquad 3.7.12a$$

$$\pi^2 E I_y / L^2 = Q \qquad\qquad 3.7.12b$$

$$\pi^2 E \Gamma / L^2 = R \qquad\qquad 3.7.12c$$

The first two expressions are obviously the Euler buckling loads for flexure in the xz and yz planes. The third is a similar formula involving the warping constant Γ, but its dimension is not that of a force since Γ

is measured in units of length raised to the sixth power.   With this notation the change in the total potential of the system can be written as

$$\delta(U+V) = \frac{\pi^2 P}{4L} \sum_{n=1}^{\infty} n^2 \left[\left(n^2 \frac{N}{P} - 1 \right) a_n^2 + \left(n^2 \frac{Q}{P} - 1 \right) b_n^2 \right.$$

$$\left. + \left(n^2 \frac{R}{P} + \frac{GC}{P} - \rho^2 \right) c_n^2 + 2y_0 a_n c_n - 2x_0 b_n c_n \right] \qquad 3.7.13$$

According to the minimal principle the original straight-line shape is a configuration of equilibrium if the change in the total potential vanishes, provided first-order small terms alone are considered.   This is obviously the case here because Eq. 3.7.13 contains only second-order small quantities in a_n, b_n, and c_n.   The equilibrium is stable as long as the second-order small terms are always positive.   The next problem is therefore to decide under what conditions the right-hand member of Eq. 3.7.13 can have only positive values for any arbitrary choice of a_n, b_n, and c_n.   The investigation is simplified if the following notation is introduced:

$$A_n = n^2(N/P) - 1 \qquad\qquad 3.7.14a$$

$$B_n = n^2(Q/P) - 1 \qquad\qquad 3.7.14b$$

$$C_n = n^2(R/P) + (GC/P) - \rho^2 \qquad\qquad 3.7.14c$$

$$X_n = A_n a_n^2 + B_n b_n^2 + C_n c_n^2 + 2y_0 a_n c_n - 2x_0 b_n c_n \qquad 3.7.15$$

Equation 3.7.13 becomes

$$\delta(U+V) = \frac{\pi^2 P}{4L} \sum_{n=1}^{\infty} n^2 X_n \qquad\qquad 3.7.16$$

The change in the total potential is positive and the equilibrium is stable if every X_n is positive for any arbitrary choice of a_n, b_n, and c_n.   To check the sign of X_n, the terms in which a_n, b_n, c_n are not raised to the second power will be eliminated.   The subscript n will be omitted for the sake of brevity.   It is observed that

$$(Cc + y_0 a - x_0 b)^2 = C^2 c^2 + 2Cy_0 ac - 2Cx_0 bc + y_0^2 a^2 + x_0^2 b^2 - 2x_0 y_0 ab$$

Hence one can write

$$CX = C(Aa^2 + Bb^2 + Cc^2 + 2y_0 ac - 2x_0 bc)$$

$$= (Cc + y_0 a - x_0 b)^2 - y_0^2 a^2 - x_0^2 b^2 + 2x_0 y_0 ab + C(Aa^2 + Bb^2)$$

$$= (Cc + y_0 a - x_0 b)^2 + (CA - y_0^2)a^2 + (CB - x_0^2)b^2 + 2x_0 y_0 ab \qquad (a)$$

Next the term containing ab will be eliminated. Obviously the identity holds

$$\left(a\sqrt{CA - y_0^2} + \frac{x_0 y_0 b}{\sqrt{CA - y_0^2}}\right)^2 = (CA - y_0^2)a^2$$

$$+ 2x_0 y_0 ab + \frac{x_0^2 y_0^2}{CA - y_0^2} b^2 \qquad (b)$$

Hence

$$(CA - y_0^2)a^2 + (CB - x_0^2)b^2 + 2x_0 y_0 ab$$
$$= \left(a\sqrt{CA - y_0^2} + \frac{x_0 y_0 b}{\sqrt{CA - y_0^2}}\right)^2$$
$$+ (CB - x_0^2)b^2 - \frac{x_0^2 y_0^2 b^2}{CA - y_0^2} \qquad (c)$$

The last two terms in the right-hand member can be written over a common denominator:

$$(CB - x_0^2)b^2 - \frac{x_0^2 y_0^2 b^2}{CA - y_0^2} = \frac{C(ABC - x_0^2 A - y_0^2 B)}{CA - y_0^2} b^2 \qquad (d)$$

Now the expression for CX can be given in the final form

$$CX = (Cc + y_0 a - x_0 b)^2 + \left(a\sqrt{CA - y_0^2} + \frac{x_0 y_0 b}{\sqrt{CA - y_0^2}}\right)^2$$
$$+ \frac{C(ABC - x_0^2 A - y_0^2 B)}{CA - y_0^2} b^2 \qquad 3.7.17$$

If the alternative expression is introduced in the perfect square of Eq. b, one obtains in the place of Eq. 3.7.17 the expression

$$CX = (Cc + y_0 a - x_0 b)^2 + \left(b\sqrt{CB - x_0^2} + \frac{x_0 y_0 a}{\sqrt{CB - x_0^2}}\right)^2$$
$$+ \frac{C(ABC - x_0^2 A - y_0^2 B)}{CB - x_0^2} a^2 \qquad 3.7.17a$$

The first two terms in the right-hand members of these equations are always positive because they are the squares of real numbers. Consequently the first two terms in the corresponding expression for X are positive if

$$C > 0 \qquad 3.7.18a$$

The last term contains four factors of which C is positive according to Eq. 3.7.18a, and b^2 (as well as a^2) is positive as a square of a real number. The remaining two factors must therefore have the same sign. It is clear from Eqs. 3.7.12 and 3.7.14 that these terms are positive when the end load P is small (and positive).   The system remains stable as long as

$$ABC - x_0^2 A - y_0^2 B > 0 \qquad\qquad 3.7.18b$$

$$CA - y_0^2 > 0 \qquad\qquad 3.7.18c$$

$$CB - x_0^2 > 0 \qquad\qquad 3.7.18d$$

Because of Eq. 3.7.18a the last two equations imply the requirements

$$A > 0 \qquad\qquad 3.7.18e$$

$$B > 0 \qquad\qquad 3.7.18f$$

As the end load P increases, the values of A, B, and C decrease according to Eqs. 3.7.14.   The left-hand member of Eq. 3.7.18b reaches the value zero before A, B, or C becomes zero, because it is obviously negative when one of the three quantities A, B, and C is zero while the other two are still positive.   Consequently neutral equilibrium is reached with increasing values of P when the equation

$$ABC - x_0^2 A - y_0^2 B = 0 \qquad\qquad 3.7.19$$

is first satisfied.   The equation can be rewritten after substitution from Eqs. 3.7.14, and with the aid of the new symbol

$$T_n = (1/\rho^2)[GC + (n^2\pi^2 E\Gamma/L^2)] = (1/\rho^2)(GC + n^2 R) \qquad 3.7.20$$

one obtains

$$[(n^2 N/P) - 1)][(n^2 Q/P) - 1]\rho^2[(T_n/P) - 1]$$
$$- x_0^2[(n^2 N/P) - 1] - y_0^2[(n^2 Q/P) - 1] = 0 \qquad 3.7.21a$$

Equation 3.7.21a is first satisfied as P increases from zero when the value of n in the equation is unity.   Consequently the condition for the lowest buckling load is

$$[(N/P) - 1][(Q/P) - 1][(T/P) - 1] - (x_0/\rho)^2[(N/P) - 1]$$
$$- (y_0/\rho)^2[(Q/P) - 1] = 0 \qquad 3.7.21b$$

where

$$N = \pi^2 E I_x/L^2 \qquad\qquad 3.7.22a$$

$$Q = \pi^2 E I_y/L^2 \qquad\qquad 3.7.22b$$

$$T = (1/\rho^2)[GC + (\pi^2 E\Gamma/L^2)] \qquad\qquad 3.7.22c$$

Effect of the Shape of the Cross Section on the Type of Buckling

A discussion of Eq. 3.7.21b reveals a number of interesting facts about torsional buckling.   When the centroid and the shear center of the section coincide, $x_0 = y_0 = 0$, and the equation reduces to

$$[(N/P) - 1][(Q/P) - 1][(T/P) - 1] = 0 \qquad 3.7.23$$

This equation is obviously satisfied if any one of the following conditions is fulfilled

$$P = N \qquad\qquad 3.7.23a$$

$$P = Q \qquad\qquad 3.7.23b$$

$$P = T \qquad\qquad 3.7.23c$$

It can be seen from Eq. 3.7.13 that the first condition corresponds to the critical load when $a_1 \neq 0$, $a_n = 0$ for all positive integral values of n except unity, and $b_n = c_n = 0$.   This means that buckling can occur according to the Euler pattern with flexure in the xz plane.   Similarly Eq. 3.7.23b is the Euler buckling load for flexure in the yz plane.   In this case $b_1 \neq 0$, $b_n = 0$ for all positive integral values except unity, and $a_n = c_n = 0$.   In both these cases the total length of the column is the half-wave length of the sine function.

The buckling load is $P = T$ when $c_1 \neq 0$, $c_n = 0$ for all positive integral values of n except unity, and $a_n = b_n = 0$.   The column may therefore fail in a purely torsional mode with the half-wave length equal to the length of the column.   This type of buckling, in which sections of the column rotate about the shear center, was discovered by Herbert Wagner. The buckling load T is often referred to as the Wagner load.

It can be concluded that columns whose centroid and shear center coincide can fail either according to the pure Euler pattern in which the column is bent about one of the principal axes of inertia, or according to the pure Wagner pattern in which the column is twisted about the axis passing through the shear centers of the sections.   The corresponding buckling loads are given in Eqs. 3.7.22.   Naturally the type of buckling observed in an experiment is the one whose buckling load is the smallest. Sections with coincident centroid and shear center are the doubly symmetric sections such as the I, and the centrally symmetric ones such as the Z.

When the section has a single axis of symmetry, that axis is also a principal axis of inertia and the shear center is situated on it.   Let the system of coordinates be chosen in such a manner that the section is symmetric with respect to the x axis.   Then obviously $y_0 = 0$, and Eq. 3.7.21b simplifies to read

$$[(N/P) - 1]\{[(Q/P) - 1][(T/P) - 1] - (x_0/\rho)^2\} = 0 \qquad 3.7.24$$

This equation is satisfied if either

$$P = N \qquad\qquad 3.7.25a$$

or

$$[(Q/P) - 1][(T/P) - 1] - (x_0/\rho)^2 = 0 \qquad\qquad 3.7.25b$$

The first condition again corresponds to pure Euler buckling with flexure in the xz plane of symmetry. The second equation is satisfied when with increasing P the values of the two expressions in brackets become sufficiently small to have a product equal to $(x_0/\rho)^2$. This will happen before either of the bracketed expressions vanishes. Consequently Eq. 3.7.25b is satisfied in the loading process before either $P = Q$ or $P = T$. This means that the buckling load defined by Eq. 3.7.25b is smaller than either Q or T. The deflections are a combination of flexure in the yz plane and torsion about the shear center. According to a theorem in geometry, they can be produced by rotations about some point on the axis of symmetry. An analysis of Eq. 3.7.13 shows that c is large compared to b when Q is large compared to T. In such a case the sections twist about a point close to the shear center. When T is large compared to Q, buckling occurs with a value of b much greater than that of c. Then the point of rotation is close to the point at infinity along the axis of symmetry, and the deformations are almost purely flexural.

The buckling load can be calculated through the solution of Eq. 3.7.25b which is a quadratic in $1/P$. One obtains

$$\frac{1}{P} = \frac{1}{2QT}\left[Q + T \pm \sqrt{(Q-T)^2 + 4\left(\frac{x_0}{\rho}\right)^2 QT}\right] \qquad 3.7.26a$$

or

$$P = \frac{1}{2[1 - (x_0/\rho)^2]}\left[Q + T \pm \sqrt{(Q-T)^2 + 4\left(\frac{x_0}{\rho}\right)^2 QT}\right] \qquad 3.7.26b$$

The quadratic has two solutions. Only the smaller value of P is of practical interest. This is the one discussed in detail and it is always smaller than Q or T.

A typical section having one axis of symmetry is the channel section. When a channel section column, or any other column whose section has one axis of symmetry, is tested in compression, buckling can occur according to the symmetric pattern corresponding to the buckling load given in Eq. 3.7.25a, or according to the antisymmetric pattern whose buckling load is presented in Eqs. 3.7.26. The buckling pattern actually observed in the test is symmetric if N is smaller than P in Eqs. 3.7.26, and antisymmetric when the opposite is true.

The most general case is represented by sections without symmetry. For these neither x_0 nor y_0 vanishes, and the buckling load has to be calculated from Eq. 3.7.21 without the possibility of further simplifications. As the equation is a cubic, three solutions are obtained. It is easy to show that the smallest of the solutions is less than the value of N, Q, or T: The triple product in Eq. 3.7.21 is very large and positive when P is small. (Only positive values of P need be considered in this discussion because compressive forces alone are of interest and P is positive when it is compressive.) Each one of its factors is large and of the same order of magnitude. Since $(x_0/\rho)^2$ and $(y_0/\rho)^2$ are both positive and, as a rule, smaller than unity, the two negative terms in the left-hand member of the equation are small compared to the positive term. With increasing P the expressions in brackets decrease in magnitude, and the triple product certainly becomes equal to the absolute value of the sum of the two negative terms before any of the bracketed expressions vanishes. This proves that the smallest root of the cubic in P is less than the value that makes any of the bracketed expressions vanish. Hence P is less than N, Q, and T.

Conclusions

The most important conclusion to be drawn from the derivations presented in this article is that the familiar Euler type of buckling is just a particular case of a more general type of instability in which the cross sections of the column are twisted as well as translated. Pure flexural buckling is possible only when the section is symmetric. Of course, most column tests have been carried out with solid rectangular sections which satisfy this requirement, and that explains why the torsional buckling phenomenon is not better known. When the section is unsymmetric, torsional and flexural deformations are mixed in the ratio that corresponds to the lowest buckling load, and the torsional pattern is noticeable only when the torsional rigidity of the section is small. This is true of the thin-walled open sections widely used in airplane structures and recently adopted in general structural engineering. With solid sections, closed thin-walled sections, and thick-walled open sections, torsional deformations are unimportant.

It might be mentioned here that the inclusion of the words *thin-walled open section* in the title of this article was necessary only because strain-energy expressions for non-uniform torsion are not known for other types of sections. It is probable that Eq. 3.7.2 holds with a reasonable degree of accuracy for columns having any arbitrary section, but values of the warping constant Γ have been derived only for thin-walled sections. This means that the general conclusions regarding torsional-flexural patterns are likely to hold for all columns, but, of course, the torsional deformations are insignificant when the St.-Venant torsional rigidity of the section is substantial.

3.8 The Buckling Load of a Truss Determined by the Convergence Criterion of the Moment-Distribution Method

Introduction

It was shown in Art. 3.6 that in a truss the elastic restraint provided by the riveted or welded connections can increase the buckling load of a compressed bar considerably above the Euler load. The elastic resistance of the joints to rotation, represented by the value of $C = \Sigma S$, was calculated under simplifying assumptions, and the multiplying factor of the Euler load was presented in diagrams as a function of the relative stiffness factor Ω. This factor was defined as the bending rigidity of the compression bar

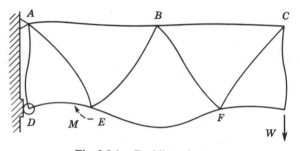

Fig. 3.8.1. Buckling of a truss.

divided by C. The disadvantage of this approach to the solution of the buckling problem is that imaginary pins must be inserted in the framework as shown in Fig. 3.6.1, in order to get definite values for the stiffnesses of the supporting members and to make sure that no supporting member is used in conjunction with more than one principal compression member. The imaginary pins naturally introduce new degrees of freedom of motion into the truss, and the resulting system has a lower buckling load than the actual one.

In reality the entire truss buckles as a single unit. Whenever one highly compressed member deflects, continuity at the rigidly connected joints requires that all the other members also deflect, as indicated in Fig. 3.8.1. In spite of the complexity of this problem, a simple procedure is available for its solution because of a peculiarity of the Hardy Cross moment-distribution method: It converges only when the loads are smaller than the critical values, and it diverges when the loads exceed the critical values.   The procedure suggested for evaluating the buckling load of the truss as a whole is as follows:

First a guess must be made regarding the magnitude of the load W under which the truss buckles.   Next the axial loads are calculated in the members of the truss by any of the routine methods based on the assumption of pin connections.   (It was shown in Art. 1.18 that the rigid connections at the joints have no appreciable effect on the axial loads.)   The stability of the truss is then tested by assuming a moment M acting at any one of the joints, say at joint E in Fig. 3.8.1, in the plane of the truss and tending to rotate joint E clockwise so as to make the truss assume the deflected shape shown.   The Hardy Cross moment-distribution method, derived in Art. 1.17 and applied to a simple framework in Art. 1.18, permits the calculation of the internal moments acting at each of the joints of the framework. In the step-by-step approximation procedure of the moment-distribution method the effect of the end loads acting in the bars must be taken into account.   This means that the stiffness coefficients and the carry-over factors must be computed from Eqs. 2.4.14 and 2.4.17, or read from Fig. 2.4.6.

Experience has shown that the moment-distribution process converges sufficiently rapidly; that is, after a reasonably small number of cycles of the process the unbalanced moments at each joint are reduced to small enough quantities to be disregarded in engineering applications.   However, this statement is true only when the truss is in a stable state of equilibrium under the load W.   When the equilibrium is unstable, the unbalanced moments increase in magnitude beyond all limits as the balancing process is continued.   This peculiarity of the moment-distribution method can be used to find the buckling load of the truss.   One can assume a few different values for W and carry out the moment distribution with the same assumed moment M at joint E but with the different stiffness coefficients and carry-over factors corresponding to each W.   It is then easy to bracket the critical value of W between closely spaced upper and lower limits, the higher value resulting in a divergent process and the lower value leading to convergence.

The practicing structural engineer may find it strange that his well-behaved moment-distribution method is expected to become divergent under sufficiently large loads.   He should remember, however, that in ordinary applications in civil engineering the safety factors are so high that the effect of the end loads on the stiffness and the carry-over factors can be disregarded when the moment distribution is carried out for the working loads.   Under such circumstances the distribution factor is always less than unity and usually not greater than one-half, and the carry-over factor is one half for a bar of uniform cross section.   Hence the moment carried over from one joint to the other is generally smaller than one-quarter the moment balanced, and the process always converges. On the other hand, in the calculation of the actual buckling load of the truss, compressive loads equal to or exceeding the Euler load of the pin-ended column are encountered. With such high compressive loads the stiffness is greatly reduced and can even be negative while the carry-over factor may attain values much greater than unity (see Fig. 2.4.6). It is quite natural then that the process may become divergent.

Method of Proof of the Convergence Criterion

The proof of the convergence criterion can be brought in the following manner: It can be shown that in each cycle of the moment-distribution procedure, consisting of balancing, distribution, and carry-over operations, the total potential of the system decreases to the minimum value compatible with the requirement that the far ends of the bars be rigidly fixed. Moreover, it is known from Arts. 3.1 and 3.4 that the total potential of a stable system has a minimum at the value corresponding to equilibrium, while the stationary value of an unstable system is either a saddle point or a maximum. In the moment-distribution method the starting point is always the undeformed configuration in which the total potential is zero. When the system is stable, the analyst moves along on the total potential surface in such a manner that in each cycle he gets down deeper in the bowl whose bottom corresponds to equilibrium. He can approach this stationary value as closely as he desires if he is willing to undertake a sufficiently large number of steps. Hence the moment-distribution process is convergent. When the equilibrium is unstable, only by the merest chance will the calculator descend to the saddle point, and normally each cycle will lead him down deeper on the slope to negative infinity, because in no cycle of the moment-distribution method can the value of the total potential increase. In each cycle of operations he gets farther away from the conditions of moment equilibrium, and thus he finds the process divergent.

This means that, if the analyst succeeds in eliminating all the unbalanced moments in a straightforward application of the moment-distribution process, he can be sure that the system investigated is stable (except in very unusual cases to be discussed later). When he is unsuccessful in his task, the failure may be due to instability or to his incompetence. In practice the lack of convergence is hardly ever the consequence of an unsuitable sequence of moment-distribution operations because it is very easy to eliminate the unbalances when the system is stable. Similarly the divergence of the procedure is readily detected when the system is unstable. This will be evident from the numerical examples which will be given later.

The Total Potential of the System

As a preliminary to the detailed proof, the total potential of a beam column must be calculated. Values of the stiffness coefficient S and the carry-over factor C were given in Eqs. 2.4.14 and 2.4.17. From the definition of these fundamental quantities of the Hardy Cross method, angle α_1 is one radian at the left support of the beam column shown in Fig. 3.8.2a when the moment at support 1 is S in.-lb and the carry-over

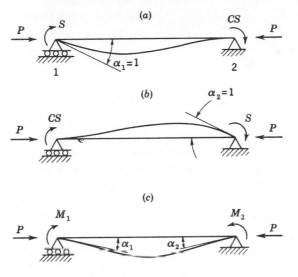

Fig. 3.8.2.   Beam column.

moment at the rigidly fixed far end (support 2) is CS in.-lb. Figure 3.8.2b shows the situation when the beam column is maintained in its original horizontal position at support 1 and the angle α_2 at support 2 is 1 radian. In the figures the positive sense of the rotations is indicated in accordance with the rigid-frame sign convention. According to this convention clockwise moments and rotations are positive. In Fig. 3.8.2c the preceding two equilibrium configurations are combined after multiplication of the quantities in the first configuration by α_1 and of those in the second by $-\alpha_2$. The resultant moments at the two supports are:

$$M_1 = S_{1,2}\alpha_1 + (CS)_{1,2}\alpha_2$$

$$M_2 = (CS)_{1,2}\alpha_1 + S_{1,2}\alpha_2 \qquad 3.8.1$$

The strain energy stored in the beam column in consequence of the application of the end moments is equal to the work done by the end moments and by the end loads during the deformations (it is assumed that the end loads $P_{1,2}$ are present when the end moments are applied):

$$U_{1,2} = \tfrac{1}{2}(M_1\alpha_1 + M_2\alpha_2) + P_{1,2}\,\Delta L_{1,2} \qquad 3.8.2$$

where $\Delta L_{1,2}$ is the second-order small shortening of the distance between supports 1 and 2 in consequence of the transverse deflections. The first term is multiplied by the factor $\tfrac{1}{2}$ because the moments increase from zero to their final values as the deformations take place. The factor is missing before the second term since P is constant during the deformation. Because of Eqs. 3.8.1 one can write Eq. 3.8.2 in the form

$$U_{1,2} = \tfrac{1}{2}S_{1,2}(\alpha_1{}^2 + a_2{}^2) + (CS)_{1,2}\alpha_1\alpha_2 + P_{1,2}\,\Delta L_{1,2} \qquad 3.8.3$$

The total strain energy stored in a truss is the sum of this quantity over all the bars contained in the truss:

$$U = \frac{1}{4}\sum_{m=1}^{N}\sum_{n=1}^{N}[S_{mn}(\alpha_m{}^2 + \alpha_n{}^2) + 2(CS)_{mn}\alpha_m\alpha_n] + \frac{1}{2}\sum_{m=1}^{N}\sum_{n=1}^{N}P_{mn}\,\Delta L_{mn} \qquad 3.8.4$$

It is assumed that the total number of joints in the truss is N; the values of S_{mn} and ΔL_{mn} are taken as zero when joint m is not connected by a bar with joint n. The factors $\tfrac{1}{4}$ and $\tfrac{1}{2}$ appear because in the summation each bar figures twice. For instance, the bar connecting joint 3 with joint 7 will contribute a term multiplied by $S_{3,7}$ and one by $S_{7,3}$.

The last term in Eq. 3.8.4 will now be evaluated. Obviously the strain energy stored in the truss is not influenced by the order in which the individual displacements are undertaken. It is permissible therefore to move each joint individually while the other joints are held fixed. It is understood that these joint displacements are caused by the transverse deflections of the bars and are second-order small, but the second-order quantities must not be neglected in stability calculations. In Fig. 3.8.3 joint n is displaced through a distance u in the x direction and through a distance v in the y direction. The increase in the distance between the end points m and n of bar mn is then according to Eq. 2.3.6

$$\Delta L_{mn} = u(L_x/L)_{mn} + v(L_y/L)_{mn} \qquad 3.8.5$$

where L_x and L_y are the x and y components, respectively, of the length L of the bar. The components are counted positive if the coordinate is larger for the moving joint than for the stationary one, and a positive ΔL_{mn} corresponds to a stretching of the bar. For the time being a positive force P_{mn} is defined as a tensile force acting from joint n on bar mn. The sum of the work done by all the forces P_{mn} during the displacement of joint n is

$$\sum_{m=1}^{N}P_{mn}\,\Delta L_{mn} = \sum_{m=1}^{N}\left[P_{mn}\left(\frac{L_x}{L}\right)_{mn}u + P_{mn}\left(\frac{L_y}{L}\right)_{mn}v\right]$$

With the sign convention adopted, $P_{mn}(L_x/L)_{mn}$ is simply the component of P_{mn} is the positive x direction, and $P_{mn}(L_y/L)_{mn}$ is the component of P_{mn} in the positive y direction.   Hence

$$\sum_{m=1}^{N} P_{mn}\, \Delta L_{mn} = R_x u + R_y v = R \cdot d \qquad\qquad 3.8.6$$

where R is the resultant of all the forces acting from joint n on the bars, R_x and R_y are its components in the positive x and y directions, d is the resultant displacement, and the dot designates the inner or scalar product.

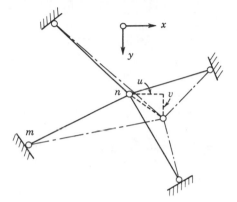

Fig. 3.8.3.   Displacement of one joint.

The scalar product is defined in vector algebra as the expression in the middle member of the chain of Eq. 3.8.6.   Hence the sum in the left-hand member of Eq. 3.8.6 is -1 times the work done by the resultant of all the forces acting from the bars on joint n during the displacement d.   A factor $\frac{1}{2}$ is not needed in the equation because the forces P_{mn} remain constant during the deformations.

Now the original sign convention is readopted and P_{mn} and ΔL_{mn} are considered positive when they correspond to compression.   Equation 3.8.5 remains unchanged if the sign convention for L_x and L_y is also inverted, and naturally the sign of the product under the summation sign in the left-hand member of Eq. 3.8.6 is not altered if both of its factors are multiplied by -1.   Consequently Eq. 3.8.4 can be written in the form

$$U = \frac{1}{4} \sum_{m=1}^{N} \sum_{n=1}^{N} [S_{mn}(\alpha_m{}^2 + a_n{}^2) + 2(CS)_{mn}\alpha_m\alpha_n] - \sum_{n=1}^{N} R_n \cdot d_n$$

$$3.8.7$$

where $R_n \cdot d_n$ is the work done by the resultant R_n of all the axial forces acting from the bars on joint n during the displacement d_n of joint n.   The sign of the last term in the right-hand member of the equation is negative because R is the resultant of the forces acting from the bars on the joint, while in Eq. 3.8.6 it designated the resultant of all the forces acting on the bars from the joint. The factor $\frac{1}{2}$ is absent because the double summation of Eq. 3.8.4 is replaced by a single summation, and thus each bar is counted only once.

The potential of all the external forces and moments is

$$V = - \sum_{n=1}^{N} M_n{}^* \alpha_n - \sum_{n=1}^{N} W_n \cdot d_n \qquad 3.8.8$$

where $M_n{}^*$ is the external moment acting on the nth joint, W_n is the external load acting on the nth joint, and $W_n \cdot d_n$ is the virtual work done by W_n during the displacement d_n. When Eqs. 3.8.7 and 3.8.8 are added up to obtain the total potential, one finds that the combination of terms

$$- \sum_{n=1}^{N} R_n \cdot d_n - \sum_{n=1}^{N} W_n \cdot d_n = - \sum_{n=1}^{N} (R_n + W_n) \cdot d_n = 0 \quad 3.8.9$$

because the sum $R_n + W_n$ of the internal and external forces acting on the nth joint is zero. Consequently the total potential of the system calculated from the undeflected state is

$$U + V = \frac{1}{4} \sum_{m=1}^{N} \sum_{n=1}^{N} [S_{mn}(\alpha_m{}^2 + \alpha_n{}^2) + 2(CS)_{mn}\alpha_m\alpha_n] - \sum_{n=1}^{N} M_n{}^*\alpha_n$$

$$3.8.10$$

Beam Column on Four Supports

The properties of the total potential will now be examined for a system having two degrees of freedom. The beam column shown in Fig. 3.8.4

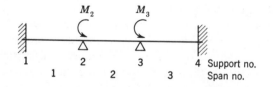

Fig. 3.8.4. Beam column on four supports.

has four supports. Since supports 1 and 4 do not permit rotation, the deflected shape of the beam column is known when the rotations at supports 2 and 3 are given. This is true, of course, only after the geometric and mechanical properties of the three spans and the axial loads acting on them are specified. The only combination in which these quantities appear in the present calculations are the stiffness S and the carry-over factor C.

For the beam column of Fig. 3.8.4 Eq. 3.8.10 reduces to

$$U + V = \tfrac{1}{2}\alpha_2^2(S_{1,2} + S_{2,3}) + \tfrac{1}{2}\alpha_3^2(S_{2,3} + S_{3,4})$$
$$+ \; \alpha_2\alpha_3(CS)_{2,3} - M_2^*\alpha_2 - M_3^*\alpha_3 \qquad 3.8.11$$

The value of $U + V$ can be laid off in the z direction of a Cartesian coordinate system whose x and y directions represent α_2 and α_3. The surface passing through the points is a quartic whose properties will now be investigated. If a small displacement is undertaken in an arbitrary direction from some point on the surface corresponding to the values α_2 and α_3, the new coordinates in the $\alpha_2\alpha_3$ plane will be $\alpha_2 + \delta\alpha_2$ and $\alpha_3 + \delta\alpha_3$. Substitution of these values in Eq. 3.8.11 and rearrangement of the terms yield

$$U + \delta U = \tfrac{1}{2}(S_{1,2} + S_{2,3})\alpha_2^2 + \tfrac{1}{2}(S_{2,3} + S_{3,4})\alpha_3^2 + (CS)_{2,3}\alpha_2\alpha_3 - M_2^*\alpha_2$$
$$- \; M_3^*\alpha_3 + (S_{1,2} + S_{2,3})\alpha_2\,\delta\alpha_2 + (S_{2,3} + S_{3,4})\alpha_3\,\delta\alpha_3$$
$$+ \; (CS)_{2,3}(\alpha_2\,\delta\alpha_3 + \alpha_3\,\delta\alpha_2) - M_2^*\,\delta\alpha_2 - M_3^*\,\delta\alpha_3$$
$$+ \; \tfrac{1}{2}(S_{1,2} + S_{2,3})\delta\alpha_2^2 + \tfrac{1}{2}(S_{2,3} + S_{3,4})\,\delta\alpha_3^2$$
$$+ \; (CS)_{2,3}\,\delta\alpha_2\,\delta\alpha_3 \qquad 3.8.12$$

The $U + V$ surface has a horizontal tangent plane, and the total potential is stationary when $U + V$ remains unchanged during the displacements as long as first-order small quantities alone are considered. Consequently the condition of a stationary value is

$$(S_{1,2} + S_{2,3})\alpha_2\,\delta\alpha_2 + (S_{2,3} + S_{3,4})\alpha_3\,\delta\alpha_3 + (CS)_{2,3}(\alpha_2\,\delta\alpha_3 + \alpha_3\,\delta\alpha_2)$$
$$- \; M_2^*\,\delta\alpha_2 - M_3^*\,\delta\alpha_3 = 0$$

Since $\delta\alpha_2$ and $\delta\alpha_3$ are entirely arbitrary, clearly the above equation is satisfied only if

$$(S_{1,2} + S_{2,3})\alpha_2 + (SC)_{2,3}\alpha_3 - M_2^* = 0 \qquad 3.8.13a$$
$$(CS)_{2,3}\alpha_2 + (S_{2,3} + S_{3,4})\alpha_3 - M_3^* = 0 \qquad 3.8.13b$$

Because of the minimal principle Eqs. 3.8.13 are also the conditions of the equilibrium of the system. The equilibrium is stable when the second-order small change in the total potential is positive. This means that for stability

$$\tfrac{1}{2}(S_{1,2} + S_{2,3})\delta\alpha_2^2 + \tfrac{1}{2}(S_{2,3} + S_{3,4})\delta\alpha_3^2 + (CS)_{2,3}\,\delta\alpha_2\,\delta\alpha_3 > 0 \quad 3.8.14$$

As Eq. 3.8.14 must hold true for any arbitrary choice of $\delta\alpha_2$ and $\delta\alpha_3$, one may choose $\delta\alpha_3 = 0$ and $\delta\alpha_2 \neq 0$. In this case the inequality reduces to

$$\tfrac{1}{2}(S_{1,2} + S_{2,3})\delta\alpha_2{}^2 > 0$$

This naturally means

$$S_{1,2} + S_{2,3} > 0 \qquad\qquad 3.8.15a$$

In a similar manner the assumption $\delta\alpha_2 = 0$ and $\delta\alpha_3 \neq 0$ leads to

$$S_{2,3} + S_{3,4} > 0 \qquad\qquad 3.8.15b$$

These two inequalities are not sufficient to insure the validity of Eq. 3.8.14 since a large negative value of $(CS)_{2,3}$ may prevail over the first two terms if they are small. A further condition can be obtained by rearranging Eq. 3.8.14. The following notation is now introduced:

$$\tfrac{1}{2}(S_{1,2} + S_{2,3}) = A \qquad\qquad (a)$$

$$(CS)_{2,3} = 2B \qquad\qquad (b)$$

$$\tfrac{1}{2}(S_{2,3} + S_{3,4}) = C \qquad\qquad (c)$$

This notation reduces Eq. 3.8.14 to

$$A\,\delta\alpha_2{}^2 + 2B\,\delta\alpha_2\,\delta\alpha_3 + C\,\delta\alpha_3{}^2 > 0$$

Moreover

$$\begin{aligned}
A\,\delta\alpha_2{}^2 + 2B\,\delta\alpha_2\,\delta\alpha_3 + C\,\delta\alpha_3{}^2 &= A[\delta\alpha_2{}^2 + 2(B/A)\,\delta\alpha_2\,\delta\alpha_3 + (C/A)\,\delta\alpha_3{}^2] \\
&= A\{[\delta\alpha_2 + (B/A)\,\delta\alpha_3]^2 - (B/A)^2\,\delta\alpha_3{}^2 \\
&\quad + (C/A)\,\delta\alpha_3{}^2\} \\
&= A\{[\delta\alpha_2 + (B/A)\,\delta\alpha_3]^2 \\
&\quad + [(AC - B^2)/A^2]\,\delta\alpha_3{}^2\} > 0 \qquad (d)
\end{aligned}$$

In the expression immediately preceding the unequal sign the first term inside the braces is always positive because it is the square of a real number. The multiplier of the expression in the second square bracket is also positive, but the expression in brackets is positive only if

$$AC - B^2 > 0 \qquad\qquad (e)$$

If this inequality is satisfied and the multiplier of the expression in braces, namely A, is also positive, the second-order change in the total potential is always positive. When $A > 0$, naturally $C > 0$ if Eq. e holds. But the conditions $A > 0$ and $C > 0$ are identical with Eqs. 3.8.15a and b because of Eqs. a and c. Equation e can be written in the form

$$(S_{1,2} + S_{2,3})(S_{2,3} + S_{3,4}) - (CS)^2{}_{2,3} > 0 \qquad\qquad 3.8.15c$$

Thus the stationary value defined by Eqs. 3.8.13 is a minimum if, and only if, Eqs. 3.8.15 are satisfied. This means that the equilibrium is stable under these conditions.

If Eqs. 3.8.15a and b are satisfied but Eq. 3.8.15c is replaced by

$$(S_{1,2} + S_{2,3})(S_{2,3} + S_{3,4}) - (CS)^2{}_{2,3} = 0 \qquad\qquad 3.8.16$$

it is easy to find displacements for which the second-order small change in the total potential vanishes.   In the next to the last member of Eq. *d* the second expression is identically zero because of Eq. 3.8.16, and the first is zero if

$$\delta\alpha_2 = -(B/A)\,\delta\alpha_3 \qquad\qquad (f)$$

This means that the value of the total potential does not change along the straight line defined by Eq. *f*, and therefore the equilibrium is neutral. Equation 3.8.16 can also be written in the form

$$(S_{1,2} + S_{2,3}) : (CS)_{2,3} = (CS)_{2,3} : (S_{2,3} + S_{3,4}) \qquad\qquad 3.8.17$$

This reveals that the equilibrium conditions expressed in Eqs. 3.8.13 are reduced to a single equation if

$$M_2 : M_3 = (S_{1,2} + S_{2,3}) : (CS)_{2,3} \qquad\qquad 3.8.17a$$

A single equation connecting two unknowns, namely α_2 and α_3, admits of an indefinitely large number of solutions.   When Eq. 3.8.17a is not satisfied, Eqs. 3.8.13a and b are contradictory and do not have any solution.

The equilibrium is unstable if in one of Eqs. 3.8.15a and b the unequal sign is replaced by an equal sign or if in Eq. 3.8.15c the left-hand side is smaller than zero.   This can be readily proved by an analysis paralleling the one just presented.

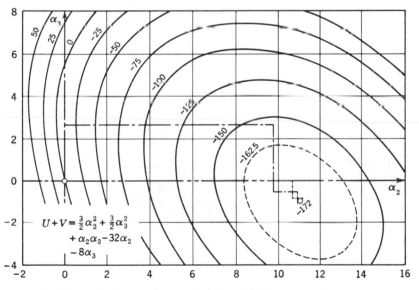

Fig. 3.8.5.   Contour map of stable total potential surface.   From author's paper in *J. Roy. Aero. Soc.*

Total Potential Surfaces

The three types of total potential surfaces are shown in Figs. 3.8.5 to 3.8.7. The figures are contour maps in which lines of constant $U + V$ are plotted just as curves of constant elevation are shown in topographic maps.

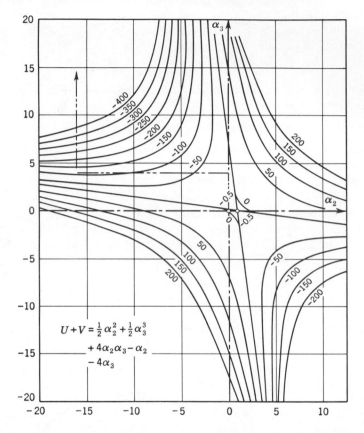

$$U+V = \tfrac{1}{2}\alpha_2^2 + \tfrac{1}{2}\alpha_3^3$$
$$+ 4\alpha_2\alpha_3 - \alpha_2$$
$$- 4\alpha_3$$

Fig. 3.8.6. Contour map of unstable total potential surface: From author's paper in *J. Roy. Aero. Soc.*

The contour lines were calculated by the methods discussed in Art. 2.3. The stable case represented by the first example shown in Fig. 3.8.5 corresponds to

$$S_{1,2} = S_{3,4} = 1, \qquad S_{2,3} = 2$$

$$C_{1,2} = C_{2,3} = C_{3,4} = 0.5$$

The surface is a paraboloid.

The second example shown in Fig. 3.8.6 characterizes the unstable condition.   The values of the constants are

$$S_{1,2} = S_{3,4} = 3, \qquad S_{2,3} = -2$$
$$C_{1,2} = C_{3,4} = 0.5, \qquad C_{2,3} = -2$$

The total potential surface is a hyperboloid with a single sheet.   Finally the limiting case between stability and instability corresponding to neutral equilibrium is shown in Fig. 3.8.7.   Its constants have the values

$$S_{1,2} = 10, \qquad S_{2,3} = -2, \qquad S_{3,4} = 4$$
$$C_{1,2} = C_{3,4} = 0.5, \qquad C_{2,3} = -2$$

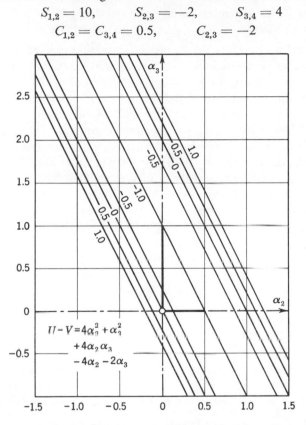

Fig. 3.8.7.   Contour map of neutral total potential surface. From author's paper in *J. Roy. Aero. Soc.*

In this case the total potential surface is a cylinder whose generators are horizontal.

The connection between the coefficients of the various terms in the expression for the total potential, the shape of the total potential surface, and the properties of the corresponding equilibrium can be discussed in a similar manner when the structure has more than two degrees of freedom of motion. This can best be done with the aid of the matrix calculus. No such more complex examples are given in this book because the large number of degrees of freedom necessitates the use of elaborate mathematics, and the mathematical complications do not contribute to a clarification of the physical ideas. Moreover, no simple geometric representation of the total potential surface is possible for systems having more than two degrees of freedom of motion.

Changes in Total Potential during a Typical Cycle of the Moment-Distribution Method

In a typical operation of the moment-distribution method all the joints of the truss except one are locked in the positions reached in the preceding operations, and the one joint that is being balanced is allowed to rotate. Its rotation can be calculated as the ratio of the total unbalanced moment acting on the joint divided by the stiffness ΣS of the joint. The balancing moment is distributed among the members common to the joint in the ratio of their stiffnesses. For instance, the share of member mn is $S_{mn}/\Sigma S$ times the balancing moment. Next the distributed moment is multiplied by the carry-over factor C, and the product is applied to the joint at the far end of the member. This carried-over moment is needed at the far end in order to maintain the tangent of the member unchanged.

In such a manner the subsystem of the truss made up of the joint to be balanced and of all the members attached to it, with the far ends of the members rigidly fixed in their positions, is completely balanced in the cycle of operations consisting of balancing, distribution, and carry-over. It is of interest to compare the cycle necessary to establish equilibrium in the subsystem by the moment-distribution method with the changes that are required to minimize the total potential of the subsystem.

Let it be assumed that α_2 and α_3 are the angles reached at a given stage of the moment-distribution process when it is applied to the system of Fig. 3.8.4. The value of the total potential of the system is that given by Eq. 3.8.11. The internal moment in beam 1–2 at joint 2 can be obtained from Eq. 3.8.1 if it is remembered that α_1 is zero because of the rigid fixation:

$$M_{2,1-2} = S_{1,2}\alpha_2$$

In the calculation of the internal moment acting in beam 2–3 at joint 2 the subscript 1 must be replaced by subscript 3 in Eq. 3.8.1:

$$M_{2,2-3} = (CS)_{2,3}\alpha_3 + S_{2,3}\alpha_2$$

As the external moment is $M_2{}^*$, the unbalanced moment remaining at joint 2 at this stage of the balancing process is

$$M_{2\text{unbal}} = M_2{}^* - (S_{1,2} + S_{2,3})\alpha_2 - (CS)_{2,3}\alpha_3 \qquad 3.8.18$$

When joint 2 is released, it rotates through an angle $\alpha_2{}^*$ whose magnitude can be determined from the minimal principle. According to the principle the change in the total potential must be zero when the angle $\alpha_2{}^*$ corresponding to the equilibrium is increased by a virtual rotation $\delta\alpha_2{}^*$. Substitution of $\alpha_2 + \alpha_2{}^* + \delta\alpha_2{}^*$ for α_2 in Eq. 3.8.11 results in

$$U + V = \tfrac{1}{2}(S_{1,2} + S_{2,3})(\alpha_2{}^2 + \alpha_2{}^{*2} + \delta\alpha_2{}^{*2} + 2\alpha_2\alpha_2{}^* + 2\alpha_2\,\delta\alpha_2{}^* + 2\alpha_2{}^*\,\delta\alpha_2{}^*)$$
$$+ \tfrac{1}{2}(S_{2,3} + S_{3,4})\,\alpha_3{}^2 + (CS)_{2,3}\alpha_3(\alpha_2 + \alpha_2{}^* + \delta\alpha_2{}^*)$$
$$- M_2{}^*(\alpha_2 + \alpha_2{}^* + \delta\alpha_2{}^*) - M_3{}^*\alpha_3 \qquad\qquad 3.8.19$$

In the application of the minimal principle, terms multiplied by the square of the virtual displacements are to be disregarded. Hence the change in the total potential is

$$\delta(U + V) = [(S_{1,2} + S_{2,3})(\alpha_2 + \alpha_2{}^*) + (SC)_{2,3}\alpha_3 - M_2{}^*]\,\delta\alpha_2{}^*$$

The equilibrium position is characterized by the requirement that $\delta(U + V)$ vanish for any arbitrary virtual rotation $\delta\alpha_2{}^*$. This means that the expression in brackets must vanish. Solution of the resulting equation for the rotation $\alpha_2{}^*$ required for equilibrium yields

$$\alpha_2{}^* - [M_2{}^* - (S_{1,2} + S_{2,3})\alpha_2 - (SC)_{2,3}\alpha_3]/(S_{1,2} + S_{2,3}) \qquad 3.8.20$$

In view of Eq. 3.8.18 this can also be written in the form

$$\alpha_2{}^* = M_{2\text{unbal}}/(S_{1,2} + S_{2,3}) \qquad\qquad 3.8.21$$

But this is exactly the amount of rotation obtained in the Hardy Cross moment-distribution method during the same operation, which shows that the application of the minimal principle leads to the same changes in the subsystem as the balancing process of the moment-distribution method.

On the other hand, it was shown in Arts. 2.2, 3.1, and 3.4 that equilibrium of a system always corresponds to a stationary value of the total potential, and the stationary value is a minimum when the system is stable. The stability depends on the sign of the second-order small changes in the total potential. In Eq. 3.8.19 the only term multiplied by $\delta\alpha_2{}^{*2}$ is $\tfrac{1}{2}(S_{1,2} + S_{2,3})$. Consequently the condition of stability for the system in which joint 2 alone is permitted to rotate, that is, for the subsystem defined earlier, is

$$(S_{1,2} + S_{2,3}) > 0 \qquad\qquad 3.8.22$$

This condition is always satisfied when the moment-distribution method is used in determining the stability of a system. When Eq. 3.8.22 is not satisfied, the subsystem is unstable even with the far ends of the members rigidly fixed. As in reality they are only elastically restrained, the subsystem, and thus the entire system, is obviously unstable, and there is no need to check the stability by moment distribution.   For this reason a moment-distribution process will be carried out only when Eq. 3.8.22, or its equivalent, is satisfied for every joint of the structure.

Graphic Representation of Changes in Total Potential during Moment Distribution

It can be concluded therefore that in every cycle of the moment distribution method the rotation of the joint that is being balanced is such that the total potential of the system attains the smallest value possible with the other joints locked.   Figure 3.8.5 illustrates this statement for the stable system of the first example.   The structure, together with the values of the stiffness coefficients and carry-over factors, is shown in Table 3.8.1.   The moment distribution is carried out in the table in two different ways, and the final moments obtained and listed under the double lines are identical within limits of accuracy sufficient for engineering purposes.   The rotation corresponding to each balancing operation was calculated and is indicated in Fig. 3.8.5.   For instance, in the first step of the second balancing succession joint 3 was balanced.   The stiffness of the joint was $\Sigma S = 3$, and, as the stiffness coefficients of beams 2–3 and 3–4 were 2 and 1, respectively, the external moment $M_3{}^* = 8$ had to be distributed in the ratio 2 to 1 between beams 2–3 and 3–4.   The rotation was

$$\alpha_3 = M_3{}^*/(\Sigma S)_3 = \tfrac{8}{3} = 2.66$$

The point $\alpha_3 = 2.66$ and $\alpha_2 = 0$ is on the dash-dotted line which indicates changes in the total potential during the second balancing succession.

It can be seen that the first balancing operation indeed changed the value of the total potential from zero at the origin ($\alpha_2 = \alpha_3 = 0$) to the lowest possible value that could be reached by proceeding from the origin of the coordinate system along the α_3 axis.   This value is -10.66.   In the second step α_3 was kept unchanged, and α_2 was changed until the lowest point of the total potential bowl was reached along the horizontal line passing through the point $\alpha_2 = 0$, $\alpha_3 = 2.66$.   In Table 3.8.1 the first balancing is followed by a carry-over of $\tfrac{1}{2}$ 5.33 = 2.66 to support 2.   This unbalanced moment was subtracted from the external moment $M_2{}^* = 32$ and the remainder distributed between beams 1–2 and 2–3 in the ratio of their stiffnesses, namely 1 to 2.   During the balancing operation the rotation at joint 2 was

$$\alpha_2 = 29.33/3 = 9.77$$

The value of the total potential reached can be calculated from the expression given in Fig. 3.8.5.   It is -153.   The dash-dotted line continues to alternate in the vertical and horizontal directions, and each segment leads to the lowest point of the valley on that particular straight-line section of the path.   The balancing process shown in Table 3.8.1 can be continued indefinitely, and the corresponding changes in the total potential make the dash-dotted line approach the bottom of the valley as closely as desired.

Table 3.8.1.   Moment Distribution—Stable Equilibrium

	$C = 0.5$ $S = 1$		$\Sigma S = 3$	$C = 0.5$ $S = 2$		$\Sigma S = 3$	$C = 0.5$ $S = 1$

$M_2^* = 32$ $M_3^* = 8$

1	No. of support	2		3		4

Support 2 balanced first:

5.33	10.66	21.33	10.66		
		−0.885	−1.77	−0.89	−0.445
0.147	0.295	0.590	0.295		
		−0.098	−0.197	−0.098	−0.049
0.016	0.032	0.066	0.033		
5.493	10.987	21.003	−0.022	−0.011	−0.006
			8.999	−0.999	−0.500

Support 3 balanced first:

		2.66	5.33	2.66	1.33
4.88	9.77	19.56	9.77		
		−3.26	−6.51	−3.26	−1.63
0.55	1.09	2.17	1.09		
		−0.363	−0.727	−0.36	−0.18
0.06	0.12	0.242	0.121		
5.49		−0.040	−0.080	−0.04	−0.02
	0.01	0.026	8.994	−1.00	−0.50
	10.99	20.995			

However, the line segments soon become so small that they cannot be indicated in the figure, and after the fifth balancing the line reaches the minimum of the total potential, namely -172, accurately enough for engineering purposes. In Table 3.8.1 eight balancing operations were carried out, but the changes in the moments after the fifth balancing were insignificant.

It is important to note that the double dash-dotted line representing the first balancing succession leads to the same minimum of the total potential, indicating that the order in which the joints are balanced does not influence the results obtained. This is always true of stable systems whose stationary value is a minimum. This minimum can be approached in various ways, but all operations lead to it because the total potential must decrease in each cycle of the moment distribution process.

The situation is entirely different when the equilibrium is unstable. This is illustrated by the second example. The balancing process is presented in Table 3.8.2, and the total potential surface is shown in

Table 3.8.2.   Moment Distribution—Unstable Equilibrium

	$C=0.5$		$C=-2$		$C=0.5$	
	$S=3$	$\Sigma S=1$	$S=-2$	$\Sigma S=1$	$S=3$	
		$M_2^*=1$		$M_3^*=4$		
1	No. of support	2		3		4

Support 2 balanced first:

	$M_2^* = 1$			$M_3^* = 4$		
1.5	3	-2	4	0		0

Support 3 balanced first:

		$M_2^* = 1$		$M_3^* = 4$		
		16	-8	12		6
-22.5	-45	30	-60			
		240	-120	180		90
-360	-720	480	-960			
.	.	.	.			.
.	.	.	.			.
.	.	.	.			.

Fig. 3.8.6. It can be seen from the figure that the stationary value of the total potential surface is a saddle point. Its coordinates are $\alpha_2 = 1$ and $\alpha_3 = 0$, and the value of the total potential is -0.5. In the NE–SW direction the surface rises abruptly, and in the NW–SE direction it descends just as rapidly.

It is of interest to mention that the saddle point is on one of the coordinate axes. This is not a coincidence, but was brought about purposely through a suitable choice of the external moments. As a consequence of the particular location of the saddle point, equilibrium was reached in Table 3.8.2 in a single operation in which a rotation $\alpha_2 = 1$ was undertaken. The path followed on the total potential surface was the short straight-line segment connecting the origin with the saddle point.

When the order of balancing was changed and the process was started at support 3, the unbalanced moments increased rapidly as may be seen from Table 3.8.2. The corresponding path on the total potential surface first reaches the deepest point along the α_3 axis and then plunges into the ravine in the second quadrant. The equilibrium of the saddle point can never be reached by this process.

Had any of the constants of the system been chosen in a slightly different manner, the stationary point would be off the α_2 axis, and no succession of balancing would ever lead to equilibrium. The balancing process would always diverge. It can be concluded therefore that no moment equilibrium can be reached, as a rule, when the moment-distribution process is applied to an unstable system. In particular cases, not likely to be encountered in practice, the unstable equilibrium may be attained through some particular succession of balancing operations, and the analyst may be misled into believing that the system is stable. The error can be discovered by changing the order in which the joints are balanced. The result is a divergent process indicating instability.

The limiting case between stability and instability is presented in Table 3.8.3 and Fig. 3.8.7. The total potential surface is a cylinder with horizontal generators. The bottom generator is the locus of points corresponding to equilibrium. It can be reached at either of two points depending on the choice of the first balancing operation. Two or more equilibrium patterns are naturally a sign of neutral equilibrium. They also indicate that the external moments satisfy Eq. 3.8.13. This is indeed the case in the example given since $S_{1,2} + S_{2,3} = 8$, $(CS)_{2,3} = 4$, and their ratio, 2, is the same as that of the two moments M_2 and M_3. When the moments are not chosen in such a manner, equilibrium is not possible, and the moment-distribution process diverges. The total potential surface is a cylinder whose generators are inclined. Consequently the total potential has no stationary value at all.

Table 3.8.3. Moment Distribution—Neutral Equilibrium

	$C = 0.5$ $S = 10$	$\Sigma S = 8$	$C = -2$ $S = -2$	$\Sigma S = 2$	$C = 0.5$ $S = 4$	
		$M_2^* = 4$		$M_3^* = 2$		
1	No. of support	2		3		4

Support 2 balanced first:

		$M_2^* = 4$		$M_3^* = 2$		
2.5		5	−1	2	0	0

Support 3 balanced at first:

		$M_2^* = 4$		$M_3^* = 2$		
0		0	4	−2	4	2

When the system has more than two degrees of freedom of motion the total potential surface cannot be shown in a contour map. However, the terminology of three-dimensional geometry can be kept, and it can be said that the total potential of a system having N degrees of freedom of motion is given in Eq. 3.8.10; the value of the total potential can be plotted in an $N + 1$-dimensional space; equilibrium corresponds to the stationary value at which an N-dimensional horizontal plane touches the $N + 1$-dimensional total potential surface; and this point corresponds to a minimum, or to a saddle point of the surface, depending on whether the equilibrium is stable or unstable. In the former case all balancing operations converge to equilibrium; in the latter case they diverge, as a rule, and only in exceptional cases can successions of balancing operations be devised that lead to unstable equilibrium. When the loading is exactly the critical one, the balancing operations diverge in general, but for particular sets of loads convergence is again possible. The final state reached, however, is not unique but depends on the order of balancing.

Numerical Evaluation of the Critical Load of a Truss

Because of its convergence properties, the moment-distribution method is most useful in determining the buckling loads of trusses. Details of the process are now given with the aid of an example. Figure 3.8.8 shows the framework proper, which is 60 in. long and 20 in. high, and the loading arm, also 60 in. long, attached to the framework by pin joints. The fixed end of the cantilever framework is supported in a statically determinate fashion by a pivot at the top and a roller at the bottom.

First it will be determined whether the framework is stable when the load W is 100 lb. The forces P caused by W in the members of the framework were calculated and listed in Table 3.8.4. The designation of the

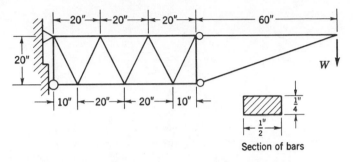

Fig. 3.8.8.   Framework used in numerical calculations.

individual members of the framework can be found in Fig. 3.8.9. The table also contains the values of $k = \sqrt{P/EI}$ and kL for each bar.

As a pin-ended column buckles when $kL = \pi$, and a rigidly fixed one at $kL = 2\pi$, a framework is obviously stable when the value $kL = \pi$ is not exceeded in any of its compression members, and obviously unstable if there is even one compression bar for which kL is greater than 2π. (The second statement is correct only if the framework is statically determinate. In a redundant framework a member can buckle without causing the entire framework to lose its load-carrying capacity.) Column 6 of Table 3.8.4 reveals that the framework has two compression members with kL values between the limits mentioned. Hence the stability of the framework must be determined by the convergence criterion of the moment distribution process.

If the ratio of gusset-plate length to total length of the bar is 0.09 for each of the members of the framework, the values of $S/(EI/L)$ and C can be read from Figs. 2.4.9 to 2.4.12. They are listed in columns 8 and 9 of Table 3.8.4. The stiffness of each member multiplied by the constant factor $20/EI$ is given in column 10. It can be seen that the sum of the stiffnesses is positive at each joint, indicating that the subsystems are all stable. If this were not true, a smaller load W would have to be assumed for the moment-distribution process since the entire structure would be unstable. (This statement holds for redundant frameworks also.)

Table 3.8.4. Stiffness Coefficients and Carry-Over Factors when $W = 100$ Lb

[1] Bar	[2] Length, L (in.)	[3] Load, P (lb)	[4] Type of Load†	[5] $k = \sqrt{P/EI}$, (in.$^{-1}$)	[6] kL	[7] $1/C$‡	[8] SL/EI	[9] C	[10] $20S/EI$
AB	20.00	550.0	T	0.28366	5.67	......	9.63	0.265	9.63
BC	20.00	450.0	T	0.25658	5.13	......	8.92	0.286	8.92
CD	20.00	350.0	T	0.22628	4.53	......	8.15	0.320	8.15
LK	10.00	600.00	C	0.29627	2.96	1.075	3.40	0.930	6.80
KJ	20.00	500.0	C	0.27046	5.41	−0.550	2.80	−1.818	−2.80
JH	20.00	400.0	C	0.24191	4.84	−0.120	−0.44	−8.330	−0.44
HG	10.00	300.0	C	0.20950	2.09	1.450	4.10	0.690	8.20
AL	20.00	0	...	0	0	......	5.078	0.5655	5.078
AK, BJ, CH	22.37	111.9	T	0.12795	2.86	......	6.38	0.426	5.70
KB, JC, HD	22.37	111.9	C	0.12795	2.86	1.130	3.48	0.885	3.11
DG	20.00	100.00	T	0.12095	2.42	......	6.02	0.456	6.02

† T signifies tension and C compression.
‡ Compression bars only.
$EI = 6835.5$ lb-in.² for all bars.

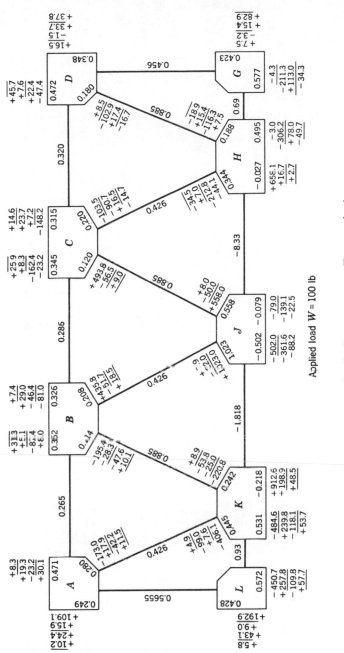

Fig. 3.8.9. Moment distribution converges. From author's paper
in *Transactions ASCE.*

Table 3.8.5.　Stiffness Coefficients and Carry-Over Factors when $W = 105$ lb

[1] Bar	[2] Length, L (in.)	[3] Load, P (lb)	[4] Type of Load†	[5] $k=\sqrt{P/EI}$ (in.$^{-1}$)	[6] kL	[7] $1/C$‡	[8] SL/EI	[9] C	[10] $20S/EI$
AB	20.00	577.5	T	0.29066	5.81	......	9.82	0.259	9.82
BC	20.00	472.5	T	0.26292	5.26	......	9.08	0.281	9.08
CD	20.00	367.5	T	0.23187	4.64	......	8.30	0.314	8.30
LK	10.00	630.0	C	0.30359	3.04	1.03	3.25	0.971	6.50
KJ	20.00	525.0	C	0.27714	5.54	-0.65	-3.65	-1.538	-3.65
JH	20.00	420.0	C	0.24788	4.96	-0.21	-0.85	-4.762	-0.85
HG	10.00	315.0	C	0.21467	2.15	1.42	4.00	0.704	8.00
AL	20.00	0	...	0	0	......	5.078	0.5655	5.078
AK, BJ, CH	22.37	117.5	T	0.13111	2.93	......	6.44	0.420	5.76
KB, JC, HD	22.37	117.5	C	0.13111	2.93	1.09	3.40	0.917	3.04
DG	20.00	105.0	T	0.12394	2.48	......	6.10	0.453	6.10

† T signifies tension and C compression.
‡ Compression bars only.
$EI = 6835.5$ lb-in.² for all bars.

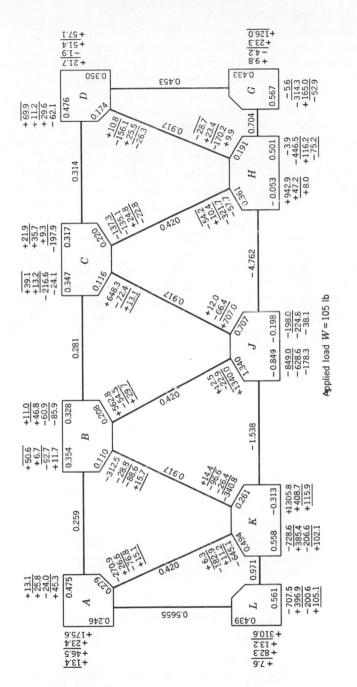

Fig. 3.8.10.   Moment distribution diverges.   From author's paper in *Transactions ASCE*

The distribution factors are written inside the polygons at each joint, and the carry-over factors are shown in the middle of each bar in Fig. 3.8.9. An external moment of 1000 in.-lb is assumed to act on joint J, the joint is balanced, and the balancing moments are distributed. The carry-over moments are calculated and applied to the far ends of the bars. Next joint K is balanced, and the procedure is continued in the usual manner except that joint J is not balanced again, but the carry-back moments are permitted to accumulate at J. When all the joints but J are balanced, the total accumulated moments are added up at joint J. In the balancing process shown in Fig. 3.8.9 the sum is 553 in.-lb. Since this is less than the 1000 in.-lb applied at the outset, one can conclude that a repetition of all the operations would reduce the unbalance even further, namely to (553/1000)553. Clearly the unbalanced moments over the whole framework can be reduced to as small values as desired if the necessary number of operations are undertaken. Hence the process converges, and by the convergence criterion the framework is stable when $W = 100$ lb.

In order to learn where the stable region ends, the calculations are now repeated for $W = 105$ lb. The constants of the problem are calculated in Table 3.8.5, and the balancing process is carried out in Fig. 3.8.10. The carry-back moments at joint J now add up to 1409 in.-lb, and, since this is larger than the initial 1000 in.-lb, the process diverges, and the structure is unstable.

It can be stated therefore that the buckling load of the framework is $W = 102.5$ lb, and the error is less than ± 2.5 lb. This accuracy is satisfactory for engineering purposes, but it is possible to improve on it by undertaking a few more moment distributions for loads between 100 and 105 lb. It should be noted that both the convergence at 100 lb and the divergence at 105 lb were very pronounced. The rapid and well-defined change from convergence to divergence made it possible to obtain the buckling load accurately enough with no great effort.

Finally it is of some interest to calculate the maximum end-fixity coefficient in the truss. Since in member KJ the average value of kL calculated from Tables 3.8.4 and 3.8.5 is about 5.47, and this is the maximum value of kL in the truss, the maximum end-fixity coefficient is $5.47/\pi^2 = 3.02$. This shows that the rigid connection between the members of the framework and the effect of the gusset plates contribute considerably to the strength of the framework.

3.9   Inelastic Buckling

The Reduced Modulus Load

All the stability investigations presented heretofore were based on the tacit assumption that the material followed Hooke's law during the buckling process. Obviously this can be true only for comparatively slender structural elements because the Euler formula derived in Art. 3.1 and presented in Eq. 3.1.31 gives stresses exceeding by far the elastic limit, and even the yield stress, of the structural materials when the length L of the column is sufficiently small. The resistance of a column to

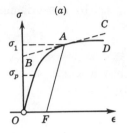

Compressive stress-strain
diagram

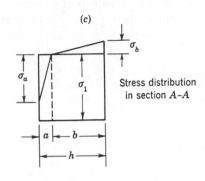

Stress distribution
in section A-A

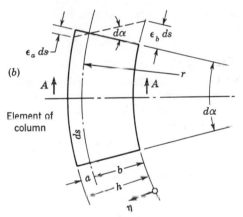

Element of
column

Fig. 3.9.1.   Inelastic buckling.

bending decreases when the compressive stress increases above the proportional limit, and this resistance must be known when the buckling load of the column is calculated.

Let it be assumed that buckling takes place at the stress σ_1 indicated in Fig. 3.9.1a. According to the definition of stability this statement has the following meaning: A perfectly straight column is subjected to a perfectly centered compressive load P. The stability of the column is investigated by imagining the midpoint of the column deflected slightly in the transverse direction and then released. If the stress P/A is different from the critical stress $\sigma_{cr} = \sigma_1$, the column is not in equilibrium in the deflected position without some transverse force.   If the load P gives rise to a stress equal to σ_1, the column is in equilibrium in the deflected position without a transverse force.

In the deflected position the fibers of the column are strained in a non-uniform manner.   If at any section the curvature is $1/r$ and the neutral axis† of the cross section is at a distance b from the most highly compressed fiber as shown in Fig. 3.9.1b, the angle $d\alpha$ subtended by two originally parallel normal sections of the column ds apart is

$$d\alpha = ds/r \qquad\qquad 3.9.1$$

after bending takes place.   The fiber at $\eta = 0$ had a length ds before bending and is $ds - \varepsilon_b\, ds$ long after bending.   From geometry

$$\varepsilon_b\, ds = b\, d\alpha = b\, ds/r$$

Similarly at $\eta = h$ the length ds changes into $ds + \varepsilon_a\, ds$ with

$$\varepsilon_a\, ds = a\, d\alpha = a\, ds/r$$

The strains in the two extreme fibers are therefore

$$\varepsilon_a = a/r, \qquad \varepsilon_b = b/r \qquad\qquad 3.9.2$$

The changes in stress can be calculated from the changes in strain when the stress-strain diagram is known.   When the compressive strain increases because of bending, as it does at $\eta = 0$, the stress follows curve AD in Fig. 3.9.1a.   When the deformations are small, the curve can be replaced with good accuracy by the tangent BAC to the curve at A.   The constant slope $d\sigma/d\varepsilon$ of this straight line is known as the tangent modulus E_t of the material corresponding to the stress σ_1.   With the aid of the tangent modulus the stress caused at $\eta = 0$ by the bending can be expressed as

$$\sigma_b = E_t\varepsilon_b = E_t b/r \qquad\qquad 3.9.3$$

At $\eta = h$ the initial compression is relieved by the bending.   When the compressive strain decreases, only the elastic part of the total deformations can be regained, and the stress decreases according to the straight line AF rather than according to the stress-strain curve AO.   As the slope of line AF is E, the original Young's modulus, the stress is

$$\sigma_a = E\varepsilon_a = Ea/r \qquad\qquad 3.9.4$$

During the application of the small transverse force the end compression P must remain constant.   Hence the total added compression caused by

---

† The neutral axis of the cross section is the line of intersection of the cross-sectional plane with the locus of those axial fibers of the column whose length remains unchanged during bending.

bending must be equal to the total tension (or relief in compression) due to bending. If the cross section of the column is of unit width, the total added compression is $\frac{1}{2}\sigma_b b$ while the relief in compression is $\frac{1}{2}\sigma_a a$. The force P remains unchanged if

$$\sigma_a a = \sigma_b b \qquad\qquad 3.9.5$$

from which it follows after substitutions that

$$Ea^2 = E_t b^2 \qquad\qquad 3.9.6$$

Since the total depth of the beam h is

$$h = a + b \qquad\qquad 3.9.7$$

the location of the neutral axis, when a small amount of bending is superimposed on a uniform state of compression at a stress σ_1, can be expressed in terms of the tangent modulus value corresponding to σ_1:

$$a = [\sqrt{E_t}/(\sqrt{E} + \sqrt{E_t})]h, \qquad b = [\sqrt{E}/(\sqrt{E} + \sqrt{E_t})]h \qquad 3.9.8$$

The couple of the forces imposed by bending is

$$M = \tfrac{1}{3}(\sigma_a a^2 + \sigma_b b^2) = (1/3r)(Ea^3 + E_t b^3) \qquad 3.9.9$$

Substitutions yield

$$M = \frac{EE_t}{(\sqrt{E} + \sqrt{E_t})^2} \frac{h^3}{3r} \qquad\qquad 3.9.10$$

Since the moment of inertia of the section is

$$I = \tfrac{1}{12}h^3 \qquad\qquad 3.9.11$$

the equation connecting the moment M with the curvature $1/r$ which it causes can be written as

$$M = E_{\text{red}} I/r \qquad\qquad 3.9.12$$

where E_{red} is the reduced modulus

$$E_{\text{red}} = 4EE_t/(\sqrt{E} + \sqrt{E_t})^2 \qquad\qquad 3.9.13$$

The reduced modulus is also known as the effective, the double, or the von Kármán modulus. In the elastic range where $E_t = E$, the formula yields $E_{\text{red}} = E$ as can be expected.

Equation 3.9.12 shows that the curvature $1/r$ caused by a moment M in a bar which is in an initial state of uniform compression is

$$1/r = M/E_{\text{red}}I \qquad\qquad 3.9.14$$

Therefore the bending rigidity under such conditions is $E_{\text{red}}I$ rather than EI. If the bending rigidity of the perfectly elastic column is replaced by this reduced bending rigidity in Eq. 3.1.11, and the equation is solved as was done in Art. 3.1, the Euler load of Eq. 3.1.29 is modified to read

$$P_{cr \cdot r} = \pi^2 E_{red} I / L^2 \qquad\qquad 3.9.15$$

and the Euler stress of Eq. 3.1.31 is replaced by

$$\sigma_{cr \cdot r} = \pi^2 E_{red} / (L/\rho)^2 \qquad\qquad 3.9.16$$

where the subscript cr · r indicates that the critical value of the load or stress was calculated with the aid of the reduced modulus.

Stress-Strain Relation and Column Curve

The critical stress of a short column, which buckles at a value of the compressive stress exceeding the proportional limit of the material, can be calculated therefore from a formula very similar to the Euler formula. The value of E_{red} can be readily found for each stress value when the tangent modulus E_t is known. This in turn can be determined if an accurate stress-strain curve for the material is available. Unfortunately the experimental determination of sufficiently accurate stress-strain curves is not an easy task, particularly if they are desired for thin sheet in compression. Data derived from tests carried out with heavy bars may differ considerably from those obtained with thin sheet or tubing because the stress-strain relationship depends greatly on the cold work to which the material was subjected prior to the test. In spite of these difficulties reliable stress-strain curves have been published for all the materials used in aircraft construction, although similar data for steels in general use in structural work and in machine design are often still lacking. Similarly plots of the tangent modulus against compressive stress can be had for a number of materials. Figure 3.9.2 shows such a plot for aluminum alloy 24S-T. Values of the reduced modulus, computed from Eq. 3.9.13, are also given.

When such an E_{red} curve is available, the buckling stress can be easily obtained with the aid of a simple graphic construction. With the particular value of the slenderness ratio L/ρ, the quantity $(1/\pi^2)(L/\rho)^2$ is computed, and the straight line $f(\sigma) = (1/\pi^2)(L/\rho)^2 \sigma$ is plotted. The intersection point of the straight line with the E_{red} curve yields the buckling stress $\sigma_{cr \cdot r}$. This procedure is simply a graphic solution of Eq. 3.9.16. The construction is indicated in Fig. 3.9.2 where the critical stress obtained is slightly above 40,000 psi. If one does not want to use the graphic construction, he can calculate the critical stress numerically by a trial-and-error process.

When critical stress values are needed for columns of different length made of the same material, it is convenient to draw the so-called column curve. Corresponding values of σ and E_{red} are read off the E_{red} curve and entered in Eq. 3.9.16 which is then solved for L/ρ. In this inverse manner corresponding values of $\sigma_{\text{cr·r}}$ and L/ρ are obtained, which are

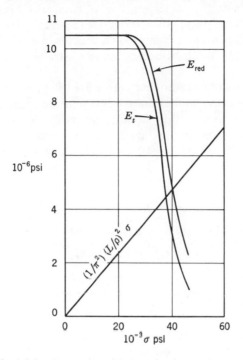

Fig. 3.9.2. Properties of 24S-T aluminum-alloy extrusions.

then plotted in a chart such as the one presented as Fig. 3.9.3. As $E_{\text{red}} = E$ when the stress is below the proportional limit, the right-hand portion of the curve is simply the Euler hyperbola. Such a column curve is very practical in structural analysis because the critical stress of any column of uniform section can be taken from it as soon as the slenderness ratio is computed. The graph is even valid for columns having elastically or rigidly fixed ends; only the geometric slenderness ratio has to be replaced in such a case by the effective slenderness ratio corresponding to the distance between inflection points.

The Effect of the Shape of the Cross Section

The E_{red} curve of Fig. 3.9.2 was derived for a column of solid rectangular section and is not strictly applicable to any other section. Fortunately the cross-sectional shape does not have much effect on the value of E_{red}. To prove this statement, the reduced modulus will be calculated for the idealized I section which represents the limiting condition when all the structural material is spaced at the greatest possible distance from the neutral axis. It is assumed that the section consists of two flanges a distance h apart, in each of which material of an area $A/2$ is concentrated. The web is infinitely thin and possesses no area; nevertheless it is perfectly rigid in shear. The uniform compressive stress σ_1 prevails in the flanges before the column is disturbed from its straight-line equilibrium configuration.

When the disturbance causes a curvature $1/r$ to appear, the increase σ_b in the compressive stress on the concave side must be equal to the relief σ_a in the stress on the convex side, because otherwise an unbalanced axial force would result. This means that the neutral axis must be located in such a manner as to yield elongations in the flanges in the ratio E_t/E. One has therefore

$$a = [E_t/(E + E_t)]h \qquad\qquad 3.9.17$$

$$b = [E/(E + E_t)]h \qquad\qquad 3.9.18$$

The strains caused by bending are then

$$\varepsilon_a = [E_t/(E + E_t)](h/r) \qquad\qquad 3.9.19$$

$$\varepsilon_b = [E/(E + E_t)](h/r) \qquad\qquad 3.9.20$$

and the stresses

$$\sigma_a = \sigma_b = [EE_t/(E + E_t)](h/r) \qquad\qquad 3.9.21$$

The bending moment corresponding to the curvature $1/r$ is the product of the force in the flange and the distance between the flanges. It can be written as

$$M = E'_{red}I/r \qquad\qquad 3.9.22$$

where

$$E'_{red} = 2EE_t/(E + E_t) \qquad\qquad 3.9.23$$

and the moment of inertia of the idealized I section is

$$I = Ah^2/4 \qquad\qquad 3.9.24$$

These formulas should be used for a column of idealized I section. For real I, channel, and other extruded, bent, or built-up structural shapes some value between E_{red} of Eq. 3.9.13 and E'_{red} of Eq. 3.9.23 is appropriate. Although there is no theoretical difficulty in deriving the proper value of the reduced modulus for any shape, the effort is not warranted because the differences in the values obtained are small. For instance in Fig. 3.9.2 the curve of E'_{red} would always plot below the E_{red} curve, but the difference would not be visible in the figure when E_{red} (or E'_{red}) is greater than 5×10^6 psi, that is, when σ is smaller than 40,000 psi. It is advisable therefore to use E_{red} for solid sections and E'_{red} for thin extrusions and bent or built-up sections.

Buckling during the Loading Process and the Tangent Modulus Load

How can these theoretical results be checked by experiment? The usual procedure is to place a column in a compression testing machine and depress the upper end of the column until the column becomes visibly and permanently bent. The maximum load read on the testing machine during the process is taken for the buckling load.

There is no obvious reason to believe that this maximum load is identical with the theoretical buckling load. It is certainly not the maximum load at which a small disturbance carries the originally perfectly straight column into a neighboring slightly curved equilibrium position. The maximum load measured depends on the initial deviations of the column from the perfect straight-line shape as was shown in Art. 3.2. Moreover the speed of loading also has an effect; it is known that a column can carry a load far in excess of its buckling load if the load is applied for only a short time. If this were not true, one could not drive a nail. It can be shown that the elasticity and the inertia of the testing machine also influence the column test.

Nevertheless the column test is a useful practical process for the determination of the load-carrying capacity of columns. In one respect it represents better the conditions under which a structural element fails than the theoretical criterion of stability; usually a structure fails while an excessively large load is being applied to it, and not under a steady load as a consequence of a transverse disturbance.

Adoption of the idea of buckling during the loading process as a criterion of column failure opens up an entirely new field of investigation. Great difficulties are encountered if dynamic effects are included in the analysis. When the load is applied and the motion of the column takes place so slowly that the mass inertia of the column can be disregarded, the following simple considerations lead to results of some usefulness:

It has just been proved that at a load less than the reduced modulus load of Eq. 3.9.15 an equilibrium state cannot be reached if the column is deflected slightly while the compressive load is maintained constant. But the bent configuration can correspond to equilibrium if simultaneously with the transverse motion the compressive load is increased by a small amount. All that is required is to insure that the increment in average compressive stress is greater than the relief in compressive stress in the extreme fiber on the convex side caused by bending. Then the compressive stresses increase all over the cross section during the process of lateral deflection. But, when the compression increases, the increment in stress is equal to the product of the increment in strain and the tangent modulus. Hence the same tangent modulus governs the bending deformations over the entire cross section and the neutral axis has to pass through the centroid of the section as in perfectly elastic bending.

The only difference between this type of bending and elastic bending is that the rate of increase of the stress with the strain is E_t rather than E. Thus in Eq. 3.1.11 the modulus E must be replaced by E_t, and not by E_{red}. The buckling load derived from this modified equation is

$$P_{cr \cdot t} = \pi^2 E_t I / L^2 \qquad\qquad 3.9.25$$

This result means that during the loading process of a perfect column a new equilibrium, without lateral loads, can be reached in a slightly bent configuration if the load increases rapidly enough to cancel all relief in the compression on the convex side, provided the end load is the tangent modulus load $P_{cr \cdot t}$ as given in Eq. 3.9.25. As E_t is always less than E_{red}, bending of the column can take place below the reduced modulus load $P_{cr \cdot r}$. Naturally bending is also possible anywhere between $P_{cr \cdot t}$ and $P_{cr \cdot r}$ if the increment in load is of suitable magnitude.

When the column starts to bend at $P_{cr \cdot t}$, the load is still on the increase. But more detailed investigations have shown that the maximum load that can be reached during such a loading process is not substantially greater than $P_{cr \cdot t}$.

The following results are now available for determining the load at which a column is likely to fail: According to the classical theory of stability a short column is in a neutral state of equilibrium under the reduced modulus $P_{cr \cdot r}$ if at the outset it was perfectly straight and was centrally loaded. Any deviation from the perfect condition results in a reduction in the load that can be reached in a quasistatic loading process. When a lateral disturbance is accompanied by a suitable simultaneous increase in the end load, deviations from the initial perfectly straight configuration can occur at a lower load, namely at any load value between $P_{cr \cdot r}$ and $P_{cr \cdot t}$. The maximum load that can be reached in this type of loading process is slightly higher than the load at which lateral deflections become possible, but again imperfections in the column and the load application tend to reduce the failing load below the theoretical maximum value.

These considerations have been somewhat crude as the stress-strain relationship was replaced by the tangent and because all inertia effects were neglected. Nevertheless one would expect that column tests would generally yield results plotting between the column curve calculated with the reduced modulus and that obtained with the tangent modulus. As the difference between these curves is not large, and the tangent modulus curve always represents the more conservative estimate, it is recommended that the latter, and thus Eq. 3.9.25, be used in structural design. Naturally this equation can also be modified by dividing its two members by the cross-sectional area. The result is

$$\sigma_{cr \cdot t} = \pi^2 E_t / (L/\rho)^2 \qquad \qquad 3.9.26$$

Values calculated from this equation for 24S-T aluminum alloy are also plotted in Fig. 3.9.3.

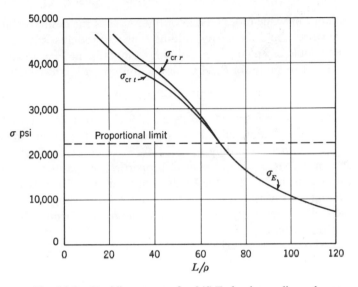

Fig. 3.9.3. Buckling stresses for 24S-T aluminum alloy columns.

Empirical Column Curves

In addition to the rational approach to inelastic column failure just presented, a purely empirical approach is also possible and has, in fact, been used widely. One can test columns of a given shape and given material, and a column curve, similar to the one shown in Fig. 3.9.3, can be established if the tests cover a sufficiently wide range of slenderness ratios. Such curves can be used conveniently in design, but it has been customary to represent them by empirical formulas. For instance, the buckling stress of SAE-1025 mild steel columns is often computed in the airplane industry in the United States from the empirical relationship

$$\sigma_{cr} = 36,000 - 1.172(L/\rho)^2 \qquad \qquad 3.9.27$$

It is of interest to note that an empirical column curve implies a uniquely determined stress-strain relationship if it can be decided which of

Eqs. 3.9.16 and 3.9.26 represents correctly the failing stress of the column. If the tangent modulus stress is chosen, the right-hand members of Eqs. 3.9.26 and 3.9.27 must be equal:

$$\pi^2 E_t/(L/\rho)^2 = 36{,}000 - 1.172(L/\rho)^2$$

Multiplication of both members of the equation by $(L/\rho)^2$ leads to a quadratic in $(L/\rho)^2$ whose solution is

$$(L/\rho)^2 = (1/2.344)[36{,}000 - \sqrt{36{,}000^2 - 4.688\pi^2 E_t}]$$

Substitution of this value in Eq. 3.9.26 or 3.9.27 yields the desired relation between the stress σ and the corresponding tangent modulus value E_t.

Another example of an empirical column formula is the one prescribed by the American Institute for Steel Construction according to which

$$\sigma_{cr} = 17{,}000 - 0.485(L/\rho)^2 \qquad\qquad 3.9.28$$

This equation is of the same form as Eq. 3.9.27, but it does not apply to exactly the same material. It also incorporates a safety factor. Columns proportioned according to Eq. 3.9.28 are considered safe even if the usual variations occur in material properties, in the dimensions of the structural elements, and in the loads, while Eq. 3.9.27 represents the actual buckling load of the column without any safety factor.

It is also well to remember that an empirical column formula is valid, as a rule, only in a limited range of the slenderness ratios. Equation 3.9.27 should be used only when L/ρ is smaller than 124, and Eq. 3.9.28 when it is smaller than 120.

Yielding and Local Buckling

When Eq. 3.9.16 is used to predict the buckling load of columns, the yield stress of the material is taken, as a rule, for the upper limit of the buckling stress. It can be seen from Fig. 3.9.3 that the theoretical critical stress increases quite rapidly when the slenderness ratio becomes very small. These high stresses cannot be realized in structural elements since a very short column can fail through yielding, even though the load acting on it is smaller than the buckling load. Similarly elastic or inelastic buckling of the thin walls of built-up, extruded, or bent sections may impose an upper limit on the buckling stress of columns. All these effects must be investigated theoretically or experimentally before the load-carrying capacity of a column can be determined with finality.

The Effect of Initial Stresses on the Buckling Stress

The load-carrying capacity of a column can be impaired by another condition, namely by initial, or locked-in, stresses which are a consequence of the process of manufacture. The region in the neighborhood of the intersection of flange and web in a hot-rolled I beam cools less rapidly than the free edges of the flange. When the longitudinal fibers along the edges contract during the cooling process and at the same time reach a completely elastic state, the fibers in the middle of the flange, near the web attachment, are still warm enough to follow the contraction of the edges plastically. When finally this middle portion also cools off, it has to contract further. But this contraction is inhibited by the fibers in the neighborhood of the edges of the flange. Because of the inability of the edge fibers to shorten substantially, the fibers in the middle portion cannot shorten as much as the drop in temperature requires; they remain

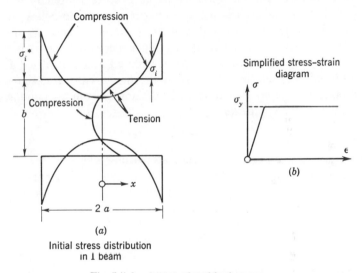

(a)
Initial stress distribution
in I beam

Fig. 3.9.4.   Effect of residual stresses.

a little longer than they would be in their natural state at room temperature as a consequence of the pull exerted on them by the fibers near the edges. The reaction of this pull is a compression applied to the edge fibers. Hence a state of initial, or internal, stress remains in the I beam as a consequence of the non-uniform cooling. The compressive stresses in the edge fibers and the tensile stresses near the junction of the flange and web are shown in Fig. 3.9.4a. The figure also shows the initial stresses in the web whose middle portion cools a little more rapidly than the regions near the flanges.

The non-uniform heating and cooling accompanying welding also causes locked-in, or initial, stresses to remain in a welded structure, and so do the cold-forming operations in the shapes so manufactured. Cold forming is possible only by producing deformations exceeding the limit of proportionality of the material; after cold forming the residual elastic stresses maintain the structural element in a shape different from the shape of the element in its original unstressed state.   As a rule bars manufactured in the mill are not perfectly straight.   They are straightened while they are cold, and thus the final product must have some residual, or locked-in, stresses.   These stresses are often designated as initial stresses.   By this term it is meant that the stresses are in the material initially, that is, before external loads are applied to the structure.

Simple reasoning will show that initial stresses reduce the buckling load of a column if the sum of the initial stress and the buckling stress exceeds the elastic limit of the material.   For the sake of simplicity it will be assumed that the material is perfectly linear and elastic below the yield stress σ_y, and it can deform indefinitely at the yield stress as shown in Fig. 3.9.4b.   Moreover, the load-carrying capacity of the web will be neglected, that is the section will be considered as an idealized I.   The maximum compressive initial stress in the flange will be designated by σ_i^* and the initial stress σ_1 in a fiber at a distance x from the middle will be represented by the expression

$$\sigma_i = -\tfrac{1}{2}\sigma_i^*[1 - 3(x/a)^2] \qquad\qquad 3.9.29$$

In this equation the positive sign indicates compression.

Let it now be assumed that the column buckles under a uniformly distributed applied stress σ_{cr}.   If buckling during the loading process is of interest, it can be assumed that the average stress in the column increases sufficiently at the time the column deflects laterally to insure that no strain reversal takes place anywhere in the section.   Under these conditions bending increases the stress in accordance with Hooke's law in every section where the sum of the initial stress and the buckling stress is less than the yield stress.   In the part of the section where algebraic addition of σ_i and σ_{cr} results in a stress greater than σ_y, naturally the stress is σ_y; according to the assumptions made the stress cannot exceed the value σ_y. Consequently, in the region where the yield stress prevails, bending cannot change the value of the stress; in fact the material does not present any resistance to increases in the deformations.

This shows that resistance to bending is only offered by that portion of the section in which the sum of the initial stress and the buckling stress is less than the yield stress of the material.   It is easy to calculate the moment of inertia of this section when the stresses are given numerically.   As an example, let it be assumed that $\sigma_y = 40{,}000$ psi, $\sigma_i^* = 30{,}000$ psi, $\sigma_{cr} = 20{,}000$ psi, $a = b = 1$ in., and $t = 0.05$ in.   At the point where $\sigma_i = 20{,}000$ psi, the sum of the initial stress and the buckling stress is equal to the yield stress.   From Eq. 3.9.29

$$(x/a)^2 = \tfrac{1}{3}[(2\sigma_i/\sigma_i^*) + 1] = \tfrac{7}{9}$$

and thus

$$x/a = 0.883$$

The area of the flange beyond $x = 0.883a$ is ineffective at buckling. Hence the effective moment of inertia of the section is only 88.3 per cent of that of the same section when initial stresses are absent, and the buckling load of the section with initial stress is 88.3 per cent of the buckling load without initial stress, provided the critical stress has the same value.   The slenderness ratio at which the buckling stress is 20,000 psi can be calculated from the Euler formula:

$$(L/\rho)^2 = \pi^2 E/\sigma_{\mathrm{cr}} = \pi^2 \times 29 \times 10^6/20 \times 10^3 = 14{,}300$$

and thus

$$L/\rho = 119.6$$

At this value the load carried by the column at buckling is 2000 lb in the absence of initial stress, and 1766 lb when the initial stresses indicated in Fig. 3.9.4 are present.

PART FOUR

Complementary Energy and Least-work Methods

4.1 The Minimum of the Complementary Energy

In this chapter the second minimal principle, or variational principle, will be derived from the first one. This second principle is complementary to the first and thus resembles it in some respects while differing from it in others. The second minimal principle has attained considerable importance in structural analysis because many problems of practical interest can be solved with its aid more conveniently than by any other method.

Let it be assumed that a structure consisting of rods in tension, columns, beams, beam columns, and bars subjected to torsion, resting on arbitrary statically determinate or indeterminate supports, some of which may be on yielding foundations, is being loaded in any arbitrarily prescribed manner. The quantity

$$dU' = dV \int_0^{\sigma_1} \varepsilon \, d\sigma + dV \int_0^{\tau_1} \gamma \, d\tau \qquad 4.1.1$$

is denoted as the complementary energy stored in the volume element dV of the structure at the time when the normal stress σ_1 and the shearing stress τ_1 are reached in the element. The total complementary energy can be obtained by integration over the volume of each member of the structure and summation of the definite integrals over all the members:

$$U' = \sum \int_V \left(\int_0^{\sigma_1} \varepsilon \, d\sigma + \int_0^{\tau_1} \gamma \, d\tau \right) dV \qquad 4.1.2$$

The symbols σ_1 and τ_1 denote the values of the normal and shearing stress, respectively, in each volume element reached simultaneously at a particular stage of the loading. In general, they are different for each volume element.

The term complementary energy is easily explained with the aid of Fig. 4.1.1. If curve OC represents the stress-strain relationship for the

332

material of a volume element dV subjected to uniaxial tension, the strain energy dU stored in the element when the tensile stress reaches the value σ_1 and the tensile strain the value ε_1 is defined as

$$dU = dV \int_0^{\varepsilon_1} \sigma \, d\varepsilon$$

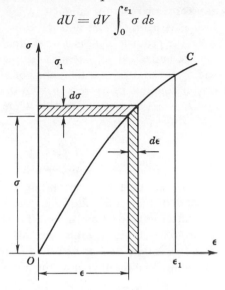

Fig. 4.1.1.   Stress-strain curve.

This was stated in Art. 2.1.   The quantity dU is dV times the area bounded by the stress-strain curve, the vertical through ε_1, and the axis of abscissas. The area complementing this area to a rectangle times dV is known as the complementary energy.   The complementary area is bounded by the stress-strain curve, the horizontal through σ_1, and the axis of ordinates.

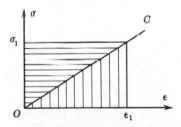

Fig. 4.1.2.   Linear stress-strain relation.

The corresponding complementary energy is expressed by the right side of Eq. 4.1.1 if the shear strain γ is set equal to zero.

When the stress-strain relationship is linear as indicated in Fig. 4.1.2,

$$\int_0^{\sigma_1} \varepsilon \, d\sigma = \int_0^{\varepsilon_1} \sigma \, d\varepsilon$$

because the area shaded horizontally is equal to the area shaded vertically. Under such conditions the complementary energy is equal numerically to the strain energy.

If the nth reaction force or reaction moment is denoted by R_n, and if the support at which it is acting yields and develops a displacement r_n in the direction of R_n, the product $R_n r_n$ is a work quantity needed in the development of the complementary-energy theorem. When several supports are on yielding foundations, or the points of attachment of some reactions are displaced for any other reason, the sum $\Sigma R_n r_n$ must be calculated for the entire structure. The quantity $-\Sigma R_n r_n$ can be referred to as the potential V' of the reactions:

$$V' = -\Sigma R_n r_n \qquad\qquad 4.1.3$$

As in general there are many different stress distributions which, together with the given external loads and the appropriate reactions, satisfy the conditions of static equilibrium, the following question can be raised: Does the system of stresses and reactions σ_1, τ_1, R_n corresponding to the actual state of equilibrium have any properties that other equilibrium systems of stresses and reactions do not have? In order to answer this question it is advantageous to determine how the quantity

$$U' - \Sigma R_n r_n = U' + V'$$

as calculated for the true state of equilibrium, changes if the values σ_1 and τ_1 of the stresses are varied while the state of deformation is maintained unchanged. During this variation the reactions R_n also undergo changes, but it is stipulated that the new stresses, together with the originally given external forces P_n, again constitute a system of equilibrium. This stipulation, of course, restricts the choice of the variation of the stresses, but it is not the intention here to consider variations of a more general type.

The variation to be undertaken is otherwise perfectly arbitrary and is not restricted at all by considerations of deformations. As a matter of fact, if one calculated the deformations corresponding to the varied state of stress, he would find that the deformed structural elements do not fit together, and under the modified state of stress the structure develops cracks. The changes in stress, as a rule, are not possible changes for the actual structure, unless the material properties or the cross-sectional dimensions of the structural elements are altered simultaneously.

This is, however, beside the point. The changes in stress need not be real changes. All that is required here is to find out how the complementary energy would change if in the calculator's imagination the state of stress were varied. Evaluation of the changes in the complementary energy during a change from the actual state of stress to a fictitious one will reveal important properties of the complementary energy of the system.

When R_n is replaced by $R_n + \delta R_n$ and the state of deformation, and consequently r_n, is maintained unchanged, the change in the potential of the reaction can be calculated from Eq. 4.1.3:

$$\delta V' = -\Sigma(R_n + \delta R_n)r_n + \Sigma R_n r_n$$
$$= -\Sigma r_n\, \delta R_n$$

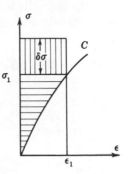

Fig. 4.1.3. Increment in complementary energy.

If at the same time the stress σ_1 in an elementary volume is increased to $\sigma_1 + \delta\sigma$ without a corresponding change in the strain, the vertically shaded area is added in the expression for the complementary energy to the horizontally shaded area in Fig. 4.1.3. The increment is then

$$\varepsilon_1\, \delta\sigma\, dV$$

With similar increments in the normal and shearing stresses in all the other elements of the various members of the structure the change in the complementary energy of the entire system is

$$\delta U' = \sum \int_V (\varepsilon_1\, \delta\sigma + \gamma_1\, \delta\tau)\, dV$$

The total change in $U' + V'$ during a variation of the state of stress is therefore

$$\delta(U' + V') = \delta U' - \sum r_n\, \delta R_n = \sum \int_V (\varepsilon_1\, \delta\sigma + \gamma_1\, \delta\tau)\, dV - \sum' r_n\, \delta R_n$$

$$4.1.4$$

It should be remembered now that, according to the specification for the variation, both the original set of stresses and reactions σ_1, τ_1, R_n, and the new set $\sigma_1 + \delta\sigma_1$, $\tau_1 + \delta\tau_1$, $R_n + \delta R_n$ are in equilibrium with the given loads P_n. Consequently the difference between the two sets, that is, the increments $\delta\sigma_1$, $\delta\tau_1$, δR_n, must constitute an equilibrium set without the external loads. If this is so, the work done by the set $\delta\sigma_1$, $\delta\tau_1$, δR_n must be zero for any virtual displacement.

One permissible set of virtual displacements (but not the only one) is defined by the finite strains ε_1 and γ_1 of the final state of the actual deformations of the structure. To these correspond the actual finite displacements of the supports r_n. If these actual displacements are chosen arbitrarily as the virtual displacements, the virtual work done by the reaction forces of the incremental set is $\Sigma r_n\, \delta R_n$. The work done by the internal forces can be calculated as follows:

If an elementary parallelepiped $dx\, dy\, dz$ is under the action of stresses

$\delta\sigma$ as shown in Fig. 4.1.4 and a virtual displacement is undertaken in such a manner that the left face of the parallelepiped is displaced to the right an amount d while the right face travels to the right a distance $d + e$, the total work dW done is

$$dW = (\delta\sigma\ dy\ dz)e = e\ \delta\sigma\ dA$$

where dA is the area of the faces under consideration.   But e is obviously

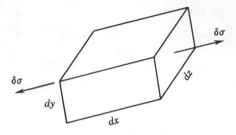

Fig. 4.1.4.   Stress increment acting on cube.

the total stretching $\varepsilon_1\ dx$ of the side dx with ε_1 denoting the final strain in the x direction.   Hence

$$dW = \delta\sigma\ dA\varepsilon_1\ dx = \varepsilon_1\ \delta\sigma\ dV$$

where dV is the volume $dx\ dy\ dz$ of the parallelepiped.   The work done by the internal forces in the parallelepiped during the virtual displacement is $-dW$ in agreement with Arts. 1.3 and 1.6.   The sign is negative because the internal stresses oppose the deformation, and thus they do negative work.   If the internal work of the shearing stresses is added, the work is integrated over the total volume of each element, and finally the individual work quantities are summed up over all the elements of the structure, the following expression is obtained for the total internal work:

$$W_i = -\sum\int_V \varepsilon_1\ \delta\sigma\ dV - \sum\int_V \gamma_1\ \delta\tau\ dV$$

The total virtual work done by the incremental set $\delta\sigma_1, \delta\tau_1, \delta R_n$ during the virtual displacements is therefore

$$-\sum\int_V \varepsilon_1\ \delta\sigma\ dV - \sum\int_V \gamma_1\ \delta\tau\ dV + \sum r_n\ \delta R_n$$

In consequence of the principle of virtual displacements this quantity must vanish, because the set of stresses and reactions $\delta\sigma, \delta\tau, \delta R_n$ is in equilibrium. After multiplication by -1 one obtains

$$\sum\int_V (\varepsilon_1\ \delta\sigma + \gamma_1\ \delta\tau)\ dV - \sum r_n\ \delta R_n = 0 \qquad 4.1.5$$

As the left-hand member of Eq. 4.1.5 is identical with the right-hand member of Eq. 4.1.4, the following property of the total complementary potential, defined as the quantity complementary energy plus the potential of the reactions, is established:

$$\delta U' - \Sigma r_n \, \delta R_n = 0 \qquad\qquad 4.1.6$$

This equation can be stated in the following form:

> When the true state of stress is varied in a system in **Theorem 14a** such a manner that the new state again constitutes equilibrium with the given set of external loads, the change in the complementary energy less the work done by the increments of the reactions while traveling through the actual displacements of the supports is zero.

By the true state is meant that state of equilibrium which prevails in reality.   In the true state not only the equilibrium conditions are satisfied but also the requirement of continuous deformations is fulfilled.

Equation 4.1.6 can be presented in the alternative form

$$\delta(U' - \Sigma R_n r_n) = 0 \qquad\qquad 4.1.7$$

or as

$$\delta(U' + V') = 0 \qquad\qquad 4.1.7a$$

which leads to the statement:

> The change in the total complementary potential   **Theorem 14b** corresponding to a variation of the state of stress is ← zero when the system is in its true state of equilibrium.

As was said earlier, only equilibrium states of stress are to be considered during the variation.   It must also be remembered that the changes must be calculated under the assumption that the state of deformation remains unchanged.   This means that, if the expression for the complementary energy is expanded into a Taylor series about a point characterized by the values of the independent variables $\sigma = \sigma_1$ and $\tau = \tau_1$, only quantities multiplied by the first power of the variations $\delta\sigma$ and $\delta\tau$ must be taken into consideration.   Terms multiplied by squares and higher powers of $\delta\sigma$ and $\delta\tau$ represent the changes due to the increments in strain caused by the increments in stress.   They must be disregarded when the stresses alone are varied and the state of deformation is maintained unchanged.

When the problem is such that there are only two independent stress quantities that can be varied, the dependent variable, namely the complementary energy plus the potential of the reactions, can be laid off in the

z direction over the point corresponding to any specified values of the two independent variables in the xy plane of a Cartesian coordinate system. The locus of all the z values thus obtained is a surface in three-dimensional space. Geometrically the requirement that the first variation of the dependent variable vanish for any arbitrary choice of the increments of the two independent variables defines the point on the surface whose tangent plane is horizontal. This point may correspond to a minimum, a saddle point, or a maximum of the values on the surface. In any case it is referred to as a stationary value. When the number of independent variables is larger, say i, visualization is of little help, but the language may be retained to say that the point defined by the complementary energy principle is a stationary value on the surface in $i + 1$ dimensions. With this in mind Theorem 14 can be restated as:

The total complementary potential has a stationary   **Theorem 15a**
value with respect to variations in stress when the
system is in its true state of equilibrium.

This is the statement of the complementary-energy principle. Just as with the first minimal principle, it is usual to modify the statement slightly because of the observation that the stationary value is a minimum when the system is stable. It can be said that

The total complementary potential is a minimum   **Theorem 15b**
with respect to variations in stress when the system is
in its true state of equilibrium.

In special cases these statements can be further simplified. For instance, when the supports are not displaced, the theorem can be expressed in the following manner:

The complementary energy is a minimum with respect to   **Corollary 1**
variations in stress when the system is in its true state of
equilibrium.

When the stress-strain relationship is linear, the complementary energy is equal to the strain energy. Then

The strain energy is a minimum with respect to variations   **Corollary 2**
in stress when the system is in its true state of equilibrium.

Corollary 2 can also be stated in the slightly modified form:

Of all possible stress distributions that satisfy the equi-   **Corollary 3**
librium conditions of an elastic body subjected to given
loads, the actual stress distribution corresponds to the
minimum of the strain energy stored.

This last statement is known as the least-work principle. It is valid only for materials following Hooke's law and for structures whose points of support do not move in the direction of the reaction. Motion perpendicular to the reaction, such as in a frictionless slide, is, of course, permissible since it does not contribute to the quantity $\Sigma R_n r_n$. Moreover, the variation in stress is meant to include only such changes as transform one equilibrium set into another equilibrium set with the external loads remaining unchanged.

It may now be of interest to recapitulate the major steps in the derivation of the complementary energy principle. The true state of stress characterized by σ_1, τ_1, R_n was compared to a fictitious one characterized by $\sigma_1 + \delta\sigma$, $\tau_1 + \delta\tau$, $R_n + \delta R_n$. Each stress system was in equilibrium with the given external loads P_n. Hence the variations $\delta\sigma$, $\delta\tau$, δR_n had to represent a self-equilibrating system. Because of the principle of virtual displacements the virtual work done by an equilibrium set is zero for any virtual displacement and therefore the work done by the system $\delta\sigma$, $\delta\tau$, δR_n must vanish, whatever choice is made for the virtual displacements. For the purposes of this investigation it was convenient to choose the transition from the unstrained state to the actual state of deformation as the virtual displacemnets. When this was done, the requirement that the virtual work vanish could be stated in the form of an equation. The left-hand member of this equation turned out to represent the change in the total complementary potential. As the change had to vanish, the quantity itself had to have a stationary value with respect to all variations of the stresses in which equilibrium was maintained.

This means that one may assume any state of stress to exist in a structure as long as it satisfies all the conditions of equilibrium. If the stresses are then modified in any arbitrary manner except for the restriction that the modified stresses again maintain equilibrium with the given external loads, the total complementary potential usually changes during the transition. This change is zero only if the assumed state of stress is the true state of stress under the given external load.

The first and second variational principles can now be compared. The first is a statement about the total potential, that is, the strain energy plus the potential of the external loads, and the second one is a statement about the total complementary potential, that is, the quantity complementary energy plus the potential of the reactions. In the first the state of deformation is varied while the stresses are maintained constant, and in the second the state of stress is changed into another statically possible one while the deformations are kept unchanged. When in an application of the first principle the external loads do not move during the variation of strain, the change in the total potential reduces to a change in the strain energy.

When the supports are fixed, the statement of the second principle reduces to a statement about the variation of the complementary energy.

Each principle states that a quantity has a stationary value, and each stationary value is a minimum when the structure is stable. For this reason the principles are known as minimal principles, although in reality they contain statements only about the first variation. The most important difference between the two principles is that the first refers to a variation of the state of deformation and the second to a variation of the state of stress. For this reason the strain energy in the first principle must be expressed in terms of strains or displacements, and the complementary energy in the second principle must be given in terms of stress or force.

In the first principle different possible patterns of deformation are compared with no regard to the possibility of equilibrium of the stresses after the deformations. The principle states that the actual pattern, that is, the one for which the equilibrium conditions are satisfied, makes the total potential a minimum. In the second principle all possible equilibrium sets of stresses and reactions are compared without any consideration of the consistency of the corresponding displacements. It is found that the actual set, that is, the one that leads to continuous deformations, corresponds to the minimum of the total complementary potential.

A restriction on the validity of the complementary-energy principle should now be removed. The derivations have been carried out for structures whose elements are essentially in a state of uniaxial stress. When a three-dimensional state of stress exists in an elastic body, Eq. 4.1.2 must be replaced by

$$U' = \sum \int_V \left(\int_0^{\sigma_{x1}} \varepsilon_x \, d\sigma_x + \int_0^{\sigma_{y1}} \varepsilon_y \, d\sigma_y + \int_0^{\sigma_{z1}} \varepsilon_z \, d\sigma_z \right.$$
$$\left. + \int_0^{\tau_{yz1}} \gamma_{yz} \, d\tau_{yz} + \int_0^{\tau_{zx1}} \gamma_{zx} \, d\tau_{zx} + \int_0^{\tau_{xy1}} \gamma_{xy} \, d\tau_{xy} \right) dV \qquad 4.1.8$$

Substitution of similar three-dimensional expressions for all the one-dimensional ones transforms the proof already given to one valid for all elastic bodies. In such elastic bodies the external loads and the reactions are often distributed over sizable areas of the surface, and thus the potential of the reactions must be found by integration of the product of pressure and displacement. The distinction between loads and reactions is straightforward: Surface forces whose magnitude and sign are specified over portions of the surface with unknown displacements are loads; surface forces of unknown magnitude acting over portions of the surface whose displacements are specified are reactions. These displacements, of course, are often specified as zero.

A final remark is of importance in practical applications. Real materials usually follow the stress-strain curve of Fig. 4.1.1 only during the loading process, and when the loads are removed the stress decreases along a straight line substantially parallel to the initial slope of the stress-strain curve. Of course, the results obtained in this article do not apply to the unloading process of such real materials. This more complex problem is treated in the theory of plasticity. The theorems just derived are valid for the loading and unloading processes of non-linearly elastic materials, such as cast iron; they can be used in the analysis of real elastoplastic materials provided no unloading takes place (see Arts. 4.2 and 4.5); and naturally the least-work principle is most useful in the solution of the statically indeterminate problems of structures in the elastic range.

4.2   The Stresses in Statically Indeterminate Structures

Castigliano's Second Theorem

In practical problems it would be difficult to investigate all the possible equilibrium stress distributions in a three-dimensional state of stress and select the one that corresponds to the minimum of the complementary energy. Fortunately experience in the last century and a half has shown that certain approximations lead to good results in engineering mechanics. Foremost among the simplifying assumptions is the one underlying the theory of the bending of beams, that plane sections perpendicular to the axis of the beam before loading remain plane and perpendicular to the axis after the loads are applied. This assumption, together with the stress-strain relationship, defines the normal stresses in any section of a beam as soon as the bending moment acting in the section is known. The contributions of the shear stresses to the total strain energy stored in a beam are usually negligible, as was shown in Art. 2.1

Consequently the stress problem of a beam is solved when the moment distribution is known. In a cantilever beam, and in a beam on two simple supports, the moment distribution can be calculated from the static equations of equilibrium alone. Such beams are called statically determinate structures. In their analysis the complementary-energy principle cannot be used because there is no more than one statically possible equilibrium stress distribution. On the other hand, the moments in a beam with one end simply supported and the other end fixed are known only after the moment of fixation or the reaction force at the simple support is obtained from considerations other than those of static equilibrium. A structure is designated as statically indeterminate if the equations of static equilibrium do not suffice for the determination of its internal forces and moments.

The so-called statically indeterminate quantity, that is, the moment of fixation or the reaction force at the simple support in the above example, can be found by enforcing the requirements of the continuity of the deformations of the structure as was done in Art. 1.13.   There the principle of virtual displacements was used to calculate the deformations.   As an alternative procedure, the magnitude of the unknown quantity can be determined from the complementary-energy principle.   The unknown forces or moments must have such values as make the total complementary potential a minimum.

Similar considerations can be offered in connection with frameworks having redundant members.   The unknown tensile or compressive force in the redundant member of a pin-jointed framework must be of such sign and magnitude as to satisfy the principle of the complementary energy.

It can be stated therefore that the analysis of statically indeterminate structures can be carried out on the basis of the following principle:

The statically indeterminate quantities make the total   **Corollary 4**
complementary potential a minimum.

When the supports do not yield and Hooke's law is valid, Corollary 4 reduces to the following statement:

The statically indeterminate quantities make the strain   **Corollary 5**
energy a minimum.

This is Castigliano's second theorem.

Beam on Three Supports

As a first example of the use of the complementary-energy principle, the bending moments will be calculated in the beam on three supports shown in Fig. 4.2.1.   It will be assumed that the supports do not yield and the material of the beam follows Hooke's law.   Then the complementary-energy principle can be used in its simplest form as given in Corollary 5.

One statically possible equilibrium configuration is characterized by the following values of the reactions:

$$R_0 = R_2 = W/2, \qquad R_1 = W \qquad\qquad 4.2.1$$

With these values the external loads are in equilibrium with the reactions, but the deformations that can be calculated from the bending moments are possible only if the continuity of the beam is disrupted.   It is easy to see that the reactions given are those that would prevail in two independent beams each of length L.   In the left span the bending moments would then be

$$M = \tfrac{1}{2}Wx, \qquad\qquad 0 \leq x \leq L/2$$

$$4.2.2$$

$$M = \tfrac{1}{2}W(L - x), \qquad L/2 \leq x \leq L$$

The bending moments in the right span are obtained if x is replaced by ξ in the expressions.   The moment diagram is shown in Fig. 4.2.1b.

Two such independent beams would deflect under the loads in the manner indicated in Fig. 4.2.1c.   The actual single beam spanning the total distance

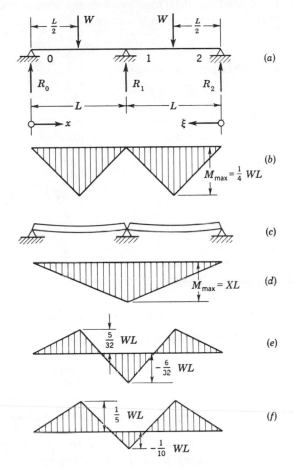

Fig. 4.2.1.   Beam on three supports.

between supports 0 and 2 would therefore have to crack over support 1. Consequently the stresses corresponding to the moments given in Eqs. 4.2.2 cannot be the true stresses of the actual system.

According to Corollary 5 the true stresses differ from any other statically possible ones inasmuch as they make the strain energy stored in the beam a minimum.   In order to obtain the combination that makes the strain energy a minimum, one has to find self-equilibrating sets of stresses and reactions to combine with those corresponding to the moments given in Eqs. 4.2.2. Each of the additional sets of stresses and reactions must be a self-contained equilibrium system because the set corresponding to Eqs. 4.2.1 and 4.2.2 already balances the given external loads. One possible set of reactions that, together with the stresses caused by it in the beam, represents equilibrium in the absence of external loads, is given by

$$R_0 = R_2 = X, \qquad R_1 = -2X \qquad\qquad 4.2.3$$

The moment in the left span is then

$$M = Xx, \qquad 0 \le x \le L \qquad\qquad 4.2.4$$

The moment in the right span is obtained through substitution of ξ for x. The moment diagram corresponding to these reactions is plotted in Fig. 4.2.1d.

The total bending moment in the left span is

$$M_1 = \tfrac{1}{2}Wx + Xx \qquad\qquad 0 \le x \le L/2$$
$$M_2 = \tfrac{1}{2}W(L - x) + Xx, \qquad L/2 \le x \le L$$

4.2.5

and the strain energy in the left span, in agreement with Eq. 2.1.12, is

$$U = (1/2EI) \int_0^{L/2} M_1{}^2 \, dx + (1/2EI) \int_{L/2}^L M_2{}^2 \, dx$$

Substitutions and the evaluation of the definite integrals yield

$$U = \tfrac{1}{96}(L^3/EI)(W^2 + 6WX + 16X^2)$$

Because of the symmetry of the bending-moment diagrams shown in Figs. 4.2.1b and d the strain energy stored in the entire structure is twice this quantity:

$$U_{\text{tot}} = \tfrac{1}{48}(L^3/EI)(W^2 + 6WX + 16X^2) \qquad\qquad 4.2.6$$

Next the stress distribution corresponding to the minimum of U_{tot} must be obtained. The bending moments according to Eqs. 4.2.2 cannot be tampered with because they are necessary for equilibrium with the external loads W. On the other hand the choice of X in Eq. 4.2.4 is entirely arbitrary as far as equilibrium is concerned. X can be chosen therefore such as to make U_{tot} a minimum, or at least a stationary value. The mathematical condition for this is that the derivative of U_{tot} with respect to X vanish. Since W is a given quantity and thus independent of X, differentiation gives

$$dU_{\text{tot}}/dX = \tfrac{1}{48}(L^3/EI)(6W + 32X) = 0$$

Omission of factors that are obviously not zero results in

$$6W + 32X = 0$$

which can be solved for the unknown reaction quantity:

$$X = -\tfrac{3}{16}W \qquad\qquad 4.2.7$$

The moment over the central support is therefore

$$M_1 = -\tfrac{3}{16}WL \qquad\qquad 4.2.8$$

The resultant moment diagram is given in Fig. 4.2.1e.

It is customary to describe the above procedure of calculation in the following manner: The statically indeterminate beam is cut over support 1 with the result that the system is reduced to a statically determinate structure consisting of two simple beams. The statically indeterminate quantity is then the unknown moment $M_1 = XL$ which in the actual system acts in the fictitious section at support 1. The strain energy is calculated from the moments caused in the determinate system under the simultaneous action of the given loads W and the unknown moment $M_1 = XL$. The strain energy is then minimized, that is, made a minimum, with respect to XL. From the minimum condition the value of the unknown moment XL (or of the unknown reaction X corresponding to $M_1 = XL$) can be calculated.

Because the system is singly redundant, that is, in it the moment distribution is completely determined by the laws of static equilibrium but for one unknown quantity, the most general type of moment distribution is the sum of one moment distribution corresponding to equilibrium with the given external loads, and another corresponding to a self-equilibrating set of reactions. The values of the moments in the latter are fixed except for an unknown constant multiplying factor which can be determined from the condition of the minimum of the strain energy. The advantage of this description of the procedure over the one contained in the preceding paragraph is that it permits a wider choice of the statically indeterminate set of moments and reactions, the only requirement being that the set constitute a self-equilibrating system.

A check of the result given in Eq. 4.2.8 against Eq. 1.13.2 shows agreement with the earlier calculations.

The example just discussed can be generalized in the following manner: Let it be assumed that after the girder of Fig. 4.2.1 is constructed the central support is observed to have settled. This will naturally modify the moment distribution. The moments can be calculated with the aid of Corollary 4. If the displacement of the support is denoted by Δ the potential of the reactions is Δ multiplied by the reaction at support 1.

The external loads yield $R_1 = W$ according to Eqs. 4.2.1 and the self-equilibrating system $R_1 = -2X$ according to Eqs. 4.2.3. The sum is $W - 2X$ acting upward. As the displacement Δ takes place in the downward direction the product of reaction and displacement is negative. However, the potential is $-\Sigma R_n r_n$, and thus in the present case

$$V' = (W - 2X)\Delta \qquad\qquad 4.2.9$$

The quantity to be minimized according to Corollary 4 is

$$U' + V' = \tfrac{1}{48}(L^3/EI)(W^2 + 6WX + 16X^2) + (W - 2X)\Delta \quad 4.2.10$$

The condition of a minimum with respect to X is

$$\tfrac{1}{48}(L^3/EI)(6W + 32X) - 2\Delta = 0$$

Consequently

$$X = -\tfrac{3}{16}W + 3(EI\Delta/L^3) \qquad\qquad 4.2.11$$

If $W = 6000$ lb, $L = 120$ in., $E = 30 \times 10^6$ psi, and $I = 10$ in.^{4}, Eq. 4.2.11 yields

$$X = -1120 + 520\Delta$$

Let it be assumed that the middle support is displaced 1 in. downward. The unknown reaction X then becomes

$$X = -600 \text{ lb} = -\tfrac{1}{10}W$$

This means that the moment over the central support is now

$$M_1 = -\tfrac{1}{10}WL$$

instead of the former value $-\tfrac{3}{16}WL$, a reduction to almost one-half the original value. The maxima in the middle of the spans, however, are increased from $\tfrac{5}{32}WL$ to $\tfrac{1}{5}WL$. The modified moment diagram is shown in Fig. 4.2.1f.

Redundant Truss beyond Elastic Limit

As a second example the forces will be determined in the three pin-jointed bars of Fig. 4.2.2a under the action of the vertical force P. If the tension in each of the slanting bars is denoted by X (the two forces must be equal because of symmetry), the force in the central bar is $P - X$ from the following considerations of equilibrium: The vertical component of X in a slanting bar is $X/2$. Hence the sum of the components in bars AB and AD added to $P - X$ in bar AC is exactly P as required. The magnitude of X will be determined from the complementary-energy principle.

It is of practical interest to calculate the forces in the bars not only under small loads but also in the neighborhood of failure. There the assumption of a linear stress-strain relationship is, of course, untenable when the material is steel or aluminum alloy.   The actual stress-strain curve can be represented analytically in good approximation by the expression

$$\varepsilon = \sigma/E + K(\sigma/E)^n \qquad\qquad 4.2.12$$

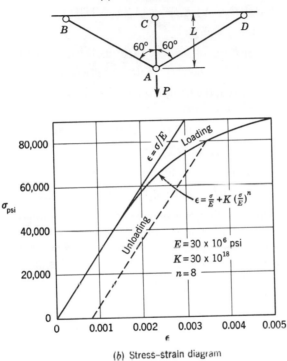

(a) Three-bar framework

(b) Stress-strain diagram

Fig. 4.2.2.   Redundant truss beyond elastic limit.

For a particular steel the values of the constants can be taken as

$$E = 30 \times 10^6, \qquad n = 8, \qquad K = 30 \times 10^{18} \qquad 4.2.13$$

The curve is shown in Fig. 4.2.2b.   The complementary energy stored in a bar of volume V is

$$U' = V \int_0^{\sigma_1} \varepsilon \, d\sigma - V \int_0^{\sigma_1} \left[\frac{\sigma}{E} + K \left(\frac{\sigma}{E} \right)^n \right] d\sigma$$

$$= EV \left[\frac{1}{2} \left(\frac{\sigma_1}{E} \right)^2 + \frac{K}{n+1} \left(\frac{\sigma_1}{E} \right)^{n+1} \right] \qquad\qquad 4.2.14$$

If all three bars have the same cross section A, and the length of bar AC is L while that of bars AB and AD is $2L$ according to Fig. 4.2.2a, the complementary energy is

In bar AB with $V = 2LA$ and the force X:

$$U'_{AB} = 2LEA\{\tfrac{1}{2}(X/EA)^2 + [K(n+1)](X/EA)^{n+1}\}$$

In bar AC with $V = LA$ and the force $P - X$:

$$U'_{AC} = LEA\{\tfrac{1}{2}[(P-X)/EA]^2 + [K/(n+1)][(P-X)/EA]^{n+1}\}$$

The total complementary energy is

$$U'_{tot} = LEA\{\tfrac{1}{2}[(P-X)/EA]^2 + [K/(n+1)][(P-X)/EA]^{n+1}\}$$
$$+ 4LEA\{\tfrac{1}{2}(X/EA)^2 + [K/(n+1)](X/EA)^{n+1}\} \qquad 4.2.15$$

Since the supports are fixed, the complementary energy must be a minimum with respect to X. The condition is

$$(dU_{tot}/dX) = 0 = -[(P-X)/EA] - K[(P-X)/EA]^n$$
$$+ 4[(X/EA) + K(X/EA)^n]$$

After rearranging the terms, one obtains

$$K\{4(X/EA)^n - [(P-X)/EA]^n\} = (P/EA) - 5(X/EA) \qquad 4.2.16$$

When P is given, this equation can best be solved by plotting the value of the left side against X and finding the intersection point of the curve obtained with the straight line representing the right side. For particular values of P the procedure is simplified further. For instance, when P is small, the left side of the equation is very small compared to the right side. With $P = 30,000$ lb and $A = 1$ sq in. the right side was found to be about one-thousand times greater than the left side in the neighborhood of the solution. The left side could therefore be neglected, and the equation

$$(P/EA) - 5(X/EA) = 0$$

solved for X:

$$X = 0.2P \qquad 4.2.17$$

The force in the vertical bar is thus 24,000 lb and that in the slanting bars 6000 lb. With $A = 1$ sq in. the stresses in all the bars are on the substantially straight portion of the stress-strain curve.

When the force is large, the right side of Eq. 4.2.16 becomes insignificant compared to the left side. For instance, at $P = 140,000$ lb, the value of the left side was found to be about one-hundred times that of the right side in the vicinity of the X value corresponding to the solution. Consequently a good approximate solution to Eq. 4.2.16 can be had if the left side of the equation is set equal to zero:

$$4(X/EA)^n - [(P-X)/EA]^n = 0$$

With $A = 1$ sq in. and the values given in Eq. 4.2.13 the above equation becomes

$$1.189X = P - X$$

and thus

$$X = 0.457P \qquad \text{4.2.18}$$

Consequently at this value of the load the central bar carries only 54.3 per cent of the load instead of 80 per cent as in the fully elastic case. The tensile force in the central bar is 76,000 lb and that in the slanting bars 64,000 lb. The stresses have numerically the same values as the forces, and thus they correspond to the noticeably curved part of the stress-strain diagram.

The example shows that the yielding of the most highly stressed member of the structure brings about a redistribution of the load. The originally lightly stressed members take over a greater share of the load and the overstressed member is relieved if the distribution is judged on a percentage basis. The actual value of the force in the central bar, of course, increases during the entire process of loading but at a progressively decreasing rate. If this were not true, Eq. 4.2.12 would not represent correctly the stress-strain relationship of the material because the curve shown in Fig. 4.2.2b is valid only for loading. When unloading takes place,only the elastic part of the strain is relieved, and the connection between stress and strain is represented in good approximation by a straight line parallel to the initial portion of the stress-strain curve. Such an unloading curve is shown dashed in the figure.

Circular Frame under Concentrated Loads

The solution of the bending-moment distribution in a circular ring frame will be discussed next. Figure 4.2.3a shows the ring and the two equal and opposite forces P acting on it. If the depth of the ring cross section is small compared to the radius r of the ring, say less than $\frac{1}{5}$, it is permissible to use the formula derived for the strain energy stored in a straight beam even though the ring frame is essentially a curved beam. The ratio depth to radius will be assumed small in this example, and the strain energy of bending will be calculated from the formula

$$U = \frac{1}{2EI} \oint M^2 \, ds = \frac{1}{2EI} \int_0^{2\pi} M^2 r \, d\phi \qquad \text{4.2.19}$$

where M is the bending moment, EI the bending rigidity, ds the element of length along the ring, and the circle crossing the integral sign indicates that the integration has to be extended around the entire circumference.

The ring is a statically indeterminate structure. It is reduced to a statically determinate one if it is cut through, say, at A. Because of the symmetry of structure and loading, the frame can be clamped at C as shown in Fig. 4.2.3b without the moment distribution being changed.   The other half of the ring, ADC, not shown in Fig. 4.2.3b, will provide the necessary clamping moment because it is subjected to forces and moments symmetric to those acting on segment ABC.

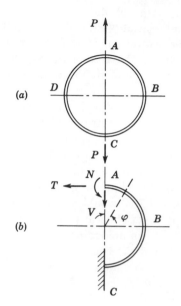

(a)

(b)

Fig. 4.2.3.   Ring subjected to equal and opposite loads.

In section A an unknown tensile force T, an unknown shear force V, and an unknown bending moment N are acting.   A shear force perpendicular to the plane of the ring, a torque, and a bending moment out of the plane cannot exist if all the loads are in the plane of the ring and the cross section of the ring is symmetric with respect to the plane of the paper.   If T, V, and N were known, the bending moment in ring segment ABC could be given as

$$M = N + Vr \sin \phi + Tr(1 - \cos \phi), \qquad 0 \leq \phi \leq \pi \qquad 4.2.20$$

The sign convention is that moments tending to increase the curvature of the ring are counted as positive.   Symmetry requires that one-half the load P be transmitted by segment ABC and the other half by segment ADC. Consequently $P/2$ is the shear force in section A.   If the directions indicated in Fig. 4.2.3b are adopted as the positive directions,

$$V = -\tfrac{1}{2}P$$

Because of the symmetry of structure and loading about axis BD, the tensile force T must be zero.   This can be understood if it is noticed that for reasons of equilibrium the shear force in section B is T.   A shear force is an antisymmetric quantity and therefore cannot be present in the plane of symmetry passing through axis BD perpendicular to the plane of the paper.

Two quantities are antisymmetric if they are mirrored images of each other with the sign reversed.   The location of the *mirror* is the axis of antisymmetry.   Thus in Fig. 4.2.4a forces F_1 and F'_1 are symmetric with respect to axis AA, while forces F_2 and F'_2 are antisymmetric with respect to the same axis.

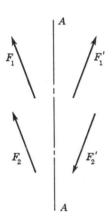

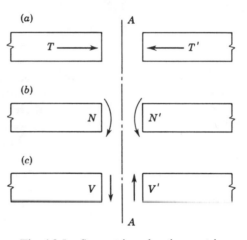

Fig. 4.2.4.   Symmetric and antisymmetric forces.

Fig. 4.2.5.   Symmetric and antisymmetric stress resultants in a section of a beam.

It is a matter of age-old experience that symmetric causes have symmetric effects and antisymmetric causes have antisymmetric effects.   In Fig. 4.2.5a the two tensile force resultants acting on opposite sides of the beam in section AA are symmetric.   They can well be present in the section if the structure and loading are symmetric with respect to axis AA. They must be absent if AA is an axis of antisymmetry.   The same reasoning holds for the internal bending moment shown in Fig. 4.2.5b.   On the contrary two shear forces acting on opposite sides of a cut are antisymmetric (see Fig. 4.2.5c); they must be absent if the cut is situated on an axis of symmetry.   It can be said therefore that

> In a plane of symmetry the shear force must vanish,   **Theorem   16** and in a plane of antisymmetry the tensile (or compressive) force and the bending moment must vanish.

In Fig. 4.2.3a, BD is an axis of symmetry, and thus the shear force in section B, namely T, is zero. Equation 4.2.20 is reduced therefore to

$$M = N - \tfrac{1}{2}Pr \sin \phi, \qquad 0 \leq \phi \leq \pi \qquad \text{4.2.21}$$

If the material of the ring follows Hooke's law, in the absence of unknown reactions the complementary-energy principle reduces to the requirement that the strain energy be a minimum with respect to N (Corollary 5). The strain-energy integral can be extended from 0 to π and multiplied by 2 because the strain energy stored in ring segment ADC must be the same as that stored in segment ABC on account of the symmetry. This is more convenient than to integrate from 0 to 2π since Eqs. 4.2.20 and 4.2.21 are not valid beyond $\phi = \pi$. Consequently

$$U = \frac{1}{2EI} 2 \int_0^\pi \left(N - \frac{1}{2} Pr \sin \phi \right)^2 r \, d\phi \qquad \text{4.2.22}$$

It should be observed, however, that the strain energy itself is not needed in the calculations, only its derivative with respect to N. Some of the work of integration can be saved therefore if the right side of Eq. 4.2.22 is differentiated with respect to N before the integration. It is shown in the calculus that the order in which the integration and the differentiation are carried out is immaterial when the functions involved and their derivatives are continuous. This condition is satisfied here, and thus

$$\partial U / \partial N = \frac{r}{EI} 2 \int_0^\pi \left(N - \frac{1}{2} Pr \sin \phi \right) d\phi$$

$$= \frac{2r}{EI} \left[N\phi + \frac{1}{2} Pr \cos \phi \right]_0^\pi$$

$$= \frac{2r}{EI}(N\pi - Pr)$$

As this quantity must vanish

$$N = Pr/\pi$$

Substitution in Eq. 4.2.21 yields

$$M = Pr[(1/\pi) - \tfrac{1}{2}\sin \phi]$$

The bending moment is shown by the shaded area in Fig. 4.2.6. The horizontal line ABC is the developed axis of the semicircular beam ABC. The maximum moment is Pr/π. This value is reached under the two loads at $\phi = 0$ and $\phi = \pi$. A secondary maximum of the absolute value of the moment exists at $\phi = \pi/2$ (and at $\phi = 3\pi/2$). There the moment is negative, namely $-[(\pi - 2)/2\pi]Pr = -0.182Pr$.   Consequently the loads tend to increase the curvature of the ring in the neighborhood of A and C, and to decrease it in the vicinity of B and D.

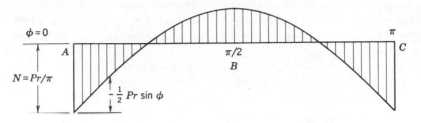

Fig. 4.2.6.   Bending moment in ring segment ABC (developed) (loading of Fig. 4.2.3).

Ring Frame Subjected to Distributed Loads

In Fig. 4.2.7 the same type of ring is subjected to a different kind of loading.   The concentrated force $2P$ at $\phi = \pi$ is balanced by a distributed load

$$q = -(2P/\pi r) \sin \phi \qquad\qquad 4.2.23$$

The quantity q is a load per inch of the circumference of the ring.   It represents the reaction shear flow transmitted from a thin circular shell (see Fig. 4.2.7b) supporting the ring, provided the shear flow is calculated from the conventional formula

$$q = \tau t = VQ/2I \qquad\qquad 4.2.24$$

whose derivation is presented in texts on airplane structures.   The negative sign indicates that the direction of the shear flow is opposed to the direction of increasing ϕ.

In the formula q is the shear flow at ϕ, τ the shear stress at ϕ, t the thickness of the shell, V the shear force resultant transmitted by the shell, Q the static moment, with respect to axis BD, of the portion of the shell cross section comprised between $-\phi$ and $+\phi$, and I is the moment of inertia of the cross section of the shell with respect to axis BD.   In the example $V = 2P$, the static moment

$$Q = \int_{-\phi}^{\phi} tr^2 \cos \phi \, d\phi = 2tr^2 \sin \phi$$

and the moment of inertia

$$I = \pi r^3 t$$

Substitution of these quantities in Eq. 4.2.24 yields Eq. 4.2.23 except for the sign.   However, Eq. 4.2.24 represents the shear flow acting on the shell while in Eq. 4.2.23 the quantity q is the shear reaction from the shell on the ring.

Equation 4.2.23 represents the shear flow distribution correctly only when the ring frame is very rigid both in its plane and perpendicularly to it. When this is not the case, a more accurate analysis is needed; this is given in Art. 4.7.

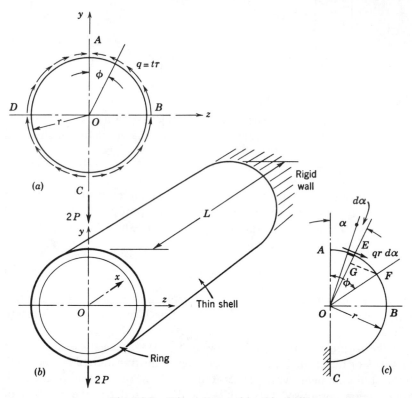

Fig. 4.2.7.   Ring supported by thin shell.

In Fig. 4.2.7b the ring is shown to have considerable depth.   This representation was chosen for the pictorial effect.   In actual thin-walled reinforced cylinders, such as those used as airplane fuselages, the ring depth is often so small that the difference between the radius of the shell and that of the median line of the ring can be disregarded.   In the following calculations the two will be assumed the same and designated by r.

If the procedure followed in the preceding calculations is adopted and the ring is again cut at A, the bending moment caused at F in Fig. 4.2.7c by the load $qr\,d\alpha$ acting at E is the load $qr\,d\alpha$ times its lever arm with respect to F. (In the figure q is shown as if it were positive.)   The perpendicular dropped from F upon OE intersects OE at G.   The distance OG is $r\cos(\phi-\alpha)$.   The lever arm, that is, the perpendicular distance of F from the line of action of the force $qr\,d\alpha$, is EG which is $r-r\cos(\phi-\alpha)$. Consequently the moment is

$$dM = -qr^2[1 - \cos(\phi - \alpha)]\,d\alpha$$

The negative sign signifies that the positive shear flow in Fig. 4.2.7c causes a negative bending moment, that is, one that tends to decrease the curvature of the ring.   The sum of the bending moments caused at F by all the distributed shear forces between A and F can be obtained by integration:

$$M = -\int_0^\phi qr^2[1 - \cos(\phi - \alpha)]\,d\alpha$$

$$= -\int_0^\phi qr^2(1 - \cos\phi\cos\alpha - \sin\phi\sin\alpha)\,d\alpha \qquad 4.2.24a$$

The value of q must now be substituted from Eq. 4.2.23.   It should be noticed, however, that the shear flow in question is that acting at $\phi = \alpha$. Consequently ϕ in Eq. 4.2.23 must be replaced by α.   One obtains

$$M = \frac{2Pr}{\pi}\int_0^\phi (\sin\alpha - \cos\phi\cos\alpha\sin\alpha - \sin\phi\sin^2\alpha)\,d\alpha$$

Since

$$\int\sin\alpha\,d\alpha = -\cos\alpha$$

$$\int\cos\alpha\sin\alpha\,d\alpha = \frac{1}{2}\int\sin2\alpha\,d\alpha = -\frac{1}{4}\cos2\alpha$$

$$\int\sin^2\alpha\,d\alpha = \frac{1}{2}\int(1 - \cos2\alpha)\,d\alpha = \frac{1}{2}\left(\alpha - \frac{1}{2}\sin2\alpha\right)$$

the moment becomes

$$M = (2Pr/\pi)[-\cos\phi + 1 + \tfrac{1}{4}\cos\phi(\cos2\phi - 1) - \tfrac{1}{2}\sin\phi(\phi - \tfrac{1}{2}\sin2\phi)]$$

$$= (2Pr/\pi)(1 - \cos\phi - \tfrac{1}{2}\sin^2\phi\cos\phi - \tfrac{1}{2}\phi\sin\phi + \tfrac{1}{2}\sin^2\phi\cos\phi)$$

or

$$M = (2Pr/\pi)(1 - \cos\phi - \tfrac{1}{2}\phi\sin\phi) \qquad 4.2.25$$

Of the statically indeterminate quantities indicated in Fig. 4.2.3b the shear force V must vanish because of the symmetry about axis AC in Fig. 4.2.7a. In this figure the loading is not symmetric with respect to the axis BD, and so a shear force can exist in section B. Hence T need not be zero. The bending moment caused by the remaining statically indeterminate quantities is therefore

$$M = N + Tr(1 - \cos \phi) \qquad 4.2.26$$

The total moment is

$$M = (2Pr/\pi)(1 - \cos \phi - \tfrac{1}{2}\phi \sin \phi) + N + Tr(1 - \cos \phi) \qquad 4.2.27$$

The strain energy stored in one-half the ring is

$$U = \frac{1}{2EI} \int_0^\pi \left[\frac{2Pr}{\pi} \left(1 - \cos \phi - \frac{1}{2} \phi \sin \phi \right) + N + Tr(1 - \cos \phi) \right]^2 r \, d\phi$$

This strain energy must be a minimum with respect to the statically indeterminate quantities T and N. Hence

$$\frac{\partial U}{\partial N} = 0 = \int_0^\pi \left[\frac{2Pr}{\pi} \left(1 - \cos \phi - \frac{1}{2} \phi \sin \phi \right) + N + Tr(1 - \cos \phi) \right] d\phi$$

$$4.2.28$$

Since

$$\int \phi \sin \phi \, d\phi = -\phi \cos \phi + \sin \phi$$

the definite integral becomes

$$\partial U/\partial N = 0 = (2Pr/\pi)(\pi - \tfrac{1}{2}\pi) + N\pi + Tr\pi$$

This equation can be simplified to read

$$\tfrac{1}{2}(2Pr/\pi) + N + Tr = 0 \qquad 4.2.29$$

Moreover

$$\frac{\partial U}{\partial T} = 0 = \int_0^\pi \left[\frac{2Pr}{\pi} \left(1 - \cos \phi - \frac{1}{2} \phi \sin \phi \right) + N \right.$$
$$\left. + Tr(1 - \cos \phi) \right] (1 - \cos \phi) \, d\phi$$

As the definite integral of the expression in brackets times $d\phi$ was set equal to zero in Eq. 4.2.29, all that remains is to equate to zero the definite integral of the expression in brackets times $\cos \phi \, d\phi$. Since

$$\int \cos^2 \phi \, d\phi = \frac{1}{2} \int (1 + \cos 2\phi) \, d\phi = \frac{1}{2} \left(\phi + \frac{1}{2} \sin 2\phi \right)$$

$$\int \phi \sin \phi \cos \phi \, d\phi = \frac{1}{2} \int \phi \sin 2\phi \, d\phi = \frac{1}{4} \left(\phi \cos 2\phi + \frac{1}{2} \sin 2\phi \right)$$

the definite integral becomes

$$\partial U / \partial T = 0 = (2Pr/\pi)[-(\pi/2) + (\pi/8)] - Tr(\pi/2)$$

This equation can be rewritten as

$$\tfrac{3}{4}(2Pr/\pi) + Tr = 0$$

Therefore

$$Tr = -\tfrac{3}{4}(2Pr/\pi) \qquad\qquad 4.2.30$$

Substitution in Eq. 4.2.29 yields

$$N = \tfrac{1}{4}(2Pr/\pi) \qquad\qquad 4.2.31$$

The expression given in Eq. 4.2.27 for the bending moment then becomes

$$M = (2Pr/\pi)[1 - \cos\phi - \tfrac{1}{2}\phi\sin\phi + \tfrac{1}{4} - \tfrac{3}{4}(1 - \cos\phi)]$$

or

$$M = (Pr/\pi)(1 - \tfrac{1}{2}\cos\phi - \phi\sin\phi) \qquad\qquad 4.2.32$$

This bending moment is shown in the curve marked *Conventional* in Fig. 4.7.4 of Art. 4.7.

4.3   The Analysis of Rigid Frames

Simple Bent Analyzed with the Aid of Castigliano's Theorem

One of the most frequently recurring tasks of the structural engineer is the analysis of building frames.   For this reason a great deal of ingenuity has been exerted to provide the analyst with tools that permit him to accomplish his task easily.   One special device developed for this purpose is the slope-deflection method discussed in Art. 1.19   Another general approach to the solution of the moment distribution problem in rigid frames is through the complementary-energy principle, or its simpler version, Castigliano's second theorem.

When the moments are sought in the rigid frame shown in Fig. 4.3.1, the symmetry of the system is not disturbed if a section is imagined through the middle of the beam.   In this section no shear force can be present in accordance with the statements made in Art. 4.2.   If the unknown moment N and the unknown tension T are introduced as in Fig. 4.3.1b, the left half of the frame shown (as well as the omitted right side) is subjected to the same forces and moments as it was in the uncut original frame, provided N and T are chosen correctly.   The magnitudes and signs of these unknown quantities will now be determined with the aid of Castigliano's second theorem.   The third unknown of the threefold redundant system, namely the shear force in the section, is known to vanish because of the symmetry.

With the sign convention adopted for the beam, the bending moment is

$$M = N - \tfrac{1}{2}wx^2, \qquad 0 \leq x \leq a \qquad\qquad 4.3.1$$

In the column, as in the beam, the bending moments will be considered positive if they cause tension in the fibers inside the rectangular frame. With this convention the moment in the column is

$$M = N - \tfrac{1}{2}wa^2 - T\xi, \qquad 0 \leq \xi \leq a \qquad\qquad 4.3.2$$

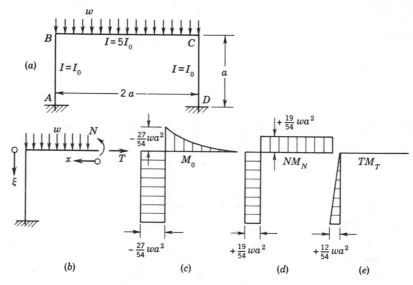

Fig. 4.3.1.   Rigid frame subjected to distributed loads.

Experience has shown that the strain energy of shear and tension or compression is very small in building frames compared to the strain energy of bending (see Art. 2.1). Similarly their rate of change with the unknown forces and moments can be neglected compared with the rate of change of the bending strain energy. For the purposes of this analysis the total strain energy stored can be given therefore as

$$U = \frac{1}{5EI_0} \int_0^a \left(N - \frac{1}{2}wx^2\right)^2 dx + \frac{1}{EI_0} \int_0^a \left(N - \frac{1}{2}wa^2 - T\xi\right)^2 d\xi \qquad 4.3.3$$

The factor $\tfrac{1}{2}$ is absent from the strain-energy expressions because the integration is indicated over half the structure, but the strain energy given is that stored in the entire structure. By virtue of Castigliano's second theorem the derivatives of U with respect to N and T must vanish. In most cases one can save work by differentiating under the integral sign and carrying out the integration afterward. If this procedure is followed the result is

$$\frac{\partial U}{\partial N} = \frac{2}{5EI_0} \int_0^a \left(N - \frac{1}{2} wx^2\right) dx + \frac{2}{EI_0} \int_0^a \left(N - \frac{1}{2} wa^2 - T\xi\right) d\xi$$

$$= \frac{2}{5EI_0} \left(Na - \frac{1}{6} wa^3\right) + \frac{2}{EI_0} \left(Na - \frac{1}{2} wa^3 - \frac{1}{2} Ta^2\right) = 0$$

This equation simplifies to

$$\tfrac{6}{5}Na - \tfrac{8}{15}wa^3 - \tfrac{1}{2}Ta^2 = 0 \qquad\qquad 4.3.4$$

Similarly

$$\frac{\partial U}{\partial T} = -\frac{2}{EI_0} \int_0^a \left(N - \frac{1}{2} wa^2 - T\xi\right) \xi \, d\xi = 0$$

which becomes

$$\tfrac{1}{2}Na^2 - \tfrac{1}{4}wa^4 - \tfrac{1}{3}Ta^3 = 0 \qquad\qquad 4.3.5$$

Solution of Eqs. 4.3.4 and 5 yields

$$T = -\tfrac{12}{54}wa$$

$$N = \tfrac{19}{54}wa^2$$

$$\qquad\qquad 4.3.6$$

Substitution of these values in the moment expressions leads to

$$M = \tfrac{19}{54}wa^2 - \tfrac{1}{2}wx^2, \qquad 0 \le x \le a$$

$$M = -\tfrac{8}{54}wa^2 + \tfrac{12}{54}wa\xi, \qquad 0 \le \xi \le a$$

$$\qquad\qquad 4.3.7$$

In parts c, d, and e of Fig. 4.3.1 the individual contributions of the given load, the redundant moment N, and the redundant tensile force T to the total bending moments are shown diagrammatically. Their sum is the same bending-moment diagram as the one presented earlier as Fig. 1.19.3c. This indicates that the analyses by means of Castigliano's theorem and the slope-deflection method are equivalent. It is easy to prove that enforcing the conditions of consistent, or compatible, deformations also leads to the same results. The proof can be given in the following manner:

In the cut system if the moment caused by the external load w alone is denoted by M_0, the moment due to $N = 1$ alone by M_N, and the moment caused by $T = 1$ alone by M_T, the total moment in any section can be given as

$$M = M_0 + NM_N + TM_T \qquad\qquad 4.3.8$$

The total strain energy stored is

$$U = \int_L \frac{1}{2EI} (M_0 + NM_N + TM_T)^2 \, ds \qquad 4.3.9$$

where ds is the element of length of the frame and L under the integral sign indicates that the integration has to be extended over the total length of all the bars of the structure. The values of N and T will now be determined with the aid of Castigliano's theorem. Minimization with respect to N yields

$$\frac{\partial U}{\partial N} = \int_L \frac{M_0 M_N}{EI} \, ds + N \int_L \frac{M_N^2}{EI} \, ds + T \int_L \frac{M_N M_T}{EI} \, ds = 0 \qquad 4.3.10$$

Similarly

$$\frac{\partial U}{\partial T} = \int_L \frac{M_0 M_T}{EI} \, ds + N \int_L \frac{M_N M_T}{EI} \, ds + T \int_L \frac{M_T^2}{EI} \, ds = 0 \qquad 4.3.11$$

If $N = 1$ is considered a dummy unit moment acting in the section imagined through the plane of symmetry of the beam, the integral

$$\int_L \frac{M_N M_T}{EI} \, ds$$

is -1 times the internal work done by the stress system corresponding to N during the deformations caused by the force $T = 1$ (compare Art. 1.12). Since the external work is $N = 1$ times the relative rotation of the two sides of the section, it follows from the principle of virtual displacements that

$$\delta_{NT} = \int_L \frac{M_N M_T}{EI} \, ds \qquad 4.3.12$$

where δ_{NT} is the relative rotation, caused by $T = 1$, of the two sides of the imaginary section. Similarly

$$\delta_{NO} = \int_L \frac{M_N M_O}{EI} \, ds \qquad 4.3.13$$

is the relative rotation of the two sides of the section as caused by the given external loads w and

$$\delta_{NN} = \int_L \frac{M_N^2}{EI} \, ds \qquad 4.3.14$$

the relative rotation of the two sides of the section as caused by the moment $N = 1$. Thus Eq. 4.3.10 can be rewritten as

$$\delta_{NO} + N\delta_{NN} + T\delta_{NT} = 0 \qquad 4.3.15$$

The physical meaning of this equation is that the actual value of N multiplied by the relative rotation of the two sides of the section as caused by $N = 1$, plus the actual value of T multiplied by the relative rotation of the two sides of the section as caused by $T = 1$ must be of such a magnitude and sign that, when added to the relative rotation caused by the external load w, the imaginary cleavage in the plane of symmetry closes.   This is one condition for the calculation of the two unknown quantities N and T. The second expresses the requirement that the gap, or relative horizontal displacement between the two sides of the imaginary section in the plane of symmetry of the rigid frame, must close under the simultaneous action of the given external load w and the unknown quantities N and T.   This requirement is the physical interpretation of Eq. 4.3.11, as is evident if the equation is rewritten in the form

$$\delta_{TO} + N\delta_{TN} + T\delta_{TT} = 0 \qquad\qquad 4.3.16$$

The first subscript of the displacement quantities δ, known as the influence coefficients, indicates the kind of displacement and the second, the force quantity that caused it.

Equations 4.3.15 and 4.3.16 closely resemble Eqs. 1.8.11 which were derived in Part 1 for the calculation of the redundant forces in statically indeterminate trusses.   Both sets of equations express the requirement that under the action of the known external and unknown internal forces the structure must deform in a consistent, continuous fashion without the development of any gaps or fissures.

When the redundant quantities are calculated for a rigid frame by means of Castigliano's second theorem in the manner described, namely by differentiation under the integral sign and integration afterward, the integrals that must be evaluated are identical with those defining the influence coefficients in Eqs. 4.3.15 and 4.3.16.   Consequently it is immaterial whether the moment distribution in a rigid frame is calculated from Castigliano's second theorem or with the aid of a consistent deformation analysis as discussed in Part 1.

Use of Hinges instead of One Complete Section

Castigliano's theorem can be used conveniently in the manner just described when the moment of inertia of the elements of the frame is constant or when it is given by other expressions that permit integration in a closed form.   When the integration must be carried out numerically or graphically, difficulties arise in connection with the accuracy of the procedure.   Parts c, d, and e of Fig. 4.3.1 show that the actual bending moment is generally obtained as a comparatively small difference of large quantities.   If one of these large quantities is inaccurate, the final result may be off considerably. Let it be assumed that through an error of 10 per cent N was obtained as $(17.1/54)wa^2$ instead of $(19/54)wa^2$. The total moment at A is then

$$M_A = (wa^2/54)(-27 + 17.1 + 12) = (2.1/54)wa^2$$

As the correct value is $(4/54)wa^2$, an error of 10 per cent in N reduces the value of the moment at A to about one-half the correct value. It would be of obvious advantage therefore to devise a method of calculation in which the M_0 moment diagram is more like the final moment diagram. If this can be achieved, inaccuracies in the numerical integration which lead to inaccurate values of N and T can have only a minor influence upon the final moments.

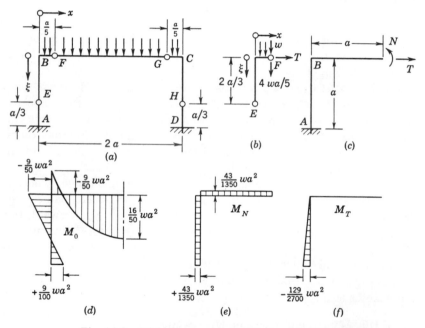

Fig. 4.3.2.   Rigid-frame analysis with the aid of hinges.

A more suitable M_0 diagram can be obtained by replacing the fictitious section through the plane of symmetry of the rigid frame with a number of fictitious hinges in the structure. The advantage of this procedure lies in the ease with which the approximate location of the inflection points in the deflected shape of the frame can be guessed in advance. At an inflection point the bending moment is zero, and thus the introduction there of a fictitious hinge does not change the moment distribution.

The qualitative deflected shape shown in Fig. 1.19.3b suggests the assumption of the fictitious hinges in the manner indicated in Fig. 4.3.2. The introduction of four hinges in a threefold redundant structure, of course, results in a mechanism rather than in a statically determinate structure. The three bars composing the top of the frame in Fig. 4.3.2a can sway to the side without any resistance. This is not objectionable as long as all the loads acting, or imagined to act, on the system are symmetric because symmetric loads cannot cause antisymmetric deformations and the sidesway is antisymmetric.

Beam FG is simply supported, and the bending moments acting on it can be determined without difficulty. One-half its total load is transmitted as a shear force to portion EF of the frame. Hence the shear force at F is $\frac{4}{5}wa$ as indicated in Fig. 4.3.2b. An unknown tensile force T also acts on EF through the hinge at F. Thus equilibrium of moments about E requires that

$$\tfrac{4}{5}wa(a/5) + T(2a/3) + (wa/5)(a/10) = 0$$

This equation can be solved for T:

$$T = -\tfrac{27}{100}wa \qquad\qquad 4.3.17$$

The bending moment at B is

$$M_B = -\tfrac{4}{5}wa(a/5) - (wa/5)(a/10) = -\tfrac{9}{50}wa^2 \qquad 4.3.18$$

As the shear force acting on the beam at B is wa, the bending moment in the beam can be expressed as

$$M_0 = -\tfrac{9}{50}wa^2 + wax - \tfrac{1}{2}wx^2 \qquad 0 \le x \le 2a \qquad 4.3.19$$

The bending moment in the column is

$$M_0 = -\tfrac{9}{50}wa^2 - T\xi = -\tfrac{9}{50}wa^2 + \tfrac{27}{100}wa\xi \qquad 0 \le \xi \le a \qquad 4.3.20$$

and at the bottom of the column

$$M_A = \tfrac{9}{100}wa^2 \qquad\qquad 4.3.21$$

These are the moments calculated for the system obtained from the actual frame through the introduction of fictitious hinges. They are shown in Fig. 4.3.2d. As it is not known whether the inflection points of the deflected shape will really coincide with these hinges, there is no justification for the assumption that the bending moment vanishes at points E, F, G, and H in the actual rigid frame. The error committed in the analysis can be rectified by assuming unknown moments to act at the fictitious hinges. It is hoped, of course, that the unknown moments will turn out to be small and will not modify materially the basic moment diagram obtained. Because of the symmetry, the moments at E and F must be equal to those at H and G, respectively, and thus the total number of redundant quantities is two and not four. This agrees with the number of unknowns in the preceding analysis where an unknown moment N and an unknown tensile force T were assumed to act in the section through the plane of symmetry of the beam.

Combinations of Redundant Moments

It would be perfectly satisfactory now to assume $M_E = M_H = 1$ and $M_F = M_G = 0$ as the first symmetric group of unknown bending moments and $M_E = M_H = 0$ and $M_F = M_G = 1$ as the second, but the establishment of equilibrium in the movable system of Fig. 4.3.2a would be time-consuming. It is much simpler to use linear combinations of these groups chosen so as to permit the satisfaction of the equilibrium conditions by inspection. It is suggested that the moments at the hinges be assigned the values that they would have if the hinges were absent and an unknown moment N and an unknown tensile force T were acting in a section through the plane of symmetry of the beam as shown in Fig. 4.3.2c. This choice represents simple symmetric linear combinations of the unknown hinge moments and agrees with the selection of the unknown quantities in the preceding solution of the problem. If the strain energy stored in the frame is a minimum with respect to these combinations, it is also a minimum with respect to the hinge moments themselves. A formal proof of this statement can be given without difficulty. It will be presented here for a system having two redundant hinge moments, but it can be easily generalized to hold for a system having any number of redundant quantities.

According to Castigliano's second theorem the strain energy U, which is a function of the unknown hinge moments M_E and M_F, must have a stationary value with respect to these unknown moments. In a mathematical formulation

$$\partial U / \partial M_E = 0$$

$$\partial U / \partial M_F = 0$$

4.3.22

If the independent linear combinations

$$p = \alpha_1 M_E + \alpha_2 M_F$$

$$q = \beta_1 M_E + \beta_2 M_F$$

4.3.23

are introduced, the strain energy can also be represented as a function of p and q. (In the present problem $\alpha_1 M_E$ is the bending moment in the plane of symmetry caused by M_E, $\alpha_2 M_F$ the bending moment in the plane of symmetry caused by M_F, $\beta_1 M_E$ the tension in the plane of symmetry caused by M_E, and $\beta_2 M_F$ the tension in the plane of symmetry caused by M_F.) Equations 4.3.22 can then be written in the form

$$(\partial U/\partial p)(\partial p/\partial M_E) + (\partial U/\partial q)(\partial q/\partial M_E) = 0$$

$$(\partial U/\partial p)(\partial p/\partial M_F) + (\partial U/\partial q)(\partial q/\partial M_F) = 0$$

4.3.24

But

$$\partial p/\partial M_E = \alpha_1, \qquad \partial p/\partial M_F = \alpha_2$$

$$\partial q/\partial M_E = \beta_1, \qquad \partial q/\partial M_F = \beta_2$$

and thus Eqs. 4.3.24 become

$$\alpha_1(\partial U/\partial p) + \beta_1(\partial U/\partial q) = 0$$

$$\alpha_2(\partial U/\partial p) + \beta_2(\partial U/\partial q) = 0$$

4.3.25

In agreement with Cramer's rule (see Art. 3.5) the only solution of these simultaneous linear equations in $\partial U/\partial p$ and $\partial U/\partial q$ is

$$\partial U/\partial p = 0 \quad \text{and} \quad \partial U/\partial q = 0 \qquad 4.3.26$$

provided that the determinant of the coefficients does not vanish

$$\begin{vmatrix} \alpha_1 & \beta_1 \\ \alpha_2 & \beta_2 \end{vmatrix} = \alpha_1\beta_2 - \beta_1\alpha_2 \neq 0$$

But, if this determinant vanishes, the determinant of the coefficients of Eqs. 4.3.23, namely

$$\begin{vmatrix} \alpha_1 & \alpha_2 \\ \beta_1 & \beta_2 \end{vmatrix} = \alpha_1\beta_2 - \beta_1\alpha_2$$

must also vanish. This is impossible unless q differs from p only by a multiplying factor, or in other words unless p and q are not two linearly independent combinations of M_E and M_F. Equations 4.3.26 give expression therefore to the fact that a stationary value of the strain energy with respect to two linearly independent combinations of the unknown moments insures a stationary value with respect to the two unknown moments.

This justifies the choice of N and T as the unknown quantities. The total moment in the beam becomes

$$M = (w/50)(-9a^2 + 50ax - 25x^2) + N, \qquad 0 \leq x \leq 2a \qquad 4.3.27a$$

In the column the total moment is

$$M = (wa/100)(-18a + 27\xi) + N - T\xi, \qquad 0 \le \xi \le a \qquad 4.3.27b$$

Minimization of the strain energy with respect to N yields the equation

$$-(43/3000)wa^2 + \tfrac{6}{5}N - \tfrac{1}{2}Ta = 0$$

Minimization with respect to T leads to

$$\tfrac{1}{2}Na^3 - \tfrac{1}{3}Ta^3 = 0$$

The latter can be solved for T:

$$T = \tfrac{3}{2}(N/a)$$

Substitution of this value in the former equation yields N:

$$N = (43/1350)wa^2 \qquad\qquad 4.3.28$$

which leads to

$$T = (129/2700)wa \qquad\qquad 4.3.29$$

The contributions of N and T to the moments in the frame are shown in Figs. 4.3.2e and f. They modify the basic moments of Fig. 4.3.2d only slightly which proves that the choice of the hinges was a suitable one. An error of 10 per cent in the value of N causes a change of $(43/13,500)wa^2 = 0.00318wa^2$ in the correct value of $0.074wa^2$ of the moment at the bottom of the column. This amounts to only 4.3 per cent. Hence inaccuracies of a numerical or graphic integration are not likely to cause trouble when the basic moments M_0 are calculated in the system with fictitious hinges.

How to Select the Unknown Quantities

The choice of hinges in strategic locations rather than a complete section through the beam has one more practical advantage. The moments in a statically indeterminate frame depend on the bending-rigidity distribution, and the analysis has to be repeated if the bending rigidity is changed anywhere in the structure. It is of importance to the designer to select the proper bending rigidities in the preliminary design phase of his work before he spends much time on the analysis. If the M_0 diagram itself is a good approximation to the final moment diagram, the bending rigidities can be chosen properly, and the results of the analysis of the statically indeterminate system are not likely to necessitate alterations in the original dimensions. This saves the work involved in one or more repetitions of the analysis.

When the frame is more complex, the manner in which it is reduced to a simple structure by means of fictitious hinges and sections is of even greater importance. The basic rules governing the choice of the redundant quantities can be stated in the following manner:

1. The simple structure obtained through the insertion of fictitious hinges and sections should have a moment diagram under the given loads that resembles as closely as possible the final moment diagram of the redundant structure.

2. The effect of each unknown quantity should be restricted to the smallest possible portion of the structure.

Fulfillment of the second requirement shortens the work of calculating the influence coefficients and at the same time reduces the interaction between the various unknowns. This simplifies the solution of the simultaneous linear equations from which the unknowns are determined. Nevertheless the work involved in the solution of the problem of the moments in a multistory building frame is very great and sufficiently intricate to deter most structural engineers from attempting it by means of Castigliano's second theorem. A much simpler solution is possible with the aid of the Hardy Cross moment-distribution method.

Frame Analysis by Means of the Hardy Cross Method

This method was explained in Art. 1.17 and was applied to the calculation of secondary stresses in a framework in Art. 1.18. Here it will be used first to recalculate the moments in the frame of Fig. 4.3.1.

The stiffness S_c of the column is

$$S_c = 4EI/L = 4EI_0/a$$

The stiffness of the beam is

$$S_c = 4EI/L = 20EI_0/2a = 10EI_0/a$$

Consequently at joints B and C the distribution factors are

$$D_b = 10(EI_0/a)/[10(EI_0/a) + 4(EI_0/a)] = \tfrac{5}{7} = 0.714 \qquad 4.3.30$$

for the beam and

$$D_c = 2/7 = 0.286 \qquad 4.3.31$$

for the column. The fixed-end moment is

$$M_{\text{fe}} = \tfrac{1}{12}wL^2 = \tfrac{1}{3}wa^2$$

Since the numerical values of w and a are not given, the most convenient procedure is to take

$$M_{\text{fe}} = \tfrac{1}{3}wa^2 = 1 \qquad\qquad 4.3.32$$

in the numerical work.   Details of the computations are given in Fig. 4.3.3.

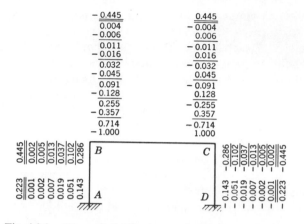

Fig. 4.3.3.   Moment-distribution analysis of frame subjected to symmetric loads.

The fixed-end moment acting on the beam at B is -1 while that acting on the beam at C is $+1$ in accordance with the frame convention.   The reaction of the former moment, namely a moment of $+1$ in.-lb, is acting on joint B.   The joint is balanced by a moment of -0.714 in.-lb acting on it from the beam and one of -0.286 from the column.   The reactions of these moments, namely $+0.714$ and $+0.286$, act on beam and column, respectively, and are listed in the figure near the end of the bar to which they are applied.   The arrangement of the figure is similar to that used in Art. 1.18.

A horizontal line drawn above these figures signifies that joint B is now balanced.   Because of the symmetry of the system it is advantageous to undertake simultaneously the same operations on the right half of the frame as on the left half.   When both B and C are balanced and the balancing moments are distributed in proportion to the distribution factors, the moments that are required to keep the far ends from rotating are carried over to the far ends of the bars.   At A and D the rigid-end fixation can take care of all the carry-over moments, and no moments are carried back to B and C.

The carry-over from C to B is one-half the distributed moment at C, that is, -0.357.   This is now the unbalanced moment at B, and it must be balanced in the same way as the original unbalanced moment. This time the distributed moment is 0.255 for the beam and 0.102 for the column. The balancing and distribution are again followed by a carry-over operation, and the sequence is repeated until the remaining unbalanced moments become negligibly small. This occurs after the sixth balancing operation in the example of Fig. 4.3.3. The carry-over moment disregarded is 0.002, that is two tenths of a per cent of the original fixed-end moment. The moments are added up, and the sums are written above double lines. The bending moment obtained at B is $-0.445\,\tfrac{1}{3}wa^2 = -0.148wa^2$ which is the same as the value of $-\tfrac{8}{54}wa^2$ obtained earlier and shown in Fig. 1.19.3.

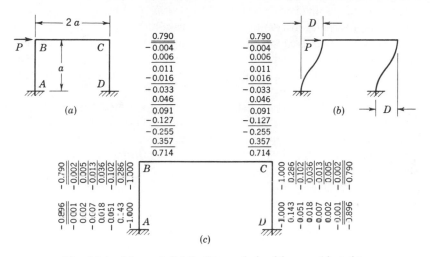

Fig. 4.3.4.   Moment-distribution analysis of frame subjected to lateral load.

The moment-distribution procedure can also be used when the frame is subjected to a lateral load, but then the sidesway must be calculated separately. Let it be assumed that the frame of Fig. 4.3.4a is moved laterally a distance D under the action of load P. For the time being the beam is supposed to remain horizontal and undeformed as shown in Fig. 4.3.4b. Calculations carried out in Art. 1.18 demonstrated that bars deform in this fashion if they are subjected to end moments.

$$M_{\text{fe}} = 6EID/L^2$$

(see Fig. 1.18.2) and shear forces

$$V = 2M_{\text{fe}}/L$$

where L is the length of the bar. In the present example the total shear force P (see Fig. 4.3.4) is evenly distributed between the two columns of length a.   Hence

$$V = P/2 = 2M_{\text{fe}}/a = 12EI_0D/a^3$$

that is,

$$D = Pa^3/24EI_0 \qquad\qquad 4.3.33$$

and

$$M_{\text{fe}} = Pa/4 \qquad\qquad 4.3.34$$

The fixed-end moment acting on the column at B is obviously counter-clockwise and thus negative.   For the sake of convenience its numerical value will be assumed to be unity.   Because of the antisymmetry of the system the fixed-end moment acting on the column at C is also negative. With these two negative moments as the unbalanced quantities, the moment distribution can be carried out in the same manner as before.   Details of it can be seen in Fig. 4.3.4c.   Again operations are undertaken simultaneously on the two sides of the frame.

The result of the moment distribution is positive moments $0.790(Pa/4)$ $= 0.1975Pa$ acting on the ends of the beams, negative moments of equal magnitude acting on the upper ends of the columns, and negative moments $-0.896(Pa/4) = -0.224Pa$ acting on the bottoms of the columns. During the balancing process, joints B and C were allowed to rotate, and in the final rotated configuration the shear force transmitted by the columns is not the same as it was in the configuration shown in Fig. 4.3.4b.   The magnitude of the shear in column AB is, according to Eq. 1.19.20,

$$V = (1/a)(M_A + M_B) = (1/a)(-0.224 - 0.1975)Pa = -0.4215P$$

The two columns therefore transmit a shear force of $-0.843P$ instead of the total shear force $-P$.   A shear force of $-0.157P$ remains unaccounted for.   The entire calculation can now be repeated for a shear force of $-0.157P$ instead of $-P$ in order to correct the results.   Moments amounting to 0.157 times those obtained in the first calculation will result in an unbalanced shear force of $-0.157 \times 0.157P$.   If the procedure is continued indefinitely, the moment acting on the beam at B will be obtained as the sum of the geometric series

$$M_B = 0.1975Pa[1 + 0.157 + (0.157)^2 + (0.157)^3 + \ldots]$$

The sum of this series is known to be

$$M_B = 0.1975Pa\,\frac{1}{1 - 0.157} = 1.187 \times 0.1974Pa = 0.235Pa$$

Similarly the moment acting on the column at A is

$$M_A = -1.187 \times 0.224Pa = -0.266Pa$$

These results agree with the values calculated in Art. 1.19 by the slope-deflection method.   They are shown in Fig. 1.19.4.

Analysis of a Building Frame

As a final example of the use of the Hardy Cross moment-distribution method, the moments will be calculated in the structure of Fig. 4.3.5 which is representative of small office buildings.   The numerical values of the moments of inertia and the lengths of the beams and columns are not given

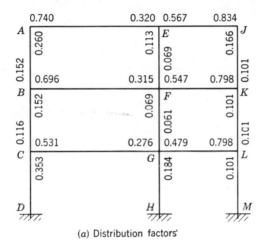

(a) Distribution factors

Fig. 4.3.5.   Moment-distribution analysis of small office building.

directly; only the distribution factors computed from these values are listed in part a of the figure.   It is of interest to observe that the distribution factors are much greater for the beams than for the columns.   This is a consequence of the fact that the columns of such office buildings are less heavily loaded, and thus have smaller sections, than the beams that have to support considerable transverse loads.   In skyscraper-type buildings, of course, the columns between the lower floors also have large sections.

The analysis will be carried out for the case when a uniformly distributed load is applied to beam BF only.   The fixed-end moments caused by it are taken as 1000.   The small distribution factors of the columns indicate that there is little interaction between the various floors.   It is advantageous therefore to balance the joints of the loaded floor rather completely before proceeding to the adjacent floors.

In Fig. 4.3.5*b* the joints of the loaded floor were balanced in the following order: *B, F, K, B, F, K,* and *B*.   At this stage of the moment-distribution process the only unbalance on floor *BFK* was 36 at joint *F*, while on floor *AEJ* the unbalanced moments amounted to 95, —55, and 22.   This

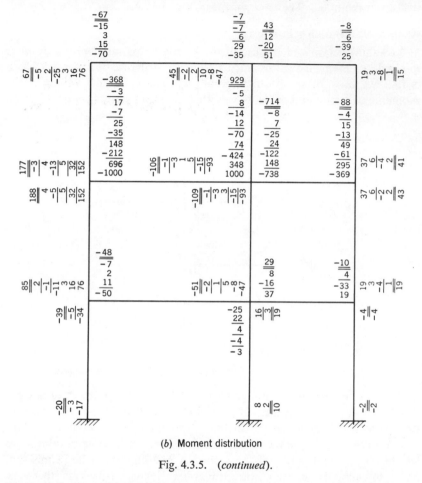

(*b*) Moment distribution

Fig. 4.3.5.   (*continued*).

floor was balanced next in the sequence *A, E, J,* and *E,* and the process was continued on the lowest floor in the sequence *C, G, L, G,* and *C*.   The largest unbalanced moment on floor *CGL* was then 4 which is less than one-half per cent of the original fixed-end moment.   On the top floor the unbalanced moments were 18 and 6 at joints *A* and *J*, respectively.   The loaded middle floor has accumulated an unbalance of —18 at *B*, and one of —6 at *K*.   At the same time the moment of 36 left at *F* was increased to 45.

It can be seen therefore that the influence of the rigidities of the adjacent floors is hardly felt at the loaded floor.   A very good engineering approximation could have been had in this example if the adjacent floors had been assumed perfectly rigid.   As this was not done, the carry-back moments as well as the original unbalanced moment on floor *BFK* were reduced further by balancing joints *F*, *K*, *B*, and *F* in turn.   Finally joint *A* was balanced once more.   After this operation the largest remaining unbalanced moment was —9 at *E*.   As this was less than 1 per cent of the original fixed-end moment, no further balancing was undertaken, and the moments were added up.   The sums are listed above or below the double lines.

The results show that the continuous beam at *F* provides almost as much restraint as a rigid wall.   The fixed-end moment was reduced from 1000 to 929.   On the other hand, at *B* the final end moment is only —368 as compared to the original —1000.   The moments in the columns are significant at *F*, and particularly at *B*.   In the top and bottom beams all the bending moments are small.

As a rule, the Hardy Cross moment-distribution method permits a much more rapid determination of the moments in a complex building than is possible with any of the analytical methods.   Its drawback is, of course, that it gives only a numerical solution of one particular problem and cannot yield formulas of general validity.

4.4   Maxwell's Reciprocal Theorem and Castigliano's First Theorem

Derivation of Maxwell's Reciprocal Theorem

In the derivation of the first and second minimal principles the true state of stress and strain was compared to slightly different states which could not possibly have existed without changes in the stress-strain law of the material.   As a matter of fact these changes would have had to vary from point to point in the structural element.   If the stress-strain law is given, a variation of the true state of displacements according to the procedure of the first minimal principle leads to unbalanced internal forces and moments in the elements, while the variation of the stresses in accordance with the second minimal principle results in cracks, that is, in incompatible deformations.   All these inconsistencies are absent only when the elastic body is in its true state of stress and strain.   The investigations presented have shown that this true state is characterized by minimal values of the total potential and the total complementary potential.

In the present article a third type of variation will be undertaken. Stresses and strains will be varied simultaneously, and the loads will also be changed in order to maintain equilibrium in the varied state.   All states compared will be possible states of equilibrium and deformations, at variance with the procedure adopted in the earlier investigations. Such a comparison of true states of stress and strain corresponding to different loads is simpler to make than the variations presented in Arts. 2.2 and 4.1. However, it does not yield principles of such fundamental nature as the first and second minimal principles. In the more limited fields of application the theorems to be derived here are often more convenient to use than the more general principles already discussed.

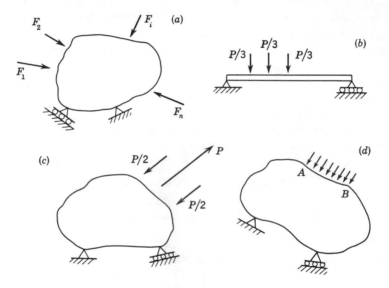

Fig. 4.4.1.   Forces and generalized forces.

Let the elastic body shown in Fig. 4.4.1a, for which Hooke's law and the principle of superposition are assumed to be valid, and which is supported in a manner that precludes rigid-body motion, be loaded with a single load F_1. The elastic body deforms under the load, and thus the point of application of F_1 moves. Since in accordance with Hooke's law displacements are proportional to the applied load, the component of the displacement of the point of application of F_1 in the direction of F_1 can be expressed by the product

$$a_{11}F_1$$

where the proportionality factor a_{11} is an influence coefficient. Its physical meaning is the displacement of the point of application of F_1 in the direction of F_1 when F_1 is equal to unity.

F_1 causes deformations in the entire elastic body. If the forces $F_2, F_3, \cdots, F_n$ are all zero, and the lines indicating them in Fig. 4.4.1a only serve the purpose of designating directions and senses at a number of points on the surface of the elastic body, the displacement of the ith point in the direction indicated by F_i caused by F_1 can be expressed in a consistent manner by the product

$$a_{i1}F_1$$

Similarly, if F_2 is the only non-zero force ($F_1 = F_3 = F_4 = \cdots = F_n = 0$), the displacement of point 1 in the direction indicated by F_1 can be written as

$$a_{12}F_2,$$

and that of point 2 in the direction of the force F_2 as

$$a_{22}F_2$$

In general, if the kth force alone is acting, the product $a_{ik}F_k$ designates the displacement of the ith point in the direction indicated by F_i as caused by the force F_k. In the notation adopted the first letter of the double subscript refers to the place where the distortion is sought, and the second to the force that caused it. If all the n forces act simultaneously, and the total displacement of the ith point in the direction indicated by F_i is denoted by d_i, the principle of superposition can be used to express the total deflections with the aid of the deflections caused by the individual forces:

$$a_{11}F_1 + a_{12}F_2 + \cdots + a_{1i}F_i + \cdots + a_{1n}F_n = d_1$$
$$a_{21}F_1 + a_{22}F_2 + \cdots + a_{2i}F_i + \cdots + a_{2n}F_n = d_2$$

$$\cdots \qquad \qquad \qquad \qquad \qquad \qquad \text{4.4.1}$$

$$a_{i1}F_1 + a_{i2}F_2 + \cdots + a_{ii}F_i + \cdots + a_{in}F_n = d_i$$

$$a_{n1}F_1 + a_{n2}F_2 + \cdots + a_{ni}F_i + \cdots + a_{nn}F_n = d_n$$

The work done by the ith force F_i during the elastic deformations can be easily calculated if it is assumed that all the n forces are applied simultaneously and are increased from zero to their final value at such a slow rate that equilibrium between loads and internal stresses is continuously maintained. The average value of F_i during this loading process is

$\frac{1}{2}F_i$, the total displacement is d_i, and thus the work done is $\frac{1}{2}F_i d_i$. The work W done by all the forces can be obtained by summation and must be equal to the strain energy U stored in the body since the body is considered perfectly elastic. Hence

$$U = W = \tfrac{1}{2}(F_1 d_1 + F_2 d_2 + \cdots + F_i d_i + \cdots + F_n d_n) \qquad 4.4.2$$

Two important theorems can be derived from Eqs. 4.4.1 and 4.4.2. First let it be assumed that F_1 alone is applied, and that it increases slowly from zero to its final value.   The deflection under F_1 is $a_{11}F_1$, and the work done $\tfrac{1}{2}a_{11}F_1^2$. Next the elastic body is loaded with F_2. As point 2 deflects through the distance $a_{22}F_2$, the work done by F_2 is $\tfrac{1}{2}a_{22}F_2^2$. But at the same time the point of application of F_1 is also displaced in consequence of the application of F_2, and the displacement is $a_{12}F_2$. Since during this second displacement of its point of application F_1 remains constant, the work done by F_1 is $a_{12}F_1 F_2$ and not $\tfrac{1}{2}a_{12}F_1 F_2$. Hence the total work W done by the applied loads, that is, the strain energy U stored in the body is

$$U = W = \tfrac{1}{2}a_{11}F_1^2 + a_{12}F_1 F_2 + \tfrac{1}{2}a_{22}F_2^2 \qquad 4.4.3$$

If the same body is loaded in the inverse order, namely, first with F_2, then with F_1, the work done by the forces can be calculated in an analogous manner.   In the first step the work done by F_2 is $\tfrac{1}{2}a_{22}F_2^2$.   In the second step F_1 does the work $\tfrac{1}{2}a_{11}F_1^2$, and at the same time F_2 performs the work $a_{21}F_1 F_2$.   The total is

$$U = W = \tfrac{1}{2}a_{11}F_1^2 + a_{21}F_1 F_2 + \tfrac{1}{2}a_{22}F_2^2 \qquad 4.4.4$$

The strain energy is independent of the order in which the loads are applied, because the principle of superposition is assumed to hold. Consequently the right-hand members of Eqs. 4.4.3 and 4.4.4 must be identical, from which it follows that

$$a_{12} = a_{21} \qquad 4.4.5$$

In the same manner an analogous statement can be proved for any pair of forces F_i and F_k:

$$a_{ik} = a_{ki} \qquad 4.4.6$$

The relation expressed by Eq. 4.4.6 is known as Maxwell's reciprocal theorem.   It can be worded as follows:

If an elastic body, for which Hooke's law and the **Theorem 17** principle of superposition hold, is supported in a manner that precludes rigid-body motion, and at each of two points of the surface of the body a direction is designated, then the displacement of point 1 in the direction there designated, caused by a unit force applied at point 2 in the direction there designated is equal to the displacement at point 2 caused by a unit force at point 1, both in the directions there designated.

Castigliano's First Theorem

The second important theorem is obtained if Eq. 4.4.2 is differentiated with respect to any, say the kth, force:

$$\partial U/\partial F_k = \tfrac{1}{2}[F_1(\partial d_1/\partial F_k) + F_2(\partial d_2/\partial F_k) + \cdots + F_i(\partial d_i/\partial F_k) + \cdots$$
$$+ F_k(\partial d_k/\partial F_k) + d_k + F_{k+1}(\partial d_{k+1}/\partial F_k) + \cdots + F_n(\partial d_n/\partial F_k)]$$

From Eqs. 4.4.1 it follows that

$$\partial d_i/\partial F_k = a_{ik}$$

Consequently

$$\partial U/\partial F_k = \tfrac{1}{2}(F_1 a_{1k} + F_2 a_{2k} + \cdots + F_i a_{ik} + \cdots$$
$$+ F_k a_{kk} + d_k + F_{k+1} a_{k(k+1)} + \cdots + F_n a_{kn})$$

If the right-hand member of this equation is rearranged, and use is made of Eqs. 4.4.6 and 4.4.1, one obtains

$$\partial U/\partial F_k = \tfrac{1}{2}(d_k + F_1 a_{k1} + F_2 a_{k2} + \cdots + F_i a_{ki} + \cdots + F_n a_{kn})$$
$$= \tfrac{1}{2}(d_k + d_k) = d_k$$

The statement contained in the equation

$$\partial U/\partial F_k = d_k \qquad\qquad 4.4.7$$

is known as Castigliano's first theorem. It can be expressed in the following manner:

Let an elastic body, for which Hooke's law and the principle of superposition hold, be supported in a manner that precludes rigid-body motion. If this body is loaded with concentrated forces, and the strain energy stored in the body is calculated and expressed as a function of the applied forces, then the partial derivative of the strain energy with respect to any one applied force is equal to the displacement of the point of application of that force in the direction of the force. **Theorem 18**

In the derivation of Castigliano's theorem it was tacitly assumed that the derivative of the strain-energy function existed. In other words, prerequisites for the validity of the theorem are that the strain energy be a continuous function of the applied forces, and have continuous derivatives with respect to all the forces. Fulfillment of this requirement is an axiom of the mathematical theory of elasticity. The engineer has no reason to expect structural materials to behave differently.

Figure 4.4.1 shows the elastic body as if it were two-dimensional, and the supports represent three constraints as required for the prevention of a rigid-body motion in a two-dimensional system. But in the development of the theorem every argument used would hold in three-dimensional space as well, as long as the necessary three-dimensional supports of the body are available. Consequently Castigliano's theorem as stated is valid for three-dimensional bodies also.

Generalized Forces and Displacements

The version of Castigliano's first theorem just presented can be generalized considerably. If one or more of the F in Eq. 4.4.1 are concentrated moments (couples) rather than forces, the equations remain valid provided the definition of the corresponding influence coefficients is suitably modified. For instance, if F_i represents a force at the ith point while F_k is a moment at the kth point, then a_{ik} is the displacement of point i in the direction of F_i caused by a unit moment applied at point k in the direction of F_k. Similarly, a_{ki} is the rotation of the elastic body at point k in the direction of F_k caused by a unit force applied at point i in the direction of F_i. It can be easily verified that with these generalized concepts of force and displacement every step in the proof remains unchanged. Consequently Maxwell's reciprocal theorem and Castigliano's first theorem hold true also for generalized forces and displacements.

As a matter of fact, the theorems must hold regardless of the nature of the generalized force, as long as the displacement is chosen to correspond with it so that Eqs. 4.4.1 and 4.4.2 are valid. For instance, the *generalized force P* in Fig. 4.4.1*b* consists of the three equal and equidistant forces $P/3$. The corresponding displacement is the average of the displacements of the points of application of the three individual forces. In Fig. 4.4.1*c* the generalized force is the group of three forces consisting of P and its balancing forces $P/2$. Here the displacement corresponding to the generalized force is the relative displacement of the point of application of P with respect to the line connecting the points of application of the two forces of magnitude $P/2$. In Fig. 4.4.1*d* the distributed load can be considered a generalized force to which the average displacement along line AB in the direction of the load corresponds as the generalized displacement. The warping group of unit forces discussed in Arts. 1.10 and 1.11 and the warping itself are also examples of a generalized force and the corresponding displacement.

Castigliano's first theorem can be used to advantage in deriving formulas for displacements and generalized displacements. For instance, all the expressions obtained in Part 1 by means of the principle of virtual displacements could also have been derived by Castigliano's first theorem. One application of the theorem is of special interest: The deflection of a beam on three non-yielding supports can be calculated formally at one of the supports by differentiating the strain energy stored in the beam with respect to that support reaction. This deflection is known to be zero. If the derivative is set equal to zero, an equation is obtained from which the unknown support reaction can be calculated. The considerations just presented can be generalized in the statement of the following lemma:

> The partial derivative of the strain energy stored in an elastic **Lemma**
> body with respect to a redundant reaction at a non-yielding
> support is zero.

From this lemma the following theorem can be deduced immediately:

> The statically indeterminate reactions make the strain **Theorem 19**
> energy a minimum.

The lemma is a necessary but not a sufficient condition for the existence of a minimum, but the stationary value of the strain energy is known to be a minimum when the system is in a stable state of equilibrium. A further generalization to a body with internal redundant quantities is Castigliano's second theorem which was derived in a different manner in Art. 4.2 as Corollary 5 to the complementary-energy principle (Theorem 15).

Examples

To show that Castigliano's first theorem yields the same results as the principle of virtual displacements, two examples given in Art. 1.12 will be recalculated. Item 3 of Fig. 1.12.8 is a beam on two simple supports loaded at its right end by a concentrated moment B. If x denotes the distance from the left end, the moment M in the beam is

$$M = (x/L)B$$

Consequently the strain energy stored in the beam is

$$U = \frac{1}{2EI} \int_0^L \left(\frac{x}{L}\right)^2 B^2\, dx = \frac{1}{6}\left(\frac{B^2 L}{EI}\right)$$

Castigliano's first theorem yields for the end rotation α at the right end

$$\alpha = \partial U/\partial B = \tfrac{1}{3}(BL/EI)$$

which agrees with the result given in the figure.

The rotation at the left end can be evaluated if a dummy moment C is assumed to act at the left end in addition to the actual moment B at the right end. The total moment in the beam is then

$$M = (x/L)B + [(L - x)/L]C$$

The strain energy is

$$U = \frac{1}{2EI} \int_0^L M^2 \, dx$$

and the rotation α at the left end, from Castigliano's theorem,

$$\alpha = \frac{\partial U}{\partial C} = \frac{1}{EI} \int_0^L M \frac{\partial M}{\partial C} \, dx$$

$$= \frac{1}{EI} \int_0^L \left(\frac{x}{L} B + \frac{L - x}{L} C \right) \frac{L - x}{L} \, dx$$

The angle of rotation in this equation is caused by a moment B acting at the right end and a moment C acting at the left end. Since the latter was introduced only for the purpose of calculation and is not part of the actual problem, C must be set equal to zero. Integration then yields

$$\alpha = \frac{B}{EI} \int_0^L \left[\frac{x}{L} - \left(\frac{x}{L} \right)^2 \right] dx = \frac{1}{6} \left(\frac{BL}{EI} \right)$$

which is the same result as that given in Fig. 1.12.8.

As an illustration of Maxwell's reciprocal theorem two deflection quantities calculated in Art. 1.12 may be mentioned. According to Fig. 1.12.6, item 1, the end deflection of a cantilever beam on whose end a concentrated moment B is acting is $\frac{1}{2}BL^2/EI$. Figure 1.12.8, item 1, shows that the end rotation of a cantilever beam is $\frac{1}{2}WL^2/EI$ if the load is a concentrated force W at the free end of the cantilever. The two expressions are obviously equal if B is a unit moment, W a unit force, and displacement and rotation are measured in consistent units. Another example was given at the end of Art. 1.8.

4.5   The Ultimate Load Carried by a Clamped Beam

Stress-Strain Relation

Structures are usually analyzed on the assumption that Hooke's law is valid for their material of construction. This assumption is satisfactory only as long as the stresses are sought under the working loads and the limit of proportionality of the material is not exceeded. When one wants to find the load under which the structure would fail, deformations considerably larger than those prevailing at the proportional limit must generally be taken into consideration. At these large deformations the deviation of the actual stress-strain curve from Hooke's straight line cannot be disregarded.

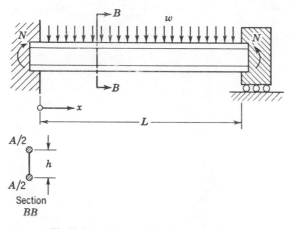

Fig. 4.5.1.   Beam with clamped ends.

The failing load must be known when the true safety of the structure is to be established because the safety factor is defined in the simplest manner as the ratio of the failing load to the maximum load to which the structure is likely to be subjected. The failing load can be determined by experiment or by calculation. In the latter case naturally the actual mechanical properties of the material must be taken into account. With most materials the properties differ considerably from the linear connection between stress and strain assumed in the theory of elasticity.

Fortunately the minimal principles are also valid for non-linear stress-strain relationships. The principle of complementary energy will now be used to examine the behavior of a beam beyond the elastic limit when the two ends of the beam are rigidly fixed (see Fig. 4.5.1). The end fixation means that the end tangents of the beam must remain horizontal during the loading process, but end point $x = L$ is permitted to approach end point $x = 0$ if large deflections of the beam give rise to such motion. With the distance fixed, tensile forces would arise in the beam which are not taken into account in this analysis.

In the calculations an analytical connection between stress and strain is needed. The stress-strain curve differs from material to material and is influenced by heat treatment and cold work previous to loading. An accurate analytical representation of the curve is complicated, and it leads to mathematical difficulties.   When the material is ordinary structural steel, a diagram consisting of two straight lines is a satisfactory approximation for the purposes of this article.

Such an approximation is shown in Fig. 4.5.2.   The initial portion has a slope $d\sigma/d\epsilon = E$ as in the theory of elasticity.   When the stress exceeds a critical value σ_y, which can be considered as the yield stress of the diagram, the slope decreases by a factor C.   This portion of the diagram is to be used when the load is increasing.   When loading takes place up to point P

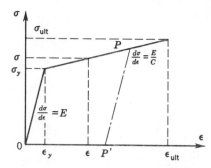

Fig. 4.5.2.   Simplified stress-strain curve.

and afterward the loads are taken off, only the elastic part of the strain is recovered and line PP', parallel to the initial portion of the diagram, represents the stress-strain relationship for unloading.   It is assumed further that the material behaves in the same manner in tension and in compression.

Calculation of the Complementary Energy

To facilitate the calculations, the cross-sectional shape of the beam is idealized, as shown in section BB of Fig. 4.5.1.   The total cross-sectional area A is assumed to be concentrated in the two flanges which are spaced a distance h apart.   They are connected by a web of negligibly small area but infinite shearing rigidity.   This idealization represents a satisfactory approximation to the actual conditions prevailing with an efficient I beam which has a thin web and two heavy flanges spaced far apart.

The moment of inertia of the idealized section is

$$I = \tfrac{1}{4}Ah^2 \qquad\qquad 4.5.1$$

and its section modulus

$$Z = \tfrac{1}{2}Ah \qquad\qquad 4.5.2$$

Consequently the tensile stress caused by a positive moment M in one flange is

$$\sigma = 2M/Ah = Mh/2I \qquad 4.5.3$$

The compressive stress in the other flange is equal in magnitude and opposed in sense. When the stress is smaller than the yield stress, the complementary energy dU' stored in a portion of the beam extending from x to $x + dx$ is

$$dU' = \tfrac{1}{2}(\sigma^2/E)A\,dx \qquad 4.5.4$$

where $A\,dx$ is the volume of the two flanges in the portion of the beam considered. Substitution of the last member of Eq. 4.5.3 for σ yields

$$dU' = \tfrac{1}{8}(M^2h^2/EI^2)A\,dx$$

which can also be written, because of Eq. 4.5.1, as

$$dU' = \tfrac{1}{2}(M^2/EI)\,dx \qquad 4.5.5$$

This equation is valid if the stresses caused by M are in the elastic range.

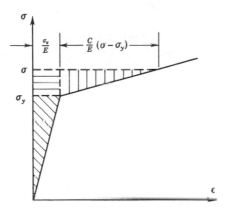

Fig. 4.5.3. Complementary energy.

When σ exceeds σ_y (which is the yield stress in tension), the complementary energy is represented by the shaded area in Fig. 4.5.3. It can be given as

$$dU' = [\tfrac{1}{2}\sigma_y(\sigma_y/E) + (\sigma - \sigma_y)(\sigma_y/E) + \tfrac{1}{2}(\sigma - \sigma_y)(C/E)(\sigma - \sigma_y)]A\,dx$$

Manipulations yield

$$dU' = \tfrac{1}{2}(A/E)[(C - 1)(\sigma - \sigma_y)^2 + \sigma^2]\,dx$$

Because of Eq. 4.5.3 one has finally

$$dU' = (1/2EI)[(C - 1)(M - M_y)^2 + M^2]\,dx, \qquad M > 0 \qquad 4.5.6$$

where M_y is the absolute value of the moment under which the beam begins to yield.   From Eq. 4.5.3 one can calculate M_y as

$$M_y = 2I\sigma_y/h, \qquad \sigma_y > 0 \qquad\qquad 4.5.7$$

where σ_y is the yield stress in tension.   When σ_y is the compressive yield stress and is considered negative, one must write

$$M_y = -2I\sigma_y/h, \qquad \sigma_y < 0 \qquad\qquad 4.5.7a$$

In the derivation of Eq. 4.5.6, M was assumed to be positive.   When M is negative, the yield stress in compression must be introduced in the calculations and M_y must be replaced by $-M_y$.   Equation 4.5.6 then becomes

$$dU' = (1/2EI)[(C - 1)(M + M_y{}^2) + M^2]\,dx, \qquad M < 0 \quad 4.5.6a$$

Equations 4.5.6 and 4.5.6a are valid in the inelastic range.   They reduce to Eq. 4.5.5 when $C = 1$, that is, when the broken line in Figs. 4.5.2 and 4.5.3 is replaced by a straight line having the slope of the initial portion of the stress-strain diagram.

At the beginning of the loading process, when w is small, the stresses are in the elastic range throughout the beam.   Then the bending-moment distribution is governed by the laws of elasticity, and the end moment N is given by

$$N = -\tfrac{1}{12}wL^2 \qquad\qquad 4.5.8$$

as was found in Art. 1.13.   At the same time the moment M_m in the middle of the beam is

$$M_m = \tfrac{1}{24}wL^2 \qquad\qquad 4.5.9$$

As w is increased, the ordinates of the bending-moment diagram are increased in the same proportion, and the shape of the diagram remains unchanged.   The yield stress is first reached at the two ends of the beam. The value of the load at which this occurs is designated by w_y.   When w exceeds w_y, the elastic solution loses its validity, and the end moment N must be calculated from the complementary-energy principle.

Let it be assumed that at some value of w greater than w_y the bending-moment diagram is represented by Fig. 4.5.4.   The absolute value of the end moment N is greater than M_y, and the bending moment M is equal to $-M_y$ at a distance p from the fixed end.   As M is negative between 0 and p, the complementary energy of the system is obtained by integrating the expression given in Eq. 4.5.6a from 0 to p, adding to the result the integral of the expression in Eq. 4.5.5 evaluated between the limits p and $L/2$, and multiplying the sum by 2:

$$U' = \frac{1}{EI} \int_0^p [(C-1)(M+M_y)^2 + M^2]dx + \frac{1}{EI}\int_p^{L/2} M^2\,dx$$

or

$$U' = \frac{1}{EI} \int_0^p (C-1)(M+M_y)^2\,dx + \frac{1}{EI}\int_0^{L/2} M^2\,dx$$

If both sides of the equation are multiplied by EI/M_y^2 and the notation

$$M/M_y = m \qquad\qquad 4.5.10$$

is introduced, one obtains

$$\frac{EIU'}{M_y^2} = \int_0^p (C-1)(m+1)^2\,dx + \int_0^{L/2} m^2\,dx \qquad 4.5.11$$

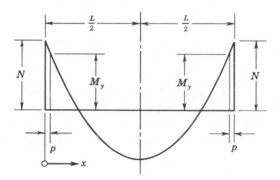

Fig. 4.5.4.   Bending-moment diagram after onset of yielding.

From considerations of static equilibrium the bending moment acting in a section x of the beam in Fig. 4.5.1 can be given as

$$M = (wL/2)x - w(x^2/2) + N$$
$$= \tfrac{1}{2}wL^2(x/L) - \tfrac{1}{2}wL^2(x/L)^2 + N$$

With the notation

$$R = w/w_y \qquad\qquad 4.5.12$$

where R is the load per unit length of the beam divided by the value of the load per unit length at the moment when the yield stress is reached at the fixed ends, the expression for the bending moment can be written in the form

$$M = 6(wL^2/12)[(x/L) - (x/L)^2] + N$$
$$= 6R(w_y L^2/12)[(x/L) - (x/L)^2] + N$$

It should now be remembered that in the elastic range the end moment N is given by Eq. 4.5.8. This equation is naturally still valid when the yield stress is reached, but is not exceeded, at the rigid support. At that time

$$N = -\tfrac{1}{12}w_y L^2$$

The absolute value of the moment under which yielding begins is denoted by M_y. Consequently

$$M_y = \tfrac{1}{12}w_y L^2 \qquad\qquad 4.5.13$$

and

$$M = 6RM_y[(x/L) - (x/L)^2] + N$$

Division by M_y and use of the symbol

$$n = N/M_y \qquad\qquad 4.5.14$$

lead to

$$m = 6R[(x/L) - (x/L)^2] + n \qquad\qquad 4.5.15$$

This is the value of m in Eq. 4.5.11.

Minimization

According to the principles derived in Art. 4.1 the complementary energy must be a minimum with respect to the unknown reaction moment N because the ends are rigidly fixed and N can do no work (Corollary 1 to Theorem 15). Since n differs from N only by a constant factor, U' is also a minimum with respect to n. Consequently the derivative of U', as given in Eq. 4.5.11, with respect to n must vanish. In the expression for U' the quantity $m = M/M_y$ is a function of n, as is evident from Eq. 4.5.15. Moreover p, the upper limit of the first definite integral, also depends on n; whenever n, and thus the end moment $N = nM_y$, is changed, the distance p changes at which the bending moment is equal to $-M_y$. The dependence of the limit of integration on the parameter n with respect to which the minimization is carried out introduces a slight complication in the differentiation.

It is shown in the calculus that the derivative of the definite integral

$$\int_{a(\alpha)}^{b(\alpha)} f(x, \alpha)\, dx$$

with respect to the parameter α is given by

$$\frac{\partial}{\partial \alpha} \int_{a(\alpha)}^{b(\alpha)} f(x, \alpha)\, dx = \int_{a(\alpha)}^{b(\alpha)} \frac{\partial f(x, \alpha)}{\partial \alpha}\, dx + f(b, \alpha)\frac{db}{d\alpha} - f(a, \alpha)\frac{da}{d\alpha}$$

$$4.5.16$$

In this equation $f(x, \alpha)$ designates a function of the independent variable x and of the parameter α; and the limits $a(\alpha)$ and $b(\alpha)$ are functions of the parameter α only. The last two terms in the right-hand member of Eq. 4.5.16 naturally vanish when the limits of integration a and b are not functions of the parameter but fixed constants, since then $da/d\alpha = db/d\alpha = 0$. Equation 4.5.16 holds if the function f as well as its derivative with respect to α is continuous between the limits a and b. The symbols $f(b, \alpha)$ and $f(a, \alpha)$ represent the values of the expression under the integral sign at the upper and lower limits.

This equation is known as Leibnitz' rule. When it is applied to the problem of minimizing U' in Eq. 4.5.11, the constant multiplier $EI/M_y{}^2$ can be disregarded. In the first integral of the right-hand member, $(m + 1)^2$ is a function of the independent variable x and of the parameter n as is evident from Eq. 4.5.15. If the parameter n of Eq. 4.5.15 is identified with the parameter α of Eq. 4.5.16, the quantity under the first integral sign can be written as

$$f(x, \alpha) = (C - 1)\{6R[(x/L) - (x/L)^2] + n + 1\}^2$$

Its derivative with respect to n is

$$(\partial/\partial\alpha)f(x, \alpha) = 2(C - 1)\{6R[(x/L) - (x/L)^2] + n + 1\}$$

If the expression is integrated with respect to x, one obtains

$$\int \frac{\partial}{\partial\alpha}f(x, \alpha)\,dx = 2(C - 1)\left\{6RL\left[\frac{1}{2}\left(\frac{x}{L}\right)^2 - \frac{1}{3}\left(\frac{x}{L}\right)^3\right] + (n + 1)x\right\}$$
$$+ \text{ constant}$$

Substitutions of the limits yields

$$\int_0^p \frac{\partial}{\partial\alpha}f(x, \alpha)\,dx = 2(C - 1)\left\{6RL\left[\frac{1}{2}\left(\frac{p}{L}\right)^2 - \frac{1}{2}\left(\frac{p}{L}\right)^3\right] + (n + 1)p\right\}$$

$$4.5.17$$

In the second term of the right-hand member of Eq. 4.5.16

$$f(b, \alpha) = [(C - 1)(m + 1)^2]_{x=p} = 0$$

because the value of $m = M/M_y$ at $x = p$ is -1 from the definition of p. The third term in the right-hand member of Eq. 4.5.16 vanishes because

$$\partial a/\partial\alpha = 0$$

as the lower limit a is a constant (zero).

In the second integral of Eq. 4.5.11 the limits are constant. Hence

$$\frac{\partial}{\partial n}\int_0^{L/2} m^2\,dx = 2\int_0^{L/2} m\,\frac{\partial m}{\partial n}\,dx = 2\int_0^{L/2}\left\{6R\left[\frac{x}{L} - \left(\frac{x}{L}\right)^2\right] + n\right\}dx$$
$$= (R + n)L$$

$$4.5.18$$

Consequently, the expression given in Eq. 4.5.11 for $EIU'/M_y{}^2$ can be a minimum only if

$$(C-1)\{6R[\tfrac{1}{2}(p/L)^2 - \tfrac{1}{3}(p/L)^3] + (n+1)(p/L)\} + \tfrac{1}{2}R + \tfrac{1}{2}n = 0 \qquad 4.5.19$$

This is the minimal condition connecting n and p with R. Another condition available for the calculation of p and n is the requirement that at $x = p$ the moment M be equal to $-M_y$, or the non-dimensional quantity $m = M/M_y$ be equal to negative unity. Substitution of -1 for m and of p for x in Eq. 4.5.15 yields

$$6R[(p/L) - (p/L)^2] + n = -1 \qquad 4.5.20$$

Equations 4.5.19 and 4.5.20 define the values of p and n for any given value of the load coefficient $R = w/w_y$ greater than unity provided inelastic deformations do not take place in the middle of the beam. In the complete absence of inelastic deformations ($R < 1$), the elastic solution given by Eqs. 4.5.8 and 4.5.9 holds. When the yield stress of the material is also exceeded in the middle of the beam, more complex calculations are required. They will be presented later.

Numerical Evaluation of Results

The computation of n and p from the simultaneous equations Eqs. 4.5.19 and 4.5.20 for a given value of R is possible but inconvenient. For this

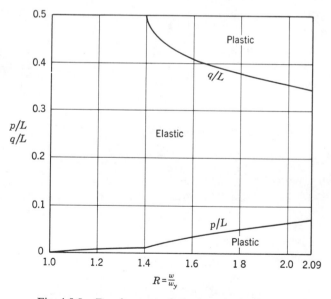

Fig. 4.5.5. Development of plastic regions. From author's brief note in *J. Appl. Mech.*

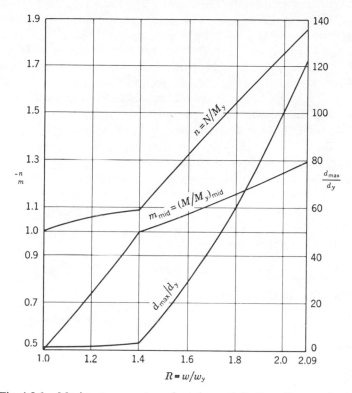

Fig. 4.5.6. Maximum moments and maximum deflection. From author's brief note in *J. Appl. Mech.*

reason n is calculated from Eq. 4.5.20 and substituted in Eq. 4.5.19. Solution for R yields

$$1/R = 1 - 6(p/L) - 6(C - 2)(p/L)^2 + 8(C - 1)(p/L)^3 \qquad 4.5.21$$

A suitable value can now be assumed for p/L and substituted in Eq. 4.5.21 to obtain R.   Equation 4.5.20 can be rewritten to read

$$-n = 6R[(p/L) - (p/L)^2] + 1 \qquad 4.5.22$$

Hence n can be computed by substituting corresponding values of p/L and R in this equation.

Computations were carried out for a beam of 12-ft length whose material properties were assumed in the following manner:

$$E = 29 \times 10^6 \text{ psi}, \qquad \sigma_y = 38,000 \text{ psi}, \qquad \sigma_{\text{ult}} = 70,000 \text{ psi}$$

The value of C was taken as 290 which, in conjunction with the values quoted, resulted in a total elongation of 32.1 per cent at the ultimate (or failing) stress of 70,000 psi. The beam section was chosen so as to represent as closely as practicable a wide-flanged I beam measuring 12 by 10 in. and weighing 64 lb per ft. The area of this section is 18.83 sq in., and its moment of inertia is 528.3 in.[4] These two values are retained for the idealized shape replacing the real section. They are compatible with Eq. 4.5.1 if h is taken as 10.6 in. The small difference between 10.6 in., the depth of the idealized section, and 12 in., which is the depth of the actual section, indicates that the idealized section is a good approximation to the real I beam. The material properties selected correspond reasonably closely to those of structural steel.

The results of the calculations are presented in the left-hand portions of Figs. 4.5.5 and 4.5.6. In addition to n and p/L, the values of the moment ratio $m = M/M_y$ at the middle of the beam, designated as m_{mid}, are plotted against the load ratio $R = w/w_y$. The expression for m_{mid} is obtained from Eq. 4.5.15 through substitution of $L/2$ for x:

$$m_{mid} = \tfrac{3}{2}R + n \qquad\qquad 4.5.23$$

Behavior of Beam after Midsection Yields

It can be seen from Fig. 4.5.6 that m_{mid} reaches the value unity at about $R = 1.4$. Under this load therefore $M = M_y$ at $x = L/2$, and the beam begins to yield at the midpoint. With the loads further increased there are three inelastic regions in the beam, one at each of the two rigidly fixed ends and one in the middle. The calculations presented must therefore be modified to take into consideration the inelastic deformations in the middle of the beam.

When the expressions are set up for the complementary energy, the integration must be carried out separately for three regions. The first extends from $x = 0$ to $x = p$ and is the inelastic range at the end of the beam. The second is the elastic region between $x = p$ and $x = q$, and the third is again an inelastic region situated between $x = q$ and $x = L/2$ (see Figs. 4.5.5 and 4.5.7). Because of the symmetry it is not necessary to integrate beyond $x = L/2$; it is simpler to multiply by 2 the value obtained for the complementary energy stored in one half of the beam.

In the first two regions the integrals to be evaluated are the same as those given in Eq. 4.5.11 with the exception that the upper limit $L/2$ is replaced by q. In the third region the elastic limit is exceeded because the moment, which is positive here, is greater than $M_y = \tfrac{1}{12}w_yL^2$. From Eq. 4.5.6 it follows that the complementary energy is

$$dU' = (1/2EI)[(C - 1)(M - M_y)^2 + M^2]\,dx \qquad\qquad 4.5.24$$

This is valid from $x = q$ to $x = L/2$. The following equation now represents the total complementary energy:

$$U' = \frac{1}{EI} \int_0^p (C - 1)(M + M_y)^2 \, dx + \frac{1}{EI} \int_q^{L/2} (C - 1)(M - M_y)^2 \, dx$$

$$+ \frac{1}{EI} \int_0^{L/2} M^2 \, dx \qquad\qquad 4.5.25$$

Division by M_y^2/EI and introduction of the symbol $m = M/M_y$ yield

$$\frac{EIU'}{M_y^2} = \int_0^p (C - 1)(m + 1)^2 \, dx + \int_q^{L/2} (C - 1)(m - 1)^2 \, dx + \int_0^{L/2} m^2 \, dx$$

$$4.5.26$$

This expression has to be minimized with respect to n; and m, p, and q are all functions of n.

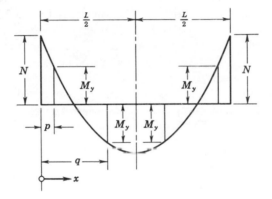

Fig. 4.5.7. Bending-moment diagram in final stages of loading.

In the process of differentiation the last two terms appearing in the right-hand member of Eq. 4.5.16 will again make no contributions to the derivative because in Eq. 4.5.26 the expression $(m + 1)$ vanishes at $x = p$ and the expression $(m - 1)$ vanishes at $x = q$. With $\partial m/\partial n = 1$ the minimal condition becomes

$$(C - 1) \int_0^p (m + 1) \, dx + (C - 1) \int_q^{L/2} (m - 1) \, dx + \int_0^{L/2} m \, dx = 0$$

$$4.5.27$$

Evaluation of the definite integrals results in

$$R\{6(q/L)^2 - 4(q/L)^3 - 6(p/L)^2 + 4(p/L)^3 - [C/(C - 1)]\}$$

$$-n\{2(p/L) - 2(q/L) + [C/(C - 1)]\} - 2(p/L) - 2(q/L) + 1 = 0$$

$$4.5.28$$

This is one of the conditions necessary for the calculation of the unknown quantities n, p, and q. A second condition is the requirement that M be equal to $-M_y$ at $x = p$. The condition was formulated mathematically in Eq. 4.5.22. A third condition is that M must take on the value M_y at $x = q$, which can be expressed as $m = 1$ at $x = q$. Substitution in Eq. 4.5.15 leads to

$$-n = 6R[(q/L) - (q/L)^2] - 1 \qquad 4.5.29$$

Equations 4.5.22, 4.5.28, and 4.5.29 suffice for the calculation of n, p, and q when R is given, but the computations are time-consuming. Numerical values were obtained therefore by the following scheme:

If p/L and q/L are known, Eqs. 4.5.22 and 4.5.29 can be solved for $1/R$. One obtains

$$1/R = 3[(q/L) - (q/L)^2 - (p/L) + (p/L)^2] \qquad 4.5.30$$

For each fixed value of p/L a number of likely values were tentatively assumed for q/L. Substitution of the values in Eqs. 4.5.29 and 4.5.30 yielded tentative values for n and R. Entering Eq. 4.5.28 with the values assumed and with those obtained, the calculator found that the left-hand member of the equation had some positive or negative value rather than zero. A plot of this error against the assumed value of q/L was a curve whose intersection with the axis indicated the correct value of q/L.

The numerical values computed are shown in the right-hand portions of Figs. 4.5.5 and 4.5.6. The figures also contain values of m_{mid} which were again calculated from Eq. 4.5.23.

Discussion of Results

As R is the ratio of the load to that value of the load under which yielding just begins at the fixed ends of the beam, Figs. 4.5.5 and 4.5.6 represent the history of the loading process from the beginning of yielding. At the outset the region in which inelastic deformations takes place increases very slowly. At $R = 1.4$, when the middle of the beam starts to yield, the end zone of yielding extends only over 1.12 per cent of the beam ($p/L = 0.0112$). The statically indeterminate end moment N also increases very slowly, its value at $R = 1.4$ being only about 8 per cent higher than that prevailing at $R = 1$. The beam therefore behaves almost as if it were hinged at the fixed ends after yielding begins. The bending moment at the middle of the beam increases from $0.5M_y = \frac{1}{24}w_yL^2$ to $M_y = \frac{1}{12}w_yL^2$. With an increase in load amounting to $0.4w_y$ the increase in the maximum moment of a simply supported beam would be $\frac{1}{8}(\frac{4}{10})w_yL^2 = \frac{1}{20}w_yL^2$ which differs little from the actual increment of $\frac{1}{24}w_yL^2$.

The situation changes radically when the middle of the beam begins to yield. The elastic load-carrying capacity of the central portion of the beam is exhausted, and the minimum of the complementary energy corresponds to rapidly expanding regions of yielding as the load is increased. When the load is doubled ($R = 2$), the end moment is 84 per cent, and the moment at the middle 158 per cent higher than their values at $R = 1$. Regions amounting to 6.5 per cent of the length have yielded in the vicinity of the fixed ends ($p/L = 0.065$), and 29 per cent of the length shows inelastic deformations in the middle ($q/L = 0.355$). The rapid growth of the yielded portions is shown graphically in Fig. 4.5.5. The region between the top horizontal line and the line representing q/L has yielded; the region between the q/L and p/L curves is still elastic; and the region between the p/L line and the bottom horizontal is again a region of yielding.

The stress at the beginning of yielding is 38,000 psi in the present numerical example. The ultimate stress of 70,000 psi is reached therefore when n attains the value $70,000/38,000 = 1.84$. Figure 4.5.6 shows that this is the case at $R = 2.09$. The beam of the present example fails under 2.09 times the load at which it begins to yield. This failure occurs by rupture of the material, and the calculations carried out naturally cannot give any indication whether buckling would take place at a smaller load.

Midpoint Deflection

The beam can also become useless for practical applications before failure occurs by rupture or instability, if its deflections reach excessively large values. For this reason the midpoint deflection of the beam is now calculated. This calculation is important for one more reason: The analysis of this article is based on the assumption of small deformations. Excessively large deformations would infringe upon this assumption and render doubtful or invalid some of the results obtained.

In the fully elastic range the midpoint deflection is given by Eq. 1.15.7:

$$d_{\max \text{ el}} = \tfrac{1}{384}(wL^4/EI) \tag{a}$$

Substitution of the numerical values yields when $L = 12$ ft

$$d_{\max \text{ el}} = (w/384)(1.44^4 \times 10^8)/(29 \times 10^6 \times 528.3) = 7.32 \times 10^{-5}w$$

if w is given in pounds per inch. From Eq. 4.5.7

$$M_y = (2I/h)\sigma_y$$

and from Eq. 4.5.13

$$M_y = \tfrac{1}{12}w_yL^2$$

Consequently

$$w_y = 24(I/hL^2)\sigma_y \tag{4.5.31}$$

where σ_y is the yield stress in tension.   With the numerical values given

$$w_y = [(24 \times 528.3)/(10.6 \times 1.44^2 \times 10^4)]38{,}000 = 2190 \text{ lb per in.}$$

if the yield stress of 38,000 psi is substituted in Eq. 4.5.31.   The maximum deflection d_y under the load at which yielding begins is then

$$d_y = 7.32 \times 2.19 \times 10^{-2} = 0.16 \text{ in.}$$

A deformation of such magnitude is certainly permissible with almost every beam of 12-ft length.

The midpoint deflection of a beam loaded beyond the yield stress can be calculated by the dummy load method discussed in Part 1, and the dummy load of 1 lb can be imagined to act in the middle of a beam whose ends are simply supported.   This change in the end conditions is permissible, as was shown in Art. 1.15.

In the left-hand half of the beam the dummy moment M_1 is then

$$M_1 = x/2 \qquad\qquad 4.5.32$$

From Eq. 4.5.15 the actual moment M_0 can be taken as

$$M_0 = 6RM_y[(x/L) - (x/L)^2] + N \qquad\qquad 4.5.33$$

The relative angle of rotation $d\phi$ of two sections a distance dx apart is, for reasons of geometry,

$$d\phi = 2\varepsilon_0 \, dx/h \qquad\qquad 4.5.34$$

where ε_0 is the strain caused in a flange by the actual moment M_0.   As the region of the beam extending from $x = p$ to $x = q$ is elastic, the strain ε_0 prevailing there can be calculated from Hooke's law as $M_0 h/2EI$.   Substitutions yield

$$d\phi = (1/EI)\{6RM_y[(x/L) - (x/L)^2] + N\} \, dx \qquad\qquad 4.5.35$$

The contribution d_1 of one elastic region to the total midpoint deflection of the beam can be calculated with the aid of the principle of virtual displacements:

$$d_1 = \int_p^q M_1 \, d\phi = \frac{L}{2EI} \int_p^q \left\{ 6RM_y \left[\left(\frac{x}{L}\right)^2 - \left(\frac{x}{L}\right)^3 \right] + N\frac{x}{L} \right\} dx \qquad 4.5.36$$

In the inelastic region of the middle of the beam the stress-strain relationship is (see Fig. 4.5.2)

$$\varepsilon_0 = (\sigma_y/E) + (\sigma - \sigma_y)(C/E) = C(\sigma/E) - (C - 1)(\sigma_y/E) \qquad 4.5.37$$

As from Eq. 4.5.3 the stress $\sigma = M_0 h/2I$, the contribution d_2 of one-half the central region to the deflection of the midpoint is

$$d_2 = \frac{L}{2EI} \int_q^{L/2} C \left\{ 6RM_y \left[\left(\frac{x}{L}\right)^2 - \left(\frac{x}{L}\right)^3 \right] + N\frac{x}{L} \right\} dx$$

$$- \frac{L}{2EI} \int_q^{L/2} (C-1)M_y \frac{x}{L} dx \qquad 4.5.38$$

In the inelastic region at the fixed end the bending moment is negative. The stress-strain law to be used is therefore

$$\epsilon_0 = -(\sigma_y/E) + (\sigma + \sigma_y)(C/E) = C(\sigma/E) + (C-1)(\sigma_y/E) \quad 4.5.39$$

The contribution d_3 of this region to the midpoint deflection is

$$d_3 = \frac{L}{2EI} \int_0^p C \left\{ 6RM_y \left[\left(\frac{x}{L}\right)^2 - \left(\frac{x}{L}\right)^3 \right] + N\frac{x}{L} \right\} dx$$

$$+ \frac{L}{2EI} \int_0^p (C-1)M_y \frac{x}{L} dx \qquad 4.5.40$$

The total midpoint deflection is

$$d_{max} = 2(d_1 + d_2 + d_3) \qquad 4.5.41$$

Because of Eqs. 4.5.31 and (g) one has at the start of yielding

$$d_y = \tfrac{1}{384}(w_y L^4/EI) = \tfrac{1}{32}(M_y L^2/EI) \qquad 4.5.42$$

After the integrations indicated in Eqs. 4.5.36, 4.5.38, and 4.5.40 are carried out and the limits are substituted, algebraic manipulations yield

$$d_{max}/d_y = -16R(C-1)[4(q/L)^3 - 3(q/L)^4 - 4(p/L)^3 + 3(p/L)^4]$$

$$- 16n(C-1)[(q/L)^2 - (p/L)^2] + 16(C-1)[(q/L)^2 + (p/L)^2]$$

$$+ 5RC + 4nC - 4(C-1) \qquad 4.5.43$$

When the middle portion is entirely elastic, $q = L/2$, and Eq. 4.5.43 reduces to

$$d_{max}/d_y = 5R + 4n + 16(n+1)(C-1)(p/L)^2$$

$$+ 16R(C-1)[4(p/L)^3 - 3(p/L)^4] \qquad 4.5.44$$

When in addition the inelastic range at the fixed end vanishes, $p/L = 0$, and Eq. 4.5.44 becomes

$$d_{max}/d_y = 5R + 4n \qquad 4.5.45$$

At the moment when yielding just begins at the fixed end, $R = w/w_y = 1$ and $n = N/M_y = -1$. Substitution of these values gives $d_{max} = d_y$, which serves as a check on the correctness of the calculations.

Numerical values computed from Eqs. 4.5.43 and 4.5.44 are shown in Fig. 4.5.6.   It can be seen that during the first phase of inelastic bending, when the middle of the beam is still perfectly elastic, the deformations increase relatively slowly, although much more rapidly than before yielding started at the fixed ends.   The increase in load from w_y to $1.4w_y$ gives rise to an increase in deflections from d_y to $2.62d_y$.   The midpoint deflection at $R = 1.4$ is therefore $2.62 \times 0.16 = 0.4192$ in., which cannot be considered excessive.   In floor beams of buildings the span divided by 360 is accepted by many engineers as the maximum permissible deflection since experience has shown that the adoption of this rule of thumb prevents the cracking of plaster in ceilings.   When yielding begins in the middle of the span of the beam of this example, the midpoint deflection is almost exactly equal to $144/360 = 0.4$ in.

As yielding proceeds in the middle of the span, the deflections increase in a manner that may be described as catastrophic.   At $R = 1.61$ the midpoint deflection is $30d_y = 4.8$ in. which is likely to interfere with the proper functioning of the beam wherever it may be used.   At $R = 2.00$ the deflection is $100d_y = 16$ in.   Probably every practicing engineer considers such a deflection equivalent to failure.

It can be concluded therefore that the beam of the present example stands up satisfactorily until $R = 1.4$ when yielding begins in its middle. Beyond this stage of the loading the deformations increase so rapidly that the beam loses its usefulness even though failure through rupture occurs only at $R = 2.09$.

4.6   Limit-Load Analysis

Purpose and Significance of Limit-Load Analysis

The purpose of a limit-load analysis is the determination of the loads under which a structure collapses.   The word *limit* refers therefore to the ultimate value, or the limit, of the load-carrying capacity.   It has nothing to do with American aeronautical terminology according to which the limit load is the maximum probable or permissible value of the actual load and the ultimate load is the limit load multiplied by the factor of safety.   In the United States the failing or collapse load of a structure is therefore known as the ultimate load to the aeronautical engineer but as the limit load to the civil engineer.   In this article limit, ultimate, collapse, or failing loads are all understood to mean the same thing, namely the load under which the load-carrying capacity of the structure is exhausted.

Naturally a collapse load analysis cannot, in general, be based on the assumption that Hooke's law holds because most structural materials have stress-strain curves deviating considerably from the linear relationship when the strains are large.   It is reasonable to expect that a failing load analysis, based on a non-linear stress-strain law, is much more difficult to carry out than an elastic analysis.   Fortunately this is not so, provided certain simplifying assumptions are made concerning the behavior of the structure.   As a matter of fact, the limit-load analysis of the ordinary building frame or of the continuous beam on several supports is often less laborious than the calculation of the bending-moment distribution by the Hardy Cross method.

When applied to beams and rigid frames, the simplifying assumptions underlying this new method of structural analysis are:

1. After a cross section of a beam or column had yielded thoroughly, rotations can take place at that section indefinitely without a change in the moment acting in the section.   A section in such a state of stress is known as a yield hinge.

2. The structure collapses when the number of its sections that have become yield hinges suffices to transform the structure, or part of it, into a mechanism.

3. The deformations under loads less than the collapse load are sufficiently small so that the changes in geometry before collapse can be disregarded.

The significance of these assumptions can be appreciated if the result obtained in Art. 4.5 are reviewed.   There it was found that yielding of an idealized I beam began at the clamped ends under a load w_y.   When the ratio $w/w_y = R$ was increased from 1 to 1.4, the moment in the yielded section increased only 8 per cent.   The assumption of a constant moment in the yielded section was therefore a satisfactory approximation during this part of the loading.   At $R = 1$ yielding took place only in one mathematical section.   As R increased the yielded region grew, but at $R = 1.4$ it still extended only over 1.12 per cent of the beam.   Hence large strains developed in a very limited region where the upper flange stretched, the lower flange was compressed, and as a consequence the beam rotated as if the yielded region had developed into a hinge.   This yield hinge is not an ideal one; it can be imagined to have a great deal of dry friction, and thus it moves only if the moment acting on it has a sufficiently large value.   This value is known as the yield moment; the sense of the moment acting from the element on the fictitious hinge is always the same as the sense of the rotation, and thus the moment always does positive work during the deformations.

It was also noted in Art. 4.5 that the deflections started to increase in the beam in a catastrophic manner after the midsection yielded at $R = 1.4$. For all practical purposes, this stage of the loading could be considered therefore as the limit of the usefulness of the beam. Yielding of the mid-section introduced a third hinge in the beam. If the yield hinges had been real hinges and the portions of the beam between the hinges had been perfectly rigid bodies, the beam with the three hinges would have been an ideal mechanism. This argument justifies the second assumption, at least for this example.

Finally the third assumption is seen to be justified because the deformations of the beam at the moment when the third hinge developed were still of the same order of magnitude as the elastic deformations.

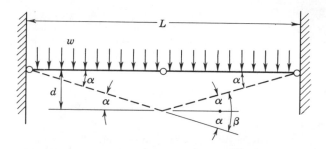

Fig. 4.6.1. Virtual displacement of beam with three yield hinges.

With the aid of the three assumptions of limit analysis it is possible to compute the collapse load without the lengthy calculations of Art. 4.5. This can be done, for instance, by means of the principle of virtual displacements. Figure 4.6.1 shows a virtual displacement of the clamped beam after the three yield hinges were inserted. Rotations take place in the yield hinges, and the segments of the beam between hinges move as if they were rigid bodies. The small average displacement of the load wL is $d/2$, and thus the virtual work done by the load is $wLd/2$. At the same time the rotation α of the end hinges is $\tan^{-1}(2d/L) \simeq 2d/L$ and that of the middle hinge $\beta = 4d/L$. As the moment acting in the hinges is the yield moment M_y, the internal work is $8M_yd/L$. As remarked earlier, the work done by the yield moments is always positive; for this reason there is no need to introduce a sign convention for moments. When the internal work is equated to the external work, the following equation is obtained:

$$8M_yd/L = wLd/2 \qquad\qquad (a)$$

Solution for the load w yields

$$w = 16M_y/L^2 \qquad\qquad 4.6.1$$

From Eq. 4.5.13

$$w_y = 12M_y/L^2 \qquad\qquad (b)$$

Comparison of Eqs. 4.6.1 and b yields

$$w/w_y = 1.33 \qquad\qquad 4.6.2$$

In this equation w is the load under which the beam loses its structural usefulness. Its value is 5 per cent below the value corresponding to $R = 1.4$ which can be considered the limit of usefulness on the basis of Art. 4.5; the 5 per cent difference is a consequence of the neglect of strain hardening in the limit analysis. In this respect, at least, the procedure of limit analysis just proposed is conservative; also, it yields results reasonably close to the values obtained from the more rigorous analysis of Art. 4.5.

The importance of determining the failing load becomes evident if the behavior of the clamped beam is compared to that of the simply supported beam. If the latter is designed with a safety factor of 2 against yielding, its midsection begins to yield when twice the design load is applied to the beam. As the beam at that moment becomes a mechanism in the sense just defined, obviously twice the design load is the ultimate load-carrying capacity. It is correct to say therefore that the beam has a safety factor of 2.

If the clamped beam is now designed on the basis of the same elastic analysis principle and for the same safety factor, it will begin to yield at the rigidly fixed supports. One-half the load causing this yielding is the design load. But in this case twice the design load does not cause failure, just yielding at the supports. The middle of the beam yields only after the load is further increased by a factor of 1.4 if Fig. 4.5.6 is accepted as correct, and by a factor of 1.33 if Eq. 4.6.2 is chosen as the standard of comparison. This means that the clamped beam fails at 2.8 (or 2.66) times its design load, and thus its true safety factor is 2.8 (or 2.66).

There is obviously no reason to discriminate against a clamped beam and to require a larger safety factor for it than for a simply supported beam unless the designer is uncertain about the effectiveness of the clamping. If this is so, however, it is much more logical to determine the actual end fixity analytically or experimentally rather than to increase arbitrarily the factor of safety.

Limit-Load Analysis of Simple Frame

The failing load of the rigid frame shown in Fig. 4.6.2 will next be determined. The moment under which a yield hinge develops in the columns is designated by M_y. It is assumed that the horizontal beam yields under three times this value of the bending moment. The horizontal and the vertical loads are of the same intensity and are denoted by w. The frame is assumed to be loaded from zero to the collapse load sw, where s

is the safety factor, in such a manner that all the loads are increased in the same proportion.   This is known as proportional loading.

According to the assumptions made, failure will occur when the yield hinges transform the redundant structure into a mechanism.   This can take place, for instance, as indicated in Fig. 4.6.2b.   Beam BC collapses in the manner discussed in connection with Fig. 4.6.1.   The yield hinges at B

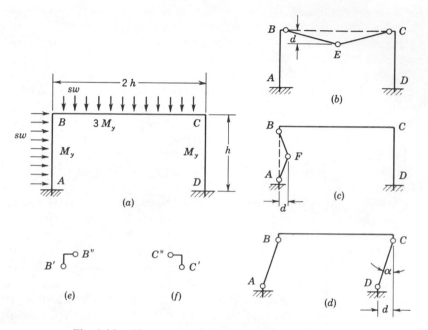

Fig. 4.6.2.   Elementary collapse mechanisms of rigid frame.

and C are in the end sections of the beam.   Naturally the external work is again the load $2hsw$ times the average displacement $(d/2)$, that is, $hswd$. The internal work is the yield moment $3M_y$ times the sum of the angles of rotation $(4d/h)$, which is $12M_yd/h$.   As the two products must be equal, the safety factor s is

$$s = 12M_y/h^2w \qquad\qquad 4.6.3$$

But this is not the only mechanism through which the structure can collapse, and thus the value of s just calculated is not necessarily the correct safety factor.   If column AB fails as shown in Fig. 4.6.2c, the external work is $swhd/2$, the internal work is $8M_yd/h$, and thus the safety factor is

$$s = 16M_y/h^2w \qquad\qquad 4.6.4$$

Which of the two values obtained for s is the correct one? Obviously the one given in Eq. 4.6.4 cannot be correct, because under three quarters of the load corresponding to it the structure would already have collapsed according to the pattern of Fig. 4.6.2b. It is not known yet, however, whether other patterns might not lead to even lower values of the safety factor. For this reason all possible mechanisms of failure must be investigated. Figure 4.6.2d represents a new possibility, namely a shear failure of the entire rigid frame as contrasted with the beam failures of its elements already discussed. In this case the external load swh has an average displacement of $d/2$, and the angle of rotation has the value $\alpha = d/h$ at each yield hinge. Hence

$$swhd/2 = 4M_y d/h \qquad\qquad (c)$$

and the safety factor is

$$s = 8M_y/h^2 w \qquad\qquad 4.6.5$$

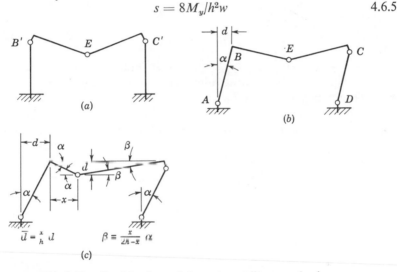

Fig. 4.6.3.   Combinations of elementary collapse mechanisms.

Apparently, with this frame, shear failure is more likely than any of the beam failures investigated. But on closer examination it is found that the beam failure illustrated in Fig. 4.6.2b is impossible for still another reason: Obviously the bending moment acting from column AB on beam BC at point B is equal in magnitude to the bending moment acting from beam BC on column AB. The moment under which the column develops a yield hinge is only one third of that necessary for the beam to develop one. Consequently a yield hinge must appear in the column before one can appear in the beam at point B, and thus the type of failure shown in Fig. 4.6.2b cannot occur.

The deformation mechanisms shown in Figs. 4.6.2e and f are introduced to provide a systematic transformation of the failure pattern of Fig. 4.6.2b into that of Fig. 4.6.3a which is more likely to occur with the frame under investigation.   The double yield hinges $B'B''$ and $C'C''$ permit rotations at points B and C which result in a transfer of the yield hinges from the beam into the columns.   The external work at failure according to Fig. 4.6.3a is the same as that according to Fig. 4.6.2b, but the amount of work done by the internal moments is different.   At point E the moment is $3M_y$, and the rotation is $2d/h$.   Thus the work done is $6M_yd/h$.   This is the same as before.   At points B and C, however, the rotation is d/h, but the yield moment is only M_y.   Thus the sum of the internal work at B and C is $2M_yd/h$, and the total internal work is $8M_yd/h$.   The requirement that the internal work be equal to the external work yields the condition

$$hswd = 8M_yd/h \qquad (d)$$

and thus the safety factor is

$$s = 8M_y/h^2w \qquad 4.6.6$$

The shift of the yield hinges from positions B'' and C'' to positions B' and C' reduced the safety factor considerably.   It is now the same for the modified beam failure and for collapse by shear.

Since the combination of pattern b with patterns e and f was so successful in reducing the value of s, the obvious conclusion is that other combinations of the elementary patterns should also be investigated.   If they yield smaller values for s, it will be expected that the structure can fail under further reduced loads.   As a first trial, pattern d of Fig. 4.6.2 will be combined with the beam failure according to Fig. 4.6.3a because these two collapse mechanisms correspond to the lowest values of s obtained heretofore.   In Fig. 4.6.3b the work done by the lateral load is $swhd/2$, and that done by the vertical load is $swhd$.   With $\alpha = d/h$ the work done at A is M_yd/h; at E it is $6M_yd/h$; at C it is $2M_yd/h$; and at D it is again M_yd/h. The principle of virtual displacements requires therefore that

$$\tfrac{3}{2}swhd = 10(M_yd/h) \qquad (e)$$

and thus

$$s = \tfrac{20}{3}(M_y/h^2w) \qquad 4.6.7$$

The new combination results in a reduction of the value of s by a factor of $\tfrac{5}{6}$.

One objection can be raised to the use of the pattern of Fig. 4.6.3b: As it is not symmetric, the yield hinge need not be at E, but may be any-where along beam BC.   The symmetry of the patterns of Figs. 4.6.2b and c, and of Fig. 4.6.3a, prescribed the hinge location, while in the new combined pattern the location of the yield hinge is unknown.   If its distance from B is denoted by x as shown in Fig. 4.6.3c, the external work W_e is

$$W_e = (swhd/2) + 2swh(x/h)(d/2) \qquad (f)$$

The internal work W_i is

$$W_i = M_y(d/h) + 3M_y(d/h) + 3M_y[x/(2h - x)](d/h)$$
$$+ M_y[x/(2h - x)](d/h) + M_y(d/h) + M_y(d/h) \qquad (g)$$

Altogether

$$W_e = (d/2)sw(h + 2x) \qquad (h)$$

and

$$W_i = M_y(d/h)\{6 + [4x/(2h - x)]\} \qquad (i)$$

Solution of the equation $W_e = W_i$ yields

$$sw = \frac{4M_y}{h} \frac{x - 6h}{2x^2 - 3hx - 2h^2} \qquad 4.6.8$$

When $x = h$, Eq. 4.6.8 reduces to Eq. 4.6.7. The yield hinge, however, should be located so as to make s a minimum. The minimum of sw is reached when the derivative of the right-hand member of Eq. 4.6.8 with respect to s vanishes. After omission of the constant factor $4M_y/h$ one obtains

$$\frac{2x^2 - 3hx - 2h^2 - (x - 6h)(4x - 3h)}{(2x^2 - 3hx - 2h^2)^2} = 0 \qquad (j)$$

This expression vanishes when the numerator is zero. This requirement leads to the quadratic equation

$$2x^2 - 24hx + 20h^2 = 0 \qquad (k)$$

whose physically significant solution is

$$x = 0.901h \qquad 4.6.9$$

Substitution in Eq. 4.6.8 yields

$$s = 6.624(M_y/h^2w) \qquad 4.6.10$$

This value is smaller than that given in Eq. 4.6.7, but the difference is so small that it is without significance in practical work. Numerous computations have shown that a shift of a yield hinge from the middle of a beam does not have, as a rule, an important effect on the safety factor of the structure. For this reason it is generally considered sufficient to investigate only those collapse mechanisms in which yield hinges occur at the ends of beams and columns, at the points of application of concentrated loads, and in the middle of beams subjected to distributed loads. Rapidly varying distributed loads may require a deviation from this practice.

Critical Review of Calculations

The following question will now have to be answered: Is it at all certain that the true value of the safety factor is $6.624(M_y/h^2w)$? The answer can be obtained if it is observed that any safety factor s calculated by the method of the preceding section is either equal to or greater than the true safety factor. This can be seen from the following reasoning:

The frame of Fig. 4.6.2 is redundant to the third degree. The four hinges indicated in Fig. 4.6.2d and in Figs. 4.6.3b and c transform it into a mechanism while the three hinges of Figs. 4.6.2b and c and of Fig. 4.6.3a transform only part of it into a mechanism. The last statement is true if the second-order small approach of the two end points of a beam corresponding to the first-order small transverse displacement of the yield hinge within the span is considered compatible with the assumptions. On the basis of practical experience this is known to be a reasonable assumption.

Because of the threefold redundancy the values of the bending moments can be prescribed at will in three arbitrary sections, and equilibrium with any prescribed loading can be established in accordance with the laws of statics. If in addition the moment is prescribed in a fourth section, the external loads cannot be freely chosen. In the examples the loads were prescribed only as far as their relative magnitudes were concerned, and the parameter specifying the absolute magnitude, namely s, was determined from the equilibrium conditions. These conditions were enforced by means of the principle of virtual displacements because this is the most convenient method to be used with mechanisms, as was already stated in Art. 1.5 of Part 1. The much more cumbersome approach of satisfying two force and one moment equilibrium conditions for each portion of the structure will be used later. The absolute magnitude of the loading was determined from the equilibrium conditions for the partial mechanisms of Figs. 4.6.2b and c and also for Fig. 4.6.3a, even though the number of moments prescribed was only three. The principle of virtual displacements made this task an easy one.

The mechanisms of Figs. 4.6.2 and 4.6.3 are therefore in equilibrium. Nevertheless they are not the real failure mechanisms if anywhere in the frame the yield moment of the structural elements is exceeded since the frame cannot support such moments. However, the safety factor calculated would be correct for the frame if its elements were sufficiently reinforced in the critical regions. Without the reinforcement the safety factor calculated is greater than the true value. A check of the correctness of the safety factor can be had therefore by determining the moments throughout the structure. Unfortunately such a check is generally rather time-consuming.

A procedure that is preferable in most cases starts out from the knowledge just gained that all safety factors calculated by means of the collapse mechanisms are too high unless they are equal to the true value. This is the minimal principle of limit analysis. Consequently the true value is the smallest value obtainable. This smallest value can be found with certainty if all possible collapse mechanisms are investigated. The total number of elementary collapse mechanisms can be ascertained from the following considerations:

In a rigid frame consisting of straight beams and columns a yield hinge can develop at each end of each beam and column because they are likely locations for the maximum bending moments. This gives six yield hinges in the frame of Fig. 4.6.2. In addition a maximum of the bending moment, and thus a yield hinge, can occur in some interior section of every element that is subjected to distributed transverse loads, such as beam BC and column AB in Fig. 4.6.2a. When concentrated transverse forces form part of the loading, their points of application can also become yield hinges. In the example of the frame of Fig. 4.6.2a the total possible number of yield hinges is eight.

At the same time the moment diagram of the entire structure is uniquely determined if the values of the moments are specified in three sections of bar AB, three sections of bar BC, and two sections of bar CD. The moment distribution is therefore known if eight quantities are given. The number of these quantities is equal to the number of possible yield hinge locations. Before yielding occurs, when the entire structure is still elastic, three moments in the structure can be considered as the statically indeterminate quantities whose magnitudes must be calculated from considerations of deformations and not from conditions of equilibrium. Hence the external loads and the eight internal moment values must be connected by $8 - 3 = 5$ independent equations of equilibrium. If this statement were not true, it would not be possible to calculate the moments in the frame after three hinges transformed it into a statically determinate structure.

The choice of any particular collapse mechanism establishes an equilibrium system. When the principle of virtual displacements is used to calculate the corresponding safety factor, an equation of equilibrium is written. The number of possible collapse mechanisms may be large, and each one furnishes an equation of equilibrium. Not all these equations are linearly independent, however, because only five independent equations can exist for the system. Any five equations can be chosen as the fundamental system, and the remainder can be considered as their linear combinations. The five mechanisms corresponding to the fundamental system of equations constitute then the fundamental set of independent collapse mechanisms. The ones selected are shown in Fig. 4.6.2. For the general building frame the number of independent collapse mechanisms is $j - n$ if j is the number of possible yield hinge locations and n is the degree of redundancy of the structure.

Naturally the investigation of the independent, or elementary, collapse mechanisms must be augmented by the analysis of their combinations, because some combination may yield a smaller safety factor than the original mechanisms. After some practice with the technique presented it

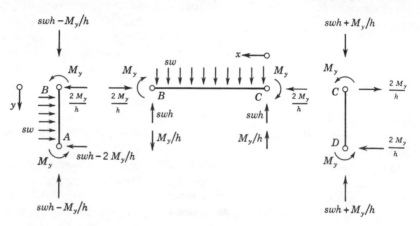

Fig. 4.6.4.   Analysis of shear collapse mechanism.

is usually not too difficult to detect those combinations that are likely to reduce the safety factor obtained for the elementary mechanisms. Before the selection of the combinations is discussed, the shear mechanism shown in Fig. 4.6.2d will be analyzed in more detail. This is the mechanism that yielded the lowest safety factor of all the elementary modes of failure. According to Eq. 4.6.5 the value was

$$s = 8M_y/h^2w \qquad (1)$$

The isolated elements are shown in their original shape in Fig. 4.6.4 because the virtual displacement is not supposed to change noticeably the geometry of the structure.

The equilibrium of column CD will be investigated first because this element is not subjected to external loads. At the time of the collapse the yield moments M_y act on the two ends of the column. Their sense is such as to resist the rotation assumed. Because both end moments are counterclockwise, they must be equilibrated by the moment of the two horizontal reaction forces indicated in the figure.

Next the horizontal beam BC is analyzed. To its right end C the reactions to the end moment and the end shear force acting at the top of column CD must be applied. At the left end M_y acts in the sense indicated because there is a yield hinge at B in the left column. The horizontal force $2M_y/h$ must be present for reasons of equilibrium. Similarly the vertical downward load $2swh$ is balanced by the two upward reaction forces swh and the two moments M_y by the moment of the reaction forces M_y/h.

The forces and moments acting on the column at B are the reactions of the corresponding quantities acting on the beam at B. The vertical and horizontal reactions at A follow then from the requirements of force

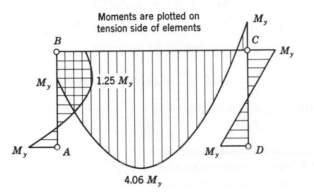

Fig. 4.6.5. Bending moments corresponding to shear collapse.

equilibrium. There remains the equilibrium of moments to satisfy for column AB which yields an equation from which s can be calculated:

$$4M_y - \tfrac{1}{2}swh^2 = 0 \tag{m}$$

and thus

$$s = 8M_y/wh^2 \tag{n}$$

This value is the same as the one obtained from the application of the principle of virtual displacements. Obviously the present calculations were much lengthier than the earlier ones. On the other hand, they permit the determination of the bending moments throughout the structure. Their distribution is shown in Fig. 4.6.5.

In column CD the moments vary linearly, and their maximum value is M_y. Since Eq. n can be written as

$$swh = 8M_y/h \tag{o}$$

the vertical reaction force at C is $9M_y/h$ and the intensity of the distributed load acting on beam BC is $8M_y/h^2$. The magnitude of the moment a distance x to the left of C is

$$M = -M_y + 9M_y(x/h) - 4M_y(x/h)^2 \qquad 4.6.11$$

A condition for a maximum of this moment is

$$9 - 8(x/h) = 0 \qquad (p)$$

from which

$$x/h = \tfrac{9}{8} \qquad 4.6.12$$

Substitution in Eq. 4.6.11 yields

$$M_{max} = \tfrac{65}{16}M_y = 4.0625M_y \qquad 4.6.13$$

Equation 4.6.11 yields $+M_y$ for the bending moment at B. If y designates the coordinate downward from B along the column, the moment in the column is

$$M = M_y + 2M_y(y/h) - 4M_y(y/h)^2 \qquad 4.6.14$$

The extremum of this moment is at

$$y/h = \tfrac{1}{4} \qquad 4.6.15$$

where

$$M_{max} = 1.25M_y \qquad 4.6.16$$

These calculations and Fig. 4.6.5 explain why the simple shear failure cannot be the true mechanism of collapse. Even though equilibrium is maintained throughout the structure and the four yield hinges do develop at the four corners of the structure, the bending moments shown in Fig. 4.6.5 cannot be supported by the structure. They exceed the yield moment over considerable portions of both beam and column. In the absence of reinforcements a yield hinge would have developed inside the span of the beam long before the value $s = 8M_y/wh^2$ was reached.

If the safety factor is divided by the ratio of the maximum moment to the yield moment of the beam, that is, by $4.0625/3 = 1.354$, the moments everywhere along the beam become equal to or less than $3M_y$ and everywhere along the column less than M_y. Such a bending-moment distribution can certainly be supported by the structure. The corresponding safety factor $s' = (8/1.353)(M_y/wh^2) = 5.91M_y/wh^2$ appears to be smaller than the true safety factor because the yield moment is reached and a yield hinge develops in only one section of the beam; thus the structure cannot collapse. The trouble is that in the absence of a sufficient number of yield hinges the conditions of consistent deformations should be, but are not likely to be, satisfied by the bending-moment distribution.† It is not difficult to prove, however, that the safety factor s' is indeed smaller than the true safety factor of the structure.

Consequently it is known that the correct numerical coefficient in the expression for the safety factor is between 5.91 and 8 in the present example. Since these bounds do not represent a close enough limitation of the value of the safety factor, it is desirable that combinations of the elementary collapse patterns be investigated. As the objective is to reduce the value obtained for the safety factor, it is advisable to choose a combination of two elementary mechanisms which eliminates the rotation in a yield hinge common to the two elementary collapse patterns. In such a manner the internal work term, which appears in the numerator of the expression from which the safety factor is computed, becomes smaller and the safety factor is reduced.

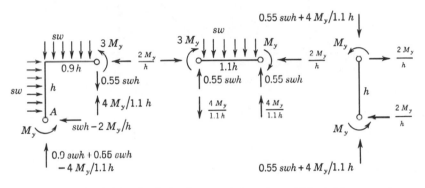

Fig. 4.6.6. Analysis of moments in system of Fig. 4.6.3c.

This procedure was followed when the patterns of Figs. 4.6.2d and 4.6.3a were combined. In the combination, shown in Fig. 4.6.3b, the beam and the column do not rotate relative to each other at B, and according to Eq. 4.6.7 the safety factor is only five sixth of the safety factors of the elementary collapse mechanisms. A further refinement in the calculations was achieved by locating hinge E at a distance x from the center of the beam as shown in Fig. 4.6.3c. The value of the safety factor corresponding to the most favorable choice of x was found to be 6.624.

A study of the collapse mechanisms showed that no other combination of the elementary patterns of Fig. 4.6.2 can yield a smaller value for the numerical coefficient of s than 6.624. Consequently this is the true value

† An elastic structure must satisfy the conditions of equilibrium and continuity. The latter are irrelevant in the presence of plastic deformations in the yield hinges because the hinges permit rotations of arbitrary magnitude under the yield moments.

of the safety factor.   In this particular case it is easy to check the correctness of this statement because the four yield hinges transform the entire structure into a mechanism and the bending moments can be obtained solely from the conditions of equilibrium.   Details of the analysis are given in Fig. 4.6.6, and the bending-moment distribution is shown in Fig. 4.6.7. The diagram shows that the moment is equal to the yield moment at each of the four yield hinges and is less everywhere else.   Hence the safety factor calculated is the true safety factor.

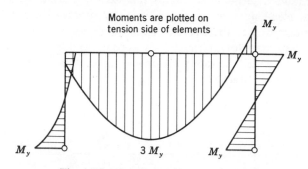

Fig. 4.6.7.   Bending moments at collapse.

Collapse of a Two-Storey Building Frame

Figure 4.6.8 shows a two-storey two-bay building frame and its loading. The yield moments of the structural elements of the first storey are double the corresponding values of the second storey, as indicated in the figure. The two loaded beams and the two loaded columns give rise to $4 \times 3 = 12$ possible yield hinge locations while the corresponding number in the remaining two beams and four columns is $6 \times 2 = 12$.   The total number j of possible yield hinge locations is therefore 24.   As long as the structure is perfectly elastic, it presents a twelvefold redundant problem of analysis. A complete section through the middle of each beam reduces the building frame to three independent statically determinate structures; as the number of statically indeterminate quantities in each section is three (one unknown moment, one unknown normal force, and one unknown shear force), the total number of unknowns is $4 \times 3 = 12$.   Consequently the number of linearly independent, elementary collapse mechanisms is $j - n = 24 - 12 = 12$.

These mechanisms can be easily found.   The two loaded beams can collapse by beam failure, and the same is true of the two columns that are subjected to transverse loading.   Figures 4.6.8b and c show these mechanisms.   In each case the safety factor listed in the figure was calculated for the element of the smaller load-carrying capacity, that is, for beam DE and column AD.

The two elementary shear failures are shown in Figs. 4.6.8*d* and *e*. These two, the four beam failures mentioned, and the rotations of the six joints *A* to *F* add up to 12 mechanisms as required.   As the beam failure of beam *DE* and the shear failure of the first storey yield the lowest

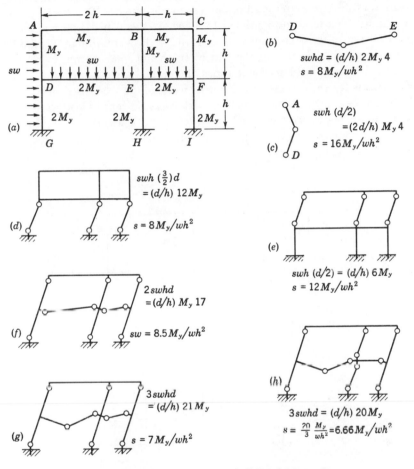

Fig. 4.6.8.   Two-storey building frame.

safety factor, their combination can be expected to result in a further decrease in the safety factor.   Unfortunately no such reduction can be realized from this combination, because the two mechanisms are entirely independent.   When the two shear failures of Figs. 4.6.8*d* and *e* are first combined into the general shear pattern of Fig. 4.6.8*f*, which in itself gives a slightly higher safety factor than the shear failure of the lower storey, and the general shear pattern is then combined with the local failure of beam *DE*, a significant reduction in the value of the safety factor results.   This combination is illustrated in Fig. 4.6.8*g*.   Its low safety factor was obtained because the clockwise rotation of the beam at *D* in pattern *b* was equal to the clockwise rotation of the column at *D* in pattern *f*, and thus the super-position of the two patterns eliminated the yield hinge at *D*.   In the trans-formation of pattern *f* into *g* the yield hinge at *D* vanished, and a new one in the middle of the beam appeared.   As the angle of rotation in the middle of the beam is twice that at the end of the beam, the internal work was increased by the yield moment times the single angle of rotation.   To compensate for the increase, the external work of the load distributed over the beam became available.   As this work was sufficient to lead to the low safety factor of pattern *b* in which the internal work was the yield moment times the quadruple angle of rotation, naturally pattern *g* had to correspond to a reduced safety factor.

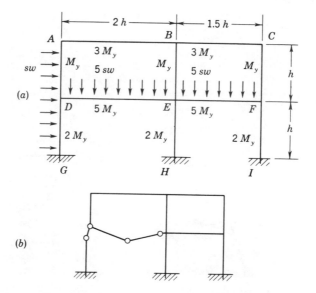

Fig. 4.6.9.   Modified two-storey building frame.

The final collapse pattern, yielding the lowest value of the safety factor, is shown in Fig. 4.6.8*h*.   It is obtained from pattern *g* by means of a joint rotation.   A careful study of all the other possible combinations of the elementary collapse mechanisms failed to produce a pattern with a lower safety factor than pattern *g*.   Hence $s = 6.66M_y/h^2w$ is considered as the correct value of the actual safety factor of the structure.   This value may still be slightly high because of the assumption that one of the yield hinges was exactly in the middle of beam DE. However, the slight correction in the value is not worth the additional work of investigating the most propitious location of the yield hinge. Similarly the values of the bending moment will not be determined all over the structure because the minimum value of s is considered to be established firmly enough without such a check. The analysis of the moments would be a lengthy process since the structure is rather complex and has one redundant quantity left after the insertion of the 10 yield hinges indicated in Fig. 4.6.8g.

It should not be concluded from this analysis that the lowest value of the safety factor is always obtained from a combination of a beam pattern with a shear pattern. If the two-storey building frame and its loading are modified as shown in Fig. 4.6.9, shear failure can take place only at much higher multiplying factors of the load w than beam failure of the highly loaded floor beam DE. In consequence the collapse pattern yielding the lowest safety factor is the one shown in Fig. 4.6.9b. The safety factor is $s = 3.6M_y/h^2w$ if the yield hinge is assumed to be in the middle of the beam.

The Fully Plastic Moment

The procedure of limit analysis just presented was based on the three assumptions listed at the beginning of this article. They appeared to be justified in view of the results obtained in Art. 4.5. But in that article an idealized I beam was analyzed which differed in its properties from real sections. In the next stage of this investigation the moment-curvature relationship of a beam of solid rectangular cross section will be established in order to examine the effect of the cross-sectional shape on the collapse load.

When the bending moment is small, all the stresses are within the elastic limit. The curvature-moment relationship is then given by the equation

$$1/\rho = M/EI \qquad\qquad 4.6.17$$

When the elastic limit is exceeded, the strain is still a linear function of the distance y, as shown in Fig. 4.6.10b, since plane sections before bending remain plane after bending. This statement is rigorously true for pure bending (under a constant bending moment) and is a good approximation when bending is accompanied by shear (and thus the moment varies along the beam).

It is assumed that the stress-strain diagram is straight up to a stress σ_y and is represented by some function $\sigma = f(\varepsilon)$ afterward. If the strain ε_y is reached at a distance e from the neutral axis, the difference in strain between the fibers at $+e$ and $-e$ is $2\varepsilon_y$, and the corresponding difference in the extension of the two fibers whose original length was dx is $2\varepsilon_y dx$. This quantity divided by the distance $2e$ between the fibers, that is, $(\varepsilon_y/e)\,dx$, is the relative angle of rotation of the sections dx apart, and thus the curvature of the beam is ε_y/e. If the curvature is designated as $1/\rho$, one has

$$1/\rho = \varepsilon_y/e = \sigma_y/Ee \qquad\qquad 4.6.18$$

The bending moment corresponding to this curvature is the static moment of the area under the stress curve in Fig. 4.6.10c with respect to the neutral axis:

$$M = \frac{2}{3}\sigma_y e^2 + 2\int_e^h \sigma_y\,dy \qquad\qquad 4.6.19$$

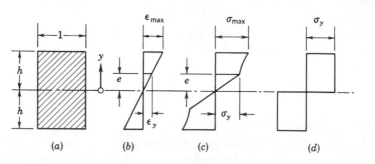

Fig. 4.6.10.   Stresses in solid rectangular cross section.

When the relationship $\sigma = f(\varepsilon)$ is known, the integration can be carried out analytically or numerically, and Eq. 4.6.19 becomes an expression for M as a function of e.   But the curvature $1/\rho$ is also a known function of e, that is, of the distance where the elastic limit is reached in the cross section. As a consequence, Eqs. 4.6.18 and 19, together with the given stress-strain relationship $\sigma = f(\varepsilon)$, represent a known relationship between the bending moment applied to the beam and the curvature caused by it.

If the stress-strain law is the same as the one used in Art. 4.5 and illustrated in Fig. 4.5.2, Eq. 4.6.19 becomes

$$M = \frac{2}{3}\,38{,}000e^2 + 2\int_e^h \left[38{,}000 + 10^5\left(\frac{y}{e} - 1\right)\frac{38}{29}\,10^{-3}\right]y\,dy \qquad 4.6.20$$

Integration yields

$$M = (1/e)(87.4h^3 + 37{,}869h^2e - 12{,}622e^3) \qquad\qquad 4.6.21$$

With

$$I = \tfrac{2}{3}h^3 \qquad\qquad 4.6.22$$

one obtains

$$M = (I/h)(h/e)[131.1 + 56{,}803(e/h) - 18{,}933(e/h)^3] \qquad 4.6.23$$

Substitution of the numerical values in Eq. 4.6.18 yields

$$1/\rho = 1.311 \times 10^{-3}/e \qquad\qquad 4.6.24$$

Corresponding values of M and $1/\rho$ were computed from Eqs. 4.6.23 and 4.6.24 with $2h = 10.6$ in. and plotted in Fig. 4.6.11.

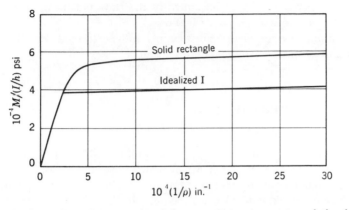

Fig. 4.6.11.   Effect of cross-sectional shape on moment-curvature relationship.

The calculations will now be repeated with the assumption that the stress remains constant when the strain corresponding to $\sigma = \sigma_y = 38{,}000$ psi is exceeded.   Under these conditions Eq. 4.6.19 becomes

$$M = \frac{2}{3} 38{,}000e^2 + 2 \int_e^h 38{,}000y \, dy$$

$$= \frac{I}{h}\frac{h}{e}\left[57{,}000\,\frac{e}{h} - 19{,}000 \left(\frac{e}{h}\right)^3 \right] \qquad 4.6.25$$

When corresponding values of M and $1/\rho$, computed from Eqs. 4.6.25 and 4.6.24, respectively, were plotted in Fig. 4.6.11, the curve obtained could not be distinguished from the curve corresponding to Eqs. 4.6.23 and 4.6.24.   Of course, this statement is true only in the range of curvature values shown in the figures.   As the curvature increases beyond all limits, the moment also increases indefinitely when the stress-strain law of Art. 4.5 is used, but it approaches a fixed value when the constant stress assumption underlying Eq. 4.6.25 is in force.   This limiting value is obtained from Eq. 4.6.25 through setting e/h equal to zero:

$$M_p = 57{,}000(I/h) \text{ in.-lb} \qquad\qquad 4.6.26$$

The subscript p signifies that the moment given in Eq. 4.6.26 is the fully plastic moment, that is, the moment acting on the cross section when $e = 0$ and the constant yield stress $|\sigma_y|$ prevails over the entire section as shown in Fig. 4.6.10d. The value obtained can be checked easily. The force acting on the upper half of the section is $\sigma_y \times 1 \times h$. As an equal and opposite force is acting on the lower half, the moment of the two resultant forces which are a distance h apart is $h^2\sigma_y$. According to Eq. 4.6.22 the moment of inertia is $\frac{2}{3}h^3$. Consequently the fully plastic moment is as given in Eq. 4.6.26 when $\sigma_y = 38,000$ psi.

At the moment when yielding just begins in the extreme fibers, the stress distribution is linear. The maximum stress is

$$\sigma_{max} = \sigma_y = 38,000 = M_y/(I/h)$$

Hence the yield moment is

$$M_y = 38,000(I/h) \qquad 4.6.27$$

It can be seen therefore that the fully plastic moment of the solid rectangular section is 1.5 times the moment M_y at which yielding begins.

For the sake of comparison the moment-curvature relationship will now be established for the idealized I beam. If the distance between the concentrated flanges is $2h$, not h as in Fig. 4.5.1, and the cross-sectional area of each flange is $\frac{1}{2}A$, the moment of inertia of the section is $I = Ah^2$ and the stress in the flange is

$$\sigma = 38,000 + 10^5[(2h/\rho) - (\tfrac{38}{29})\,10^{-3}] = 37,869 + 10^5(2h/\rho) \qquad 4.5.28$$

This is true provided the stress exceeds the yield stress. The moment acting in a cross section is

$$M = Ah\sigma = (I/h)\sigma = (I/h)[37,869 + 10^5(2h/\rho)] \qquad 4.6.29$$

With $2h = 10.6$ in. one obtains

$$M/(I/h) = 37,869 + 1.06 \times 10^6(1/\rho) \qquad 4.6.30$$

This relationship is also plotted in Fig. 4.6.11.

It was found in Art. 4.5 that the yield hinge at the middle of the beam developed at a stage of the loading when the end moment was about 8 per cent higher than the yield moment. From Eq. 4.6.30 it follows that such a moment corresponds to a curvature of approximately 0.00286. Figure 4.6.11 therefore represents the entire range of curvatures occurring in the clamped beam of Art. 4.5 until the third yield hinge develops.

It appears that the behavior of the solid rectangular section differs from that of the idealized I section in that the resistance of the former increases rather rapidly after the extreme fiber yields, and the value of the bending moment necessary to increase the curvature further soon becomes approximately equal to the fully plastic moment. With the idealized I section there is an abrupt change in the tangent to the moment-rotation curve at the yield point, and from there on the moment increases very slowly with the curvature. In the range of curvatures of importance in a limit analysis the bending moment in a beam of solid rectangular section is substantially equal to the fully plastic moment M_p, while in a beam of idealized I section it is very nearly equal to the yield moment M_y. For these reasons in place of the yield moment it is customary to use the fully plastic moment in the limit analysis of the structures made of real, not idealized, sections. Experimental evidence is available to prove that this procedure leads to structures of sufficient strength.

The fully plastic moment of a beam of arbitrary cross-sectional shape can be easily determined. Above the neutral axis the yield stress in tension is assumed to act, while below the neutral axis the yield stress in compression prevails. The two regions of uniform stress are separated by an infinitesimal region of elastic deformations. Since the yield stress in tension is assumed to be equal numerically to the yield stress in compression, the sum of the absolute values of the static moments of the two portions of the cross section multiplied by the yield stress is equal to the fully plastic moment. Computations have shown that the fully plastic moment of standard American I sections is about 1.15 times the yield moment of the section. For wide-flanged I sections the ratio is closer to 1.10. These ratios reduce the determination of the fully plastic moment to a simple multiplication, because the yield moment is the yield stress times the section modulus, and the latter is tabulated for all standard sections. For the sake of comparison it may be restated that the ratio of the fully plastic moment to the yield moment is 1.5 for the solid rectangular section and 1.0 for the idealized I section.

In closing this article it is once more emphasized that the analysis of the collapse load here presented is applicable only when the material of the structure can yield a great deal without an appreciable amount of strain hardening. This is generally true of frames built of structural steel. Moreover, the limit analysis does not give any information about the possibility of failure by buckling before the structure collapses through yielding in the hinges.

4.7 Stresses Caused by a Radial Load in a Reinforced Monocoque Airplane Fuselage

Monocoque Construction

Present-day airplanes are generally constructed according to the reinforced monocoque principle. The major portion of the fuselage, or body, of such a plane (see Fig. 4.7.1) is essentially a circular, or approximately circular, thin-walled aluminum cylinder reinforced by uniformly spaced circumferential and longitudinal stiffening elements. All the stiffeners are arranged on the inside of the cylinder to keep the exterior smooth. As a

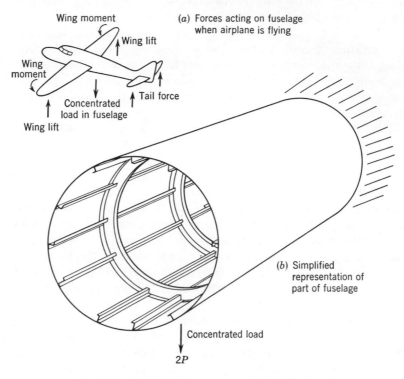

Fig. 4.7.1.   Reinforced monocoque airplane fuselage.

rule both sets of stiffeners are continuous, and the longitudinal stiffeners, known as stringers, pass through holes cut in the circumferential stiffeners called rings. The diameter of the cylinder is about 120 in. in a modern transport plane, the thickness of the sheet covering from 0.025 to 0.072 in. and the depth of the ring about 2.5 in. The stringers are about 1 in. deep. Both stringers and rings are usually Z, hat, or channel-shaped sections manufactured of 0.03 to 0.08-in.-thick strips of aluminum-alloy sheet. It is rather obvious that a cylinder whose thickness-to-radius ratio is of the order of magnitude of 1 to 1000 would buckle when subjected to small shear forces, torques, or bending moments. The addition of the two sets of stiffeners, however, transforms the cylinder into a rather sturdy fuselage structure.

Heavy pieces of equipment are always attached to the rings of the fuselage. The weight of such equipment is balanced mainly by the resultant lifting force and moment of the pressure exerted by the moving air on the wings. Any remaining small unbalanced moments can be taken care of by the pilot through adjusting the horizontal tail surfaces until the vertical tail force reaches the value necessary for moment equilibrium (see Fig. 4.7.1a). The major portion of the concentrated load is thus transmitted forward to the wing and constitutes the shear force acting on the fuselage. This shear is carried almost entirely by the sheet covering because the bending rigidity of the individual stringers is small. It gives rise to tensile forces in the top and to compressive forces in the bottom of the fuselage which are resisted by the stringers as well as by the sheet covering.

Stresses According to Engineering Theory of Bending

Until a few years ago the stresses in a reinforced monocoque type of fuselage were generally calculated from the engineering theory of the bending of beams. This beam theory is based on the assumptions of Bernoulli and Navier, according to which cross sections of the cylinder which were plane and perpendicular to the axis of the cylinder before bending remain the same after the loads are applied, and the cross-sectional shape of the cylinder does not distort in its own plane during loading. It follows from these assumptions that the normal stress caused in the cylinder by a bending moment can be calculated from the formula

$$\sigma = My/I = (Mr/I) \cos \phi \qquad 4.7.1$$

where σ is the normal stress due to bending, M the applied moment in the section, y the distance of the point where the stress is sought from the horizontal neutral axis of the circular cylinder, I the moment of inertia of the cross section with respect to the horizontal neutral axis, r the radius of the cylinder, and ϕ the angle subtended with the vertical direction by the radius passing through the point in question (Fig. 4.2.7). According to the same theory the shear stress in the sheet is

$$\tau = VQ/2It \qquad 4.7.2$$

where V is the applied shear force in the section, Q is the static moment of that portion of the section which lies above a horizontal line passing through the point at which the shear stress is sought, and t is the thickness of the cylindrical sheet.

Simplified Model of Fuselage

In most airplane fuselages the stringers are evenly spaced and so close together that no great error is made if the structure is idealized through the assumption of a uniform distribution of the stringer cross-sectional area over the entire circumference of the cylinder.   In such a manner a fictitious thickness t' of the sheet, effective in resisting the fuselage bending moments, is obtained.   It is defined as the total cross-sectional area of all the stringers divided by the circumference of the cylinder plus the actual thickness t of the sheet covering.†   The thickness t alone resists the shear if the shearing and the bending rigidities of the individual stringers are neglected as is the usual procedure.   Since $I = \pi r^3 t'$, Eq. 4.7.1 can be rewritten as

$$\sigma = (M/\pi r^2 t') \cos \phi \qquad\qquad 4.7.1a$$

Similarly, Eq. 4.7.2 becomes

$$\tau = -(2P/\pi rt) \sin \phi \qquad\qquad 4.7.2a$$

if τ is the reaction shear stress acting from the shell on the ring (compare Eq. 4.2.23) and $2P$ is the shear force in accordance with the loading shown in Figs. 4.2.7 and 4.7.1.

The physical model corresponding to this imaginary reinforced cylinder consists essentially of many closely spaced longitudinal elements carrying the normal stresses.   They are connected by a sheet which transmits only shear stresses acting parallel to the surface.   The sheet is considered a thin membrane which cannot support bending moments or shear forces perpendicular to the surface if they are of any appreciable magnitude. This clear physical picture of closely spaced stringers and sheet is slightly obscured through the assumption that the stringers are infinitely close and cover the circumference of the cylinder continuously.   Such an assumption is necessary to reduce the mathematical difficulties.   It leads to the concepts of an actual sheet of thickness t carrying the shear stresses and a fictitious sheet of thickness t' carrying the normal stresses.   However, the physical action can still be visualized best by imagining normal stress-carrying longitudinal elements connected by narrow strips of shear-carrying sheet.

Rigorous Analysis

Equations 4.7.1 and 4.7.2 represent the stresses very accurately when the section in which they are sought is far enough from the section in which the external load is applied.   This is a consequence of St.-Venant's principle which is discussed in Art. 4.8.   One of the important tasks of the airplane stress analyst is, however, the determination of the maximum moment

---

† When the sheet covering buckles, a portion of the sheet cross section rather than all of it should be added to the cross section of the stringers in accordance with the effective width theory.

in the reinforcing ring to which the load is applied, and of the maximum shear stress in the sheet covering of the fuselage in the immediate vicinity of the load.   There is not the slightest justification for the assumption that Eqs. 4.7.1 and 4.7.2 hold true so close to the load.   The assumption will be discarded therefore, and a more rigorous analysis will be carried out for a simplified model of a fuselage consisting of one ring, at which the external load is applied, and a thin reinforced cylindrical sheet of length L whose far end is attached to a rigid wall (see Fig. 4.2.7).

The assumption represented by Eq. 4.7.2a can be easily corrected.   It can be replaced by a Fourier series with unknown coefficients, and the values of the coefficients can be determined with the aid of the minimal principle.   The Fourier series is capable of representing any arbitrary physically possible shear stress distribution, and thus also the actual shear stress distribution (compare Art. 2.8).   The minimal principle to be used in this problem is the second one which compares all possible equilibrium systems; in this principle the energy is expressed in terms of stress.   As the far end of the cylinder is assumed to be rigidly fixed, the reactions cannot do any work.   If in addition the material of construction is assumed to follow Hooke's law, the principle reduces to the requirement that the total strain energy stored in the system be a minimum with respect to the parameters defining the various possible states of equilibrium.

If the shear flow $q = \tau t$ is introduced as in Art. 4.2, the assumption can be written as

$$q = a_1 \sin \phi + a_2 \sin 2\phi + a_3 \sin 3\phi + \cdots = \sum_{n=1}^{\infty} a_n \sin n\phi \qquad 4.7.3$$

Positive coefficients correspond to shear stresses directed in the sense of increasing ϕ.   Equilibrium requires that the vertical resultant of the shear flow be equal and opposite to the shear force $2P$.   At any angle α the vertical component dF_v of the shear flow $qr \, d\alpha$ acting on the length $r \, d\alpha$ of the circumference is

$$dF_v = qr \sin \alpha \, d\alpha$$

as can be seen from Fig. 4.2.7c.   The total vertical shear force transmitted by the sheet is then

$$F_v = 2 \int_0^\pi qr \sin \alpha \, d\alpha \qquad 4.7.4$$

Substitution from Eq. 4.7.3 at $\phi = \alpha$ and evaluation of the definite integral lead to terms of the type

$$\int_0^\pi \sin \alpha \sin n\alpha \, d\alpha$$

It follows from Eq. 2.7.18 that this integral vanishes for every value of n except unity. (When the upper limit L is substituted in Eq. 2.7.18, the argument of the sine function becomes π). When $n = 1$ one has

$$\int_0^\pi \sin^2 \alpha \, d\alpha = \frac{\pi}{2}$$

which follows from Eq. 2.7.5. Consequently Eq. 4.7.4 becomes

$$F_v = 2ra_1 \int_0^\pi \sin^2 \alpha \, d\alpha = \pi a_1 r$$

Finally, with $F_v = -2P$, one obtains

$$a_1 = -2P/\pi r \qquad\qquad 4.7.5$$

With the first coefficient known, the shear flow expression can be written in the form

$$q = -(2P/\pi r) \sin \phi + a_2 \sin 2\phi + a_3 \sin 3\phi + \cdots \qquad 4.7.6$$

The coefficient of the first term in the series is identical with the coefficient of $\sin \phi$ in the elementary solution. The bending moment caused by the first term in the ring is therefore identical with the moment of the elementary theory. This was worked out in the last example of Art. 4.2 and given as Eq. 4.2.32. It will be designated here as M_1, with the subscript referring to the first term of the complete solution:

$$M_1 = (Pr/\pi)(1 - \tfrac{1}{2} \cos \phi - \phi \sin \phi) \qquad 4.7.7$$

Each of the remaining terms of the series modifies the shear stress distribution without changing the vertical shear force resultant. Naturally the bending moment in the ring is also influenced by every term. The next task is therefore the calculation of the bending moment in the ring as caused by any one of the shear flow components, say by $a_n \sin n\phi$. The total bending moment can then be obtained by adding up the contributions of all the shear flow components.

Bending Moment in Elastic Ring

The bending moment at point F of Fig. 4.2.7c was given in Eq. 4.2.24a as

$$M = -r^2 \int_0^\phi q(1 - \cos \phi \cos \alpha - \sin \phi \sin \alpha) \, d\alpha$$

The shear flow at $\phi = \alpha$ is represented now by

$$q = a_n \sin n\alpha$$

where n is an integer greater than unity.   Thus the integral becomes

$$M = -a_n r^2 \int_0^\phi (\sin n\alpha - \cos \phi \cos \alpha \sin n\alpha - \sin \phi \sin \alpha \sin n\alpha)\, d\alpha \quad 4.7.8$$

or symbolically

$$M = -a_n r^2 (I_1 + I_2 + I_3) \qquad 4.7.8a$$

The first one of these integrals, I_1, is easily evaluated:

$$I_1 = \int_0^\phi \sin n\alpha \, d\alpha = -\frac{1}{n}\Big[\cos n\alpha\Big]_0^\phi = \frac{1}{n}(1 - \cos n\phi) \qquad (a)$$

In the second integral, $\cos \phi$ is a constant multiplying factor during the integration.   A known trigonometric identity yields

$$\cos \alpha \sin n\alpha = \tfrac{1}{2}[\sin (n + 1)\alpha + \sin (n - 1)\alpha]$$

The integral of this expression is

$$\int_0^\phi \cos \alpha \sin n\alpha \, d\alpha = -\frac{1}{2(n + 1)}[\cos (n + 1)\phi - 1] - \frac{1}{2(n - 1)}$$
$$\times [\cos (n - 1)\phi - 1]$$
$$= -\frac{1}{2(n^2 - 1)}[(n - 1)(\cos n\phi \cos \phi - \sin n\phi \sin \phi - 1)$$
$$+ (n + 1)(\cos n\phi \cos \phi + \sin n\phi \sin \phi - 1)]$$
$$= \frac{1}{n^2 - 1}(n \quad n \cos n\phi \cos \phi \quad \sin n\phi \sin \phi)$$

The second integral is therefore

$$I_2 = -[1/(n^2 - 1)](n \cos \phi - n \cos n\phi \cos^2 \phi - \sin n\phi \sin \phi \cos \phi) \quad (b)$$

In the third integral $\sin \phi$ is a constant factor.   Moreover

$$\sin n\alpha \sin \alpha = \tfrac{1}{2}[\cos (n - 1)\alpha - \cos (n + 1)\alpha]$$

and for this reason

$$\int_0^\phi \sin n\alpha \sin \alpha \, d\alpha = \frac{1}{2(n - 1)} \sin (n - 1)\phi - \frac{1}{2(n + 1)} \sin (n + 1)\phi$$
$$= \frac{1}{2(n^2 - 1)}[(n + 1)(\sin n\phi \cos \phi - \cos n\phi \sin \phi)$$
$$-(n - 1)(\sin n\phi \cos \phi + \cos n\phi \sin \phi)]$$
$$= \frac{1}{n^2 - 1}(\sin n\phi \cos \phi - n \cos n\phi \sin \phi)$$

With this value the third integral becomes

$$I_3 = -[1/(n^2 - 1)](\sin n\phi \cos \phi \sin \phi - n \cos n\phi \sin^2 \phi) \qquad (c)$$

Addition of Eqs. b and c yields

$$I_2 + I_3 = -[1/(n^2 - 1)][n \cos \phi - n \cos n\phi(\cos^2 \phi + \sin^2 \phi)]$$

$$= -[1/n(n^2 - 1)](n^2 \cos \phi - n^2 \cos n\phi) \qquad (d)$$

If the numerator and the denominator of the expression obtained for I_1 are multiplied by $(n^2 - 1)$ one obtains

$$I_1 = [1/n(n^2 - 1)](n^2 - 1)(1 - \cos n\phi) \qquad (e)$$

Addition of Eqs. d and e results in

$$I_1 + I_2 + I_3 = [1/2(n^2 - 1)](n^2 - 1 + \cos n\phi - n^2 \cos \phi) \qquad (f)$$

Substitution of the right-hand member of Eq. f in Eq. 4.7.8a leads to

$$M = -a_n r^2\{(1/n) + [1/n(n^2 - 1)](\cos n\phi - n^2 \cos \phi)\} \qquad 4.7.9$$

This is the bending moment caused by the nth component of the shear flow in Eq. 4.7.3. However, the moment was calculated from the static conditions of equilibrium for a ring split at point A (compare Figs. 4.2.7c and 4.2.3b). In the actual unsplit redundant ring the unknown tensile force T, the unknown shear force V, and the unknown moment N also contribute to the moment in every section. The magnitude of the unknown quantities can be calculated from the minimal principle, as was done in the last example of Art. 4.2.

Because of the symmetry of the shear flow loading of the ring, no shear force can be present in section A of the ring in Fig. 4.2.7a. Hence the moment terms caused by the unknown quantities in section A are the same as the ones given in Eq. 4.2.26:

$$M = N_n + T_n r(1 - \cos \phi)$$

where the subscript n indicates that the moment is caused by the nth component of the shear flow. The total bending moment in the ring is therefore

$$M_n = N_n + T_n r(1 - \cos \phi) - a_n r^2\{(1/n) + [1/n(n^2 - 1)](\cos n\phi - n^2 \cos \phi)\}$$
$$4.7.10$$

The values of N_n and T_n will now be obtained from the minimal principle. When the nth component of the shear flow alone is acting, the strain energy stored in a ring is given by

$$U_n = \frac{1}{2EI} \int_0^{2\pi} M_n{}^2 r \, d\phi$$

The strain energy has a stationary value (a minimum) when

$$\partial U_n / \partial N_n = 0, \qquad \partial U_n / \partial T_n = 0$$

Since

$$\partial M_n / \partial N_n = 1, \qquad \partial M_n / \partial T_n = r(1 - \cos \phi)$$

the first requirement becomes

$$\frac{\partial U_n}{\partial N_n} = \frac{1}{EI} \int_0^{2\pi} M_n \frac{\partial M_n}{\partial N_n} \, dx = 0$$

or

$$\int_0^{2\pi} M_n r \, d\phi = 0 \qquad\qquad 4.7.11$$

From the second requirement the constant factor r can be omitted when the derivative is set equal to zero.   One obtains

$$\int_0^{2\pi} M_n (1 - \cos \phi) r \, d\phi = 0$$

which can be reduced, in view of Eq. 4.7.11, to the statement

$$\int_0^{2\pi} M_n \cos \phi \, r \, d\phi = 0 \qquad\qquad 4.7.12$$

Because

$$\int_0^{2\pi} \cos \phi \, r \, d\phi = \int_0^{2\pi} \cos n\phi \, r \, d\phi = \int_0^{2\pi} \cos \phi \cos n\phi \, r \, d\phi = 0$$

and

$$\int_0^{2\pi} \cos^2 \phi \, r \, d\phi = \pi r$$

evaluation of the integral indicated in Eq. 4.7.11 yields

$$N_n \pi + T_n r\pi - a_n r^2 (\pi / n) = 0 \qquad\qquad 4.7.13$$

Similarly Eq. 4.7.12 becomes

$$-T_n r\pi + [n a_n r^2 / (n^2 - 1)]\pi = 0 \qquad\qquad 4.7.14$$

Equation 4.7.14 can be solved for T_n:

$$T_n = (n/(n^2 - 1)]a_n r \qquad 4.7.15$$

Substitution of this value in Eq. 4.7.13 leads to

$$N_n = (1/n)a_n r^2 - [n/(n^2 - 1)]a_n r^2 = -[1/n(n^2 - 1)]a_n r^2 \qquad 4.7.16$$

Substitution of N_n from Eq. 4.7.16 and T_n from Eq. 4.7.15 in Eq. 4.7.10 yields after some simplification

$$M_n = -[a_n r^2/n(n^2 - 1)] \cos n\phi \qquad 4.7.17$$

This is the bending moment caused by the nth component of the shear flow, and n is any integer greater than unity.   From the principle of superposition the bending moment due to the shear flow represented by the entire Fourier series can be obtained by summing up all the contributions from $n = 2$ to $n = \infty$ in accordance with Eq. 4.7.17 and adding to the sum the bending moment caused by the first component as given in Eq. 4.7.7. The resulting bending moment is

$$M = \frac{Pr}{\pi}\left[1 - \frac{1}{2}\cos\phi - \phi\sin\phi\right] - r^2 \sum_{n=2}^{\infty} \frac{a_n}{n(n^2 - 1)} \cos n\phi \qquad 4.7.18$$

With this expression the problem of the bending moment in the ring is solved for any arbitrary shear flow distribution if the coefficients a_n are known.   In the present case, of course, they are unknown and have to be determined from the requirement that the sum total of the strain energy stored in the entire system be a minimum.

The strain energy stored in the ring is

$$U_r = \frac{r}{EI}\int_0^\pi M^2\,d\phi \qquad 4.7.19$$

with M given in Eq. 4.7.18.

Shear Strain Energy in Sheet

The shear strain energy stored in the sheet covering can be easily calculated because the shear flow distribution around the circumference is represented by Eq. 4.7.3, and the shear flow is constant in the x direction. This latter statement is a consequence of the assumption that the sheet carries shear only, but no axial or circumferential normal stresses.

Figure 4.7.2 shows a portion of the sheet covering with the stringers attached. For the sake of simplicity the stringers are drawn as solid rectangular bars. A small element of the sheet is cut out and shown separately as Fig. 4.7.2b. A view of the same element from the x direction is presented in Fig. 4.7.2c and one from the radial direction in Fig. 4.7.2d. The axial (x) view is helpful in proving that no circumferentia lstresses σ_ϕ can exist in the sheet because their resultant would have to be resisted either by a radial internal pressure p, which was not assumed to be present, or by variable shear stresses τ acting perpendicularly to the surface of the sheet. Such shear stresses would give rise to bending moments in the thin sheet; in the state of membrane stresses assumed for the sheet such shear stresses and bending moments do not exist. A thin sheet is not capable of resisting them to any appreciable degree.

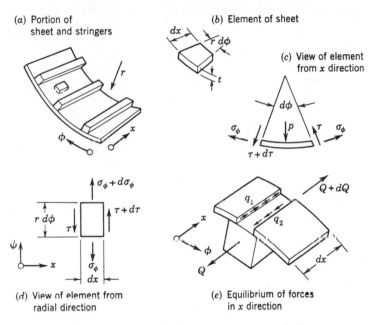

(a) Portion of sheet and stringers

(b) Element of sheet

(c) View of element from x direction

(d) View of element from radial direction

(e) Equilibrium of forces in x direction

Fig. 4.7.2.   Elements of reinforced monocoque shell.

Figure 4.7.2d is presented to prove that in the membrane stress condition the shear stress, and thus the shear flow, must be constant in the axial direction. In the figure the shear stress τ in a section x is different from the shear stress $\tau + d\tau$ in a section $x + dx$. If the increment $d\tau$ is positive, there is a resultant force $d\tau(r\,d\phi)t$ acting in the positive ϕ direction. Equilibrium is then possible only if the normal stress σ_ϕ in a section ϕ differs from the normal stress $\sigma_\phi + d\sigma_\phi$ in a section $\phi + d\phi$. The normal stress resultant $d\sigma_\phi(t\,dx)$ balances the shear force resultant if

$$d\sigma_\phi = -r(d\phi/dx)\,d\tau$$

However, such an equilibrium is impossible because the circumferential normal stress σ_ϕ is identically equal to zero for every value of ϕ, as has just been shown. It follows then that the shear flow cannot vary with x.

The shear strain energy is therefore

$$U_\tau = \frac{1}{2G} \int_V \tau^2 \, dV = \frac{1}{2G} \, tL \int_0^{2\pi} \left(\frac{q}{t}\right)^2 r \, d\phi$$

that is

$$U_\tau = \frac{Lr}{2Gt} \int_0^{2\pi} q^2 \, d\phi \qquad\qquad 4.7.20$$

with q given in Eq. 4.7.3.

Strain Energy in Stringers

When the shear flow in two adjacent panels varies (there is no restriction on the variation of q with ϕ because the σ_x stresses necessary for equilibrium can be present in the stringers), the difference between the shear forces $q_2 \, dx$ and $q_1 \, dx$ represents an axial load for the length dx of the stringer (see Fig. 4.7.2e). The total tensile force Q in a section x of the stringer is obtained by integration:

$$Q = (q_2 - q_1) \int_0^x dx$$

where $(q_2 - q_1)$ is written before the integral sign because the shear flow does not vary with x. Integration gives

$$Q = (q_2 - q_1)x + C$$

where the integration constant C is determined from the end condition that at $x = 0$ the force in the stringer must vanish. (There is no axial load applied at the free end of the cantilever cylinder.) Hence

$$Q = (q_2 - q_1)x \qquad\qquad 4.7.21$$

Even though the actual monocoque cylinder has a finite number of stringers, and thus the shear flow in its sheet covering must change by finite amounts from panel to panel, the assumption was made at the beginning of this article that the stringers were distributed continuously around the circumference and the shear flow varied continuously according to Eq. 4.7.3. The assumption simplifies the work of calculation without influencing adversely the results since there is a very large number of stringers in a modern airplane fuselage. As a consequence of this assumption the finite difference $(q_2 - q_1)$ prevailing over a distance equal to the stringer spacing must be replaced by the infinitesimal difference $(dq/d\phi) \, d\phi$ corresponding to the infinitesimal distance $r \, d\phi$. Similarly the force Q

acting on the total cross-sectional area of the stringer is replaced by $\sigma_x t' r\, d\phi$, the resultant of the normal stress σ_x over the infinitesimal area $t' r\, d\phi$. Thus Eq. 4.7.21 becomes

$$\sigma_x t' r\, d\phi = x(dq/d\phi)\, d\phi$$

from which it follows that

$$\sigma_x = (1/t'r)(dq/d\phi)\, dx \qquad 4.7.22$$

The strain energy stored in the reinforced cylinder because of the normal stresses is then

$$U_\sigma = \frac{1}{2E}\int_V \sigma_x{}^2\, dV = \frac{1}{2E}\, t'\int_0^{2\pi}\int_0^L \sigma_x{}^2 r\, dx\, d\phi \qquad 4.7.23$$

Substitution of σ_x changes this equation into

$$U_\sigma = \frac{1}{2Ert'}\int_0^{2\pi}\int_0^L \left(\frac{dq}{d\phi}\right)^2 x^2\, dx\, d\phi \qquad 4.7.24$$

As all the quantities involved except x are independent of x, the integration with respect to x can be carried out immediately. One obtains

$$U_\sigma = \frac{L^3}{6Ert'}\int_0^{2\pi}\left(\frac{dq}{d\phi}\right)^2 d\phi \qquad 4.7.25$$

The Minimum of the Strain Energy

The total strain energy stored in the reinforced cylinder is

$$U_{tot} = U_r + U_s + U_\sigma \qquad 4.7.26$$

with the three strain-energy quantities given in Eqs. 4.7.19, 4.7.20, and 4.7.25.

The strain energy is a function of the known external load $2P$ and the indefinitely large number of unknown coefficients a_n. Any arbitrary choice of these coefficients results in equilibrium for the entire system, and the actual values of the coefficients differ from all other values inasmuch as they make the strain energy a minimum. The analytical requirements for this minimum (or more correctly for a stationary value) can be written as

$$\partial U_{tot}/\partial a_i = 0 \qquad 4.7.27$$

where a_i is a typical coefficient. This represents an indefinitely large number of equations because i can be any integer greater than unity. The equations suffice for the calculation of all the unknown coefficients.

Differentiation yields

$$\frac{\partial U_{\text{tot}}}{\partial a_i} = \frac{2r}{EI} \int_0^\pi M \frac{\partial M}{\partial a_i}\, d\phi + \frac{Lr}{Gt} \int_0^{2\pi} q \frac{\partial q}{\partial a_i}\, d\phi$$

$$+ \frac{L^3}{3Ert'} \int_0^{2\pi} \left(\frac{dq}{d\phi}\right)\left(\frac{\partial}{\partial a_i}\right)\left(\frac{\partial q}{\partial\phi}\right)\, d\phi \qquad 4.7.28$$

where $(\partial/\partial a_i)(\partial q/\partial\phi)$ is the derivative of $\partial q/\partial\phi$ with respect to a_i. From Eq. 4.7.18

$$\partial M/\partial a_i = -[r^2/i(i^2 - 1)] \cos i\phi \qquad (g)$$

In the evaluation of the first term of the right-hand member of Eq. 4.7.28 the following integral is needed:

$$\int_0^\pi \phi \sin\phi \cos i\phi\, d\phi = \frac{1}{2} \int_0^\pi \phi[\sin(i+1)\phi - \sin(i-1)\phi]d\phi$$

$$= \frac{1}{2(i+1)^2}\left[\sin(i+1)\phi - (i+1)\phi \cos(i+1)\phi\right]_0^\pi$$

$$- \frac{1}{2(i-1)^2}\left[\sin(i-1)\phi - (i-1)\phi \cos(i-1)\phi\right]_0^\pi$$

The terms containing the sine function vanish at both limits. The terms containing the cosine function also vanish at the lower limit because of the factor ϕ. At the upper limit their sign depends on i. If i is even, the function $\cos(i+1)\pi$ is -1; and, if i is odd, the function has the value $+1$. The same is true of the function $\cos(i-1)\pi$. The simplest formula expressing these conditions is

$$\cos(i+1)\pi = \cos(i-1)\pi = -(-1)^i \qquad 4.7.29$$

The definite integral can now be written in the form

$$\int_0^\pi \phi \sin\phi \cos i\phi\, d\phi = \pi(-1)^i \left[\frac{1}{2(i+1)} - \frac{1}{2(i-1)}\right]$$

$$= -\frac{\pi}{i^2 - 1}(-1)^i$$

When the product of the right-hand members of Eqs. 4.7.18 and g is integrated from 0 to π, all terms multiplied by the factor Pr/π vanish with the exception of the one just calculated. Similarly the integral between the limits 0 and π of every term of the infinite series in Eq. 4.7.18 multiplied by $\cos i\phi$ vanishes with the exception of

$$\int_0^\pi \cos^2 i\phi \, d\phi = \frac{\pi}{2}$$

One obtains therefore

$$\frac{2r}{EI} \int_0^\pi M \frac{\partial M}{\partial a_i} \, d\phi = -\left(\frac{2r}{EI}\right)\left(\frac{Pr}{\pi}\right) \frac{r^2}{i(i^2-1)} \left(\frac{\pi}{i^2-1}\right)(-1)^i$$
$$+ \left(\frac{2r}{EI}\right) \frac{r^2}{i(i^2-1)} \, r^2 \frac{a_i}{i(i^2-1)} \left(\frac{\pi}{2}\right)$$

that is,

$$\frac{2r}{EI} \int_0^\pi M \frac{\partial M}{\partial a_i} \, d\phi = -\frac{2}{i(i^2-1)^2}(-1)^i \frac{Pr^4}{EI} + \frac{\pi}{i^2(i^2-1)^2}\left(\frac{a_i r^5}{EI}\right) \quad 4.7.29b$$

In the second term of the right-hand member of Eq. 4.7.28 the value of q must be substituted from Eq. 4.7.3. Differentiation with respect to a_i gives

$$\partial q/\partial a_i = \sin i\phi$$

For this reason

$$\frac{Lr}{Gt} \int_0^{2\pi} q \frac{\partial q}{\partial a_i} \, d\phi = \frac{Lr}{Gt} \pi a_i \qquad 4.7.30$$

In the third member the derivative of q with respect to ϕ is needed:

$$\partial q/\partial \phi = -(2P/\pi r) \cos \phi + 2a_2 \cos 2\phi + 3a_3 \cos 3\phi + \cdots \qquad (h)$$

The derivative of this quantity with respect to a_i is

$$(\partial/\partial a_i)(\partial q/\partial \phi) = i \cos i\phi \qquad (i)$$

When the product of the right-hand members of Eqs. h and i is integrated between the limits 0 and 2π, all integrals except that of $\cos^2 i\phi$ vanish because of the orthogonality of the cosine functions. One obtains

$$\frac{1}{3}\left(\frac{L^3}{Ert'}\right) \int_0^{2\pi} \left(\frac{\partial q}{\partial \phi}\right) \frac{\partial}{\partial a_i}\left(\frac{\partial q}{\partial \phi}\right) \, d\phi = \frac{L^3}{3Ert'} \pi i^2 a_i \qquad 4.7.31$$

The requirement that the derivative of the total strain energy (see Eq. 4.7.28) with respect to the unknown coefficient a_i vanish is now satisfied if the sum of the right-hand members of Eqs. 4.7.29b, 4.7.30, and 4.7.31 is set equal to zero:

$$-\frac{2}{i(i^2-1)^2}(-1)^i \frac{Pr^4}{EI} + \frac{\pi}{i^2(i^2-1)^2}\left(\frac{r^5}{EI}\right) a_i$$
$$+ \frac{Lr}{Gt} \pi a_i + \frac{L^3}{3Ert'} \pi i^2 a_i = 0 \qquad 4.7.32$$

The only unknown in this equation is a_i. Before the equation is solved for it, the notation

$$\gamma = 1/i^2(i^2 - 1)^2 \qquad 4.7.33$$

is introduced and the equation is multiplied by $6Ert'/L^3$:

$$-2i\gamma(-1)^i(P/r)(6r^6t'/L^3I) + \pi a_i\gamma(6r^6t'/L^3I)$$
$$+ \pi a_i 6(E/G)(r/L)^2(t'/t) + \pi a_i 2i^2 = 0$$

With the non-dimensional parameters

$$A = 6r^6t'/IL^3 \qquad 4.7.34$$

$$B = 6(E/G)(t'/t)(r/L)^2 \qquad 4.7.35$$

the equation can be written as

$$a_i(2i^2 + B + \gamma A) = 2i\gamma(-1)^i A(P/\pi r)$$

and thus the unknown coefficient is given as

$$a_i = -(1)^i \frac{i\gamma A}{2i^2 + B + \gamma A}\left(\frac{2P}{\pi r}\right) \qquad 4.7.36$$

As the parameters A and B depend only on the geometric and mechanical properties of the structure, the shear flow coefficients can be calculated from Eq. 4.7.36 without difficulty.

Numerical Example

To find out how much the actual shear flow just calculated differs from that assumed in the elementary theory, a numerical example was worked out which represents conditions that prevail in the fuselages of modern transport airplanes. Many such fuselages are characterized by the following approximate values:

$$A = 10^7, \qquad B = 500 \qquad 4.7.37$$

The coefficient a_1 of the first term in the series (Eq. 4.7.3) representing shear flow is $-2P/\pi r$ from Eq. 4.7.5. The next 15 coefficients were calculated from Eq. 4.7.36. All the coefficients are tabulated below:

i	=	1	2	3	4	5	6	7	8	9
$a_i/(2P/\pi r)$	=	-1	1.99	-2.91	3.36	-2.79	1.70	-0.90	0.48	-0.26

i	=	10	11	12	13	14	15	16
$a_i/(2P/\pi r)$	=	0.144	-0.084	0.051	-0.032	0.021	-0.014	0.009

It can be seen from these values that the actual shear flow differs greatly from that corresponding to elementary theory; the coefficients of the second through sixth terms of the Fourier series exceed the first coefficient in absolute value. The following coefficients decrease in magnitude, and the sixteenth amounts to less than 1 per cent of the absolute value of the first coefficient. Beyond the sixteenth the coefficients decrease monotonically in absolute value, and for very large values of i the value of the coefficient approaches zero. This can be seen from the following considerations:

When i is large, the difference between i^2 and $i^2 - 1$ is negligibly small. Then in good approximation

$$\gamma \cong i^{-6} \qquad\qquad 4.7.38$$

With this substitution the absolute value N_i of the numerical coefficient of a_i becomes

$$N_i \cong i\gamma A/(2i^2 + B + \gamma A) \cong (i^{-5}A)/(2i^2 + B + i^{-6}A)$$

Multiplication of both numerator and denominator by i^5 yields

$$N_i \cong A/(2i^7 + i^5B + i^{-1}A)$$

Since A and B are finite, not indefinitely large, it is always possible to choose i sufficiently large to make i^5B negligibly small compared to $2i^7$, and A/i negligibly small compared to i^5B. For instance, with $i = 100$ and the values stated in Eq. 4.7.37 the denominator becomes

$$2i^7 + i^5B + (A/i) = 2 \times 10^{14} + 5 \times 10^{12} + 10^5$$

When $i = 1000$ one obtains

$$2i^7 + i^5B + (A/i) = 2 \times 10^{21} + 5 \times 10^{17} + 10^4$$

This indicates that, for very large values of i, N_i is given in good approximation as

$$N_i \cong \tfrac{1}{2}A/i^7 \qquad\qquad 4.7.39$$

The formula clearly shows that N_i approaches zero as i approaches infinity. Moreover the sum of all the infinite number of terms is finite; that is, in mathematical language the series is convergent.

In actual numerical calculations only a limited number of coefficients are taken into account. For instance with $A = 10^7$ and $B = 500$ the 16 coefficients listed seem sufficient to define the shear flow with engineering accuracy. With the fuselage of a small fighter airplane the non-dimensional parameters are likely to have the values $A = 10^4$ and $B = 50$. In such a case six or eight coefficients are found to suffice.

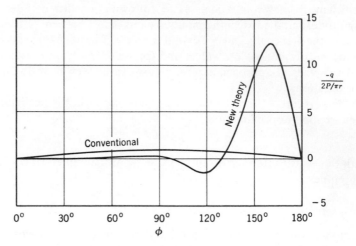

Fig. 4.7.3. Shear flow in sheet ($A = 10^7$, $B = 500$, according to new theory; also conventional curve. Both for single field case). From V. L. Salerno's doctoral thesis, Polytechnic Institute of Brooklyn, 1947.

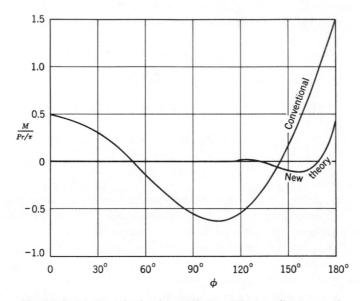

Fig. 4.7.4. Moments in ring ($A = 10^7$, $B = 500$, according to new theory; also conventional moment diagram. Both for single field case). From V. L. Salerno's doctoral thesis, Polytechnic Institute of Brooklyn, 1947.

The shear flow distribution is shown in Fig. 4.7.3.   It was calculated by adding up the first 16 terms of the series given in Eq. 4.7.6 for several values of the independent variable ϕ.   For the sake of comparison the sinusoidal shear flow curve of the elementary theory, corresponding to Eq. 4.7.2a, is also plotted.   The two curves are entirely different, and the maximum actual shear stress exceeds the conventional maximum value by a factor of 12.4.   The maximum of the conventional sinusoidal shear distribution occurs at $\phi = 90°$.   The present analysis shows that a very strong shear stress concentration occurs in the immediate vicinity of the applied load, and the upper half of the cylinder is practically free of stress.

The bending moments acting in the ring were computed from Eq. 4.7.18 with the aid of the coefficients given in Eq. 4.7.36.   They are plotted in Fig. 4.7.4, and the curve obtained is labeled *new theory*.   For comparison the conventional bending-moment distribution, calculated from Eq. 4.7.7, is also presented.   It can be seen that the maximum actual bending moment in the ring is only 27 per cent of the maximum bending moment of the elementary theory.   Large bending moments can be observed only in the immediate vicinity of the applied load, and the upper two thirds of the ring is practically free of bending moment.

Fuselages Having Many Elastic Rings

The calculations presented here were carried out for a single-field reinforced monocoque cylinder consisting of one ring and the longitudinally reinforced sheet.   In reality the concentrated load is transmitted over a relatively long portion of the fuselage, subdivided into many fields by the elastic rings, before the stresses approach the distribution represented by the elementary theory.   When the cylinder consists of many fields, an infinite series must be assumed independently for each field to represent the shear flow.   Thus the number of unknown coefficients is p times infinity if there are p fields and p elastic rings in the structure.   The work involved in determining all these coefficients from the minimal principle is naturally greater than that presented in this article.   Nevertheless it is not prohibitively large, and the stresses and moments have been calculated for many structural arrangements of interest.   A few results of such calculations are given in Fig. 4.7.5.

The most important questions to which the structural designer must have an answer are: (1) How much higher is the maximum shear stress in the neighborhood of the concentrated load than the maximum calculated from the conventional sinusoidal stress distribution? (2) How much smaller is the maximum bending moment in the ring under the concentrated load than that calculated from the elementary theory?   The answers can be given in the form of multiplying factors.   The first one, the shear stress concentration factor μ, is defined as the maximum shear stress according to the present theory divided by the maximum shear stress of the sine law. As the absolute value of the latter is from Eq. 4.7.2a

$$\tau_{\text{max elem}} = 2P/\pi rt$$

the maximum actual shear stress can be calculated from the equation

$$\tau_{\text{max}} = \mu 2P/\pi rt \qquad\qquad 4.7.40$$

provided μ is known. The value of μ depends on the non-dimensional parameters A and B. Computed values are plotted against A for two

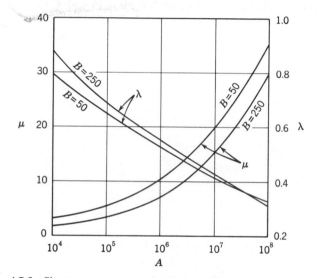

Fig. 4.7.5. Shear stress concentration factor μ for single-ring monocoque and moment reduction factor λ for semi-infinite monocoque. From author's paper in *J. Aero. Sciences.*

different values of B in Fig. 4.7.5. The figure shows that the stresses increase rapidly with A and are little influenced by B. The calculations were carried out for the single-field monocoque because the stresses decrease when the number of elastic rings increases. The curve presented corresponds therefore to the worst possible conditions.

The second quantity of interest is the moment reduction factor λ, defined as the maximum moment in the ring according to the present theory divided by the maximum moment of the elementary theory. From Eq. 4.7.7 the latter is

$$M_{\text{max elem}} = (3/4\pi)2Pr$$

The maximum actual moment can be obtained from the equation

$$M_{max} = \lambda(3/4\pi)2Pr \qquad 4.7.41$$

provided λ is known.   Values of λ are plotted against A for two different values of B in Fig. 4.7.5.   They were calculated for a cylinder extending to infinity from the ring at which the concentrated load is applied; the cylinder contained an indefinitely large number of equally spaced elastic rings.   When there are fewer rings in the fuselage, the maximum bending moment is smaller.   Proportioning the loaded ring and the sheet with the aid of Fig. 4.7.5 results therefore in conservative design.   The value of λ decreases rapidly with A and depends little on B.   Even in a relatively stubby fuselage having $A = 10^4$ the actual maximum moment is noticeably smaller than that predicted by the elementary theory.

Physical Significance of Parameters and Experimental Results

It might be added that a theory can be in error even though all the derivations in it are correct, if the basic assumptions do not represent the actual conditions with sufficient accuracy.   Whenever the accuracy of these assumptions is in doubt, they should therefore be checked by comparing the results of experiments with the predictions of the theory.   This was done for the concentrated radial load acting on a reinforced monocoque cylinder, and the results corroborated the theory.

Finally the physical significance of the non-dimensional parameters will be discussed.   With the moment of inertia of the total cross section of the shell designated as I_{tot}, one has

$$I_{tot} = \pi r^3 t'$$

The parameter A can then be written as

$$A = 1.91(\pi r^0 t'/I_{ring})(r/L)^3 = 1.91(I_{tot}/I_{ring})(r/L)^3 \qquad 4.7.42$$

where the symbol I_{ring} is used instead of I to designate the moment of inertia of the cross section of a ring. A is therefore essentially the ratio of the moment of inertia of the entire fuselage to the moment of inertia of a ring.   When the latter is relatively small, A is large, and the stress distribution in the structure deviates considerably from the stress distribution calculated from elementary considerations.   When the rings are very rigid, the value of A is small, and the stress concentration is less pronounced.

If it is observed that the total cross-sectional area A_{tot} of the fuselage is

$$A_{tot} = 2\pi r t'$$

and the cross-sectional area of the sheet carrying shear is

$$A_{\text{sheet}} = 2\pi rt$$

and, if the ratio E/G is taken as 2.6, the expression defining B can be written in the form

$$B = 15.6(A_{\text{tot}}/A_{\text{sheet}})(r/L)^2 \qquad\qquad 4.7.43$$

B is therefore essentially the ratio of the area available for carrying normal stress to that available for carrying shear stress. The effect of B on the stress concentration and stress distribution is much less pronounced than that of A. This conclusion is in perfect agreement with experimental results.

4.8   Saint-Venant's Principle

In his daily work the structural analyst repeatedly makes use of Saint-Venant's principle, even though he may not realize it. For instance, when he determines the normal and shearing stresses in a beam, his calculations are based on this principle which was stated by Love in the following form:

> According to this principle the strains that are pro-   **Theorem   20**
> duced in a body by the application, to a small part of
> its surface, of a system of forces statically equivalent to
> zero force and zero couple, are of negligible magnitude
> at distances which are large compared with the linear
> dimensions of the part.

The principle was enunciated by Saint-Venant as a fact deduced from experience. It will now be demonstrated by a strain-energy analysis of a beam, and in the course of the analysis the significance of the principle as well as the manner in which it is applied to practical problems will be discussed.

The beam of Fig. 4.8.1a is under the action of the uniformly distributed load w. The total load wL is balanced by the distributed shear $V = \frac{1}{2}wL$ applied to each of the two end sections of the beam. Let it be assumed that the shear stresses are applied to the end sections in accordance with diagram b.

If the normal and shear stresses in the beam are calculated from the engineering theory of bending, a linear normal stress distribution and a parabolic shear stress distribution are obtained. It is known that these results are only approximate, as they are consequences of the assumption that plane cross sections before bending remain plane after bending, but their accuracy has proved to be most satisfactory in all engineering applications dealing with beams of the kind shown in Fig. 4.8.1.   For this reason it will be assumed in this analysis that the linear normal stress distribution and the parabolic shear stress distribution represent the correct answer to the problem of the beam, at least at a sufficient distance from the disturbing influence of the reactions in the end sections.

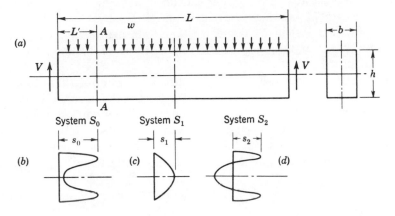

Fig. 4.8.1.   Disturbance stresses in beam.

How can the parabolic shear stress distribution calculated from the theory and shown in diagram c be reconciled with the distribution of the applied shear stresses represented by diagram b?   As a first step one can add to the system S_0 of the actual end shear stresses the system S_1 of the theoretical end shear stresses as well as the system $-S_1$, because adding and subtracting the same quantity obviously cannot change the stress distribution.   The system S_1 can then be combined with the loads w acting over the entire length of the beam; from this combination stresses result which are in perfect agreement with the engineering theory of bending everywhere in the beam, including the end sections.

The remaining systems S_0 and $-S_1$ of end shears are next grouped together and yield a shear stress distribution in the end section of the beam as illustrated in diagram d.   Naturally in this combined system S_2, the shear stresses acting upward near the extreme fibers are balanced by the shear stresses acting downward near the middle of the beam.   By virtue of Saint-Venant's principle the effect of these self-equilibrating shear stresses acting on the area bh of the end section cannot extend farther along the beam than a distance whose order of magnitude is h or b.   Only this region can be affected by the peculiar application of the end shear, and the rest of the beam must have stresses in agreement with the engineering theory of bending as calculated earlier.

In examining this statement, let it be assumed that the length of the region of local irregularities of shear is L'. As a measure of the effect one can adopt the ratio of the absolute value of the maximum shear stress in a section caused by system S_2 to the absolute value s_1 of the maximum shear stress in the end section according to engineering theory. The region of local irregularities can then be defined as that portion of the beam of length L' in which the maximum shear stress caused by the self-equilibrating system S_2 is greater than or equal to s_1/n, where n is a number chosen on the basis of the requirements of accuracy in any particular problem.

As in every section between the end section and section AA the absolute value of the maximum shear stress is s_1/n, or greater, the absolute value of the maximum shear strain is s_1/nG, where G is the shear modulus, or greater. The average of the absolute values of the shear stresses is, of course, less than s_1/n in some sections, but the total strain energy stored in the region is greater than $(\frac{1}{2})(s_1/n)(s_1/nG)bhL'$, if n is sufficiently large. This strain energy

$$U > s_1{}^2(bhL'/n^2G) \qquad\qquad 4.8.1$$

was stored because the applied shear stresses of system S_2 did work, and the strain energy so stored must be equal to the work done.

Let the absolute value of the maximum shear stress of diagram d be designated s_2. Under the action of the externally applied stresses, normal and shear stresses arise in the interior of the beam both in the direction of the shear force V and perpendicularly to it. The order of magnitude of these stresses cannot be greater than s_2, and the order of magnitude of the corresponding strain cannot be greater than s_2/G. If the base of the end section is held fixed in its location and direction, the relative displacement of the top cannot be greater than $(s_2/G)h$.[†] Part of the cross section may be displaced in the direction and the other part opposite to the direction of the local applied shear stress. But, even if all the displacements and stresses had the same direction and sense, the total work done would be smaller than $\frac{1}{2}s_2bh(s_2/G)h$. Hence the external work

$$W_e < \tfrac{1}{2}s_2{}^2(bh^2/G) \qquad\qquad 4.8.2$$

---

† In reality the rotations of the elements may cause a slightly larger maximum displacement. This can be taken into account by multiplying the right-hand member of inequality 4.8.2 by a positive number m whose order of magnitude should be unity in the present problem.

From the requirement of the principle of the conservation of energy, that is, $U = W_e$, it follows that

$$\tfrac{1}{2}s_1^2(bhL'/n^2G) < \tfrac{1}{2}s_2^2(bh^2/G)$$

and thus

$$L' < n^2(s_2/s_1)^2h \qquad\qquad 4.8.3$$

This inequality will now be discussed in some detail.

If the applied shear stresses have a different distribution from that required by engineering beam theory and s_2 is equal to s_1, the critical length becomes

$$L' < n^2h \qquad\qquad 4.8.4$$

Within a distance $L' = 4h$ the maximum disturbance stress is reduced to less than one-half its value at the end section, and within a distance of $L' = 16h$ to less than one-quarter. The expression *less than* may in reality mean *a small fraction of* in many practical applications. Hence the details of the application of the end reactions are of little importance when the stresses are sought in a beam far away from the supports. This result gives the stress analyst the right to calculate the stresses from engineering beam theory even though the end conditions required by that theory are not accurately satisfied. Many other engineering theories are also satisfactory for practical purposes only because Saint-Venant's principle holds.

Some difficulty arises in the application of the principle when loads and reactions are transmitted in a concentrated fashion, for instance through knife edges. Under such conditions forces of finite magnitude are assumed to be applied to vanishingly small areas with the result that the local stresses increase beyond all limits. In Eq. 4.8.3, s_1 remains unchanged but s_2 is indefinitely large; disturbance stresses of the magnitude of the stresses of engineering beam theory may then appear anywhere in the beam according to the inequality. This contradicts the empirical knowledge expressed in Saint-Venant's principle. The contradiction is explained if it is remembered that concentrated loads are not real; they only represent mathematical idealizations. Every real knife edge causes sufficient deformations to reduce the actual stress values to such as can be supported by materials of construction which have but a limited strength. The result obtained from the strain-energy considerations is nevertheless useful. It shows that a rigorous analysis of the details of the stress distribution caused by mathematically idealized concentrated loads may, although it need not, lead to stress values that are not in agreement with what is generally known as Saint-Venant's principle if the calculations are based on the assumption that the material behaves perfectly elastically even under indefinitely large stresses.

One of the assumptions most often encountered in engineering stress analysis is that the stress distribution in a bar subjected to tension is uniform. This assumption is obviously not true at the ends of the bar if the load is applied to them by means of rivets, bolts, or the jaws of a testing machine. At a distance from the load application equal to two or three diameters of a circular cylindrical bar, however, the stresses must be distributed uniformly if Saint-Venant's theorem holds. An analogous statement is true for strips of metal for which the validity of the principle was checked by calculations made with the aid of the Rayleigh-Ritz method.

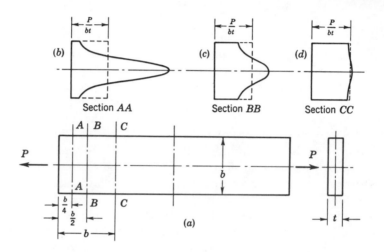

Fig. 4.8.2. Stresses caused in strip by concentrated loads. From *Theory of Elasticity*, by Timoshenko and Goodier, McGraw-Hill Book Co., 1951

In these calculations the loads were assumed to be concentrated in the middle of the shorter sides of the strip as shown in Fig. 4.8.2a. The stress distribution in three consecutive sections is given in Figs. 4.8.2b, c, and d. In section AA, at a distance from the end equal to one-quarter the width b of the strip, the stresses vary greatly and the maximum stress is considerably greater than the average stress P/bt, with t the thickness of the strip. At a distance of half the width of the strip the stress concentration is considerably less pronounced while at a distance equal to the width of the strip the maximum deviation from uniformity amounts to only 3 per cent of the average stress. The example shows that in problems of this kind the stress analyst can rely on Saint-Venant's principle.

The strain-energy derivation of the principle also indicates that the effect of a local disturbance in the shear stress distribution in one section of a thin-walled reinforced monocoque cylinder, such as the fuselage of an airplane, may have a noticeable effect even at a distance several times the diameter of the cylinder. When a concentrated radial load is applied to a reinforcing ring of such a cylinder, the load distribution differs considerably from the sinusoidally distributed shear stresses which should prevail in the section of the fuselage according to engineering beam theory (see Art. 4.7). If the stresses in the fuselage are calculated from engineering beam theory, the disturbance stresses must be superimposed on the results.

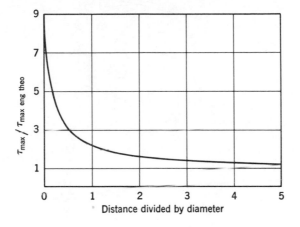

Fig. 4.8.3. Variation of maximum shear stress with distance from loaded end of reinforced monocoque cylinder. From V. L. Salerno's paper in Reissner Anniversary Volume, Edwards, 1949.

The disturbance stresses are caused by the combination of the negative of the shear stress distribution of the engineering theory in the end section and the actual shear load which is a concentrated force. This self-equilibrating system extends over an entire cross section, and thus the maximum linear dimension involved is the diameter of the fuselage.

It must also not be forgotten that in a beam of solid section, or in a three-dimensional solid body, the self-equilibrating external forces have many more paths over which they can balance one another than in a thin-walled monocoque where all the forces must pass through the thin shell. For this reason the distance over which disturbance stresses are of sizable magnitude is generally much greater in a monocoque beam than in a solid beam of the civil engineering variety. Figure 4.8.3 shows how the maximum shear stress in the sheet covering decreases with distance from the end section at which the concentrated load is applied to the reinforcing ring.   The values plotted were calculated from an extension of the theory presented in Art. 4.7.   The cylinder was assumed to extend from the loaded end section to infinity and to have dimensions like those of the fuselages of large transport airplanes.   The values of the non-dimensional parameters defined in Art. 4.7 were $A = 10^7$ and $B = 150$. It can be seen from the curve that the maximum shear stress in the end section is nine times the value obtained from engineering beam theory.

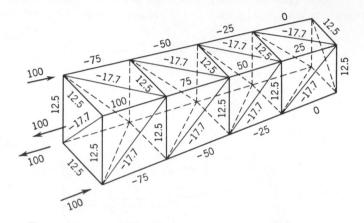

Fig. 4.8.4.   Forces in statically determinate space framework.
From authors' paper in *J. Aero. Sciences.*

This maximum decreases rapidly with distance from the loaded section and amounts to about 110 per cent of the conventional value when the distance is equal to 5 diameters.   In the neighborhood of the concentrated load the stresses are entirely different from those of the engineering beam theory, as can be seen from Figs. 4.7.3 and 4.7.4 of Art. 4.7.

After dealing with the plate and the shell it is of interest to investigate the pin-jointed framework in which the paths for force transmission are further reduced in number.   Figure 4.8.4 represents a space framework subjected to a warping group of forces of 100 lb each.   The four forces add up to zero resultant force and zero resultant moment.   According to Saint-Venant's principle, their effect should be felt only in a region whose length is of the order of magnitude of the depth or the width of the cross section.   On the other hand, the space framework is statically determinate, and thus there is only a single force distribution that satisfies the equations of equilibrium.   The numbers written along the bars are the values of the forces in the bars in pounds, with the positive sign signifying tension and

the negative sign compression.   It can be seen that at a distance of four
depths from the loaded end the force is still one quarter of the maximum
value.   As a matter of fact, in a similar structure of ten times the length,
the force in the end section would still amount to 2.5 per cent of the maximum value.   Consequently Saint-Venant's principle does not seem to
work here.

The situation improves a little when the framework is provided with
crossed surface diagonals.   The maximum load in the first bay is reduced
from 100 to 87.5 lb, and in the last bay from 25 to 12.5 lb, as a comparison
of Fig. 4.8.4 with Fig. 1.11.8 reveals.   The forces in the highly redundant
cross-braced framework were calculated in Art. 1.11.   The same article also
contains the evaluation of the forces in the space framework when in
addition crossed diagonals are provided in three transverse planes in the
interior of the structure.   The effect of these last redundant elements is
striking: The maximum force in the last bay is reduced to 0.7 lb, as may
be seen from Fig. 1.11.7.

The significant conclusion from the analysis of the space framework is
that Saint-Venant's principle does not work when it cannot work.   For a
statically determinate framework the strain-energy demonstration of the
principle ceases to be valid because the evaluation of the displacements of
the points of application of the loads loses its validity.   If the statically
determinate framework is lengthened through the addition of more bays,
the warping increases, and the increased work of the external forces
provides for the increase in the strain energy stored in the structure.
Such large displacements would cause very high stresses in the redundant
bars of the framework of Fig. 1.11.7, and for this reason in the redundant
framework the principle of the minimum of the energy leads to a modifica-
tion of the forces.   The transverse bracing elements are much more
effective in this redistribution of the forces than the ones arranged on the
surface of the framework.

A far more fundamental redistribution of the stresses takes place when
the space framework is replaced by a three-dimensional prismatic solid
whose edges are the lines previously occupied by the longitudinal bars of
the space framework.   Obviously in the solid a large amount of warping
under the action of the warping group of forces would lead to extreme
shear strains in the material.   The consequent very large strain energy
would be in contradiction to the requirements of the minimal principles.
Hence the corners of the solid are restrained from large displacements by
the shearing rigidity of the material, and the effect of the warping group
is localized.   Such a restriction of the displacements cannot exist in a
statically determinate framework in which the length of each bar can be
changed at will, and the result is no strain, just a change in geometry.

It follows from these considerations that self-equilibrating systems of loads applied to solid bodies have a noticeable effect only in regions whose linear dimensions are approximately equal to the linear dimensions of the region to which the loads are applied. In very flexible thin-walled structures the affected region may be much larger; and for statically determinate frameworks Saint-Venant's principle cannot hold.

Appendix 1

Introduction

In these pages are collected remarks that in other books are usually relegated to footnotes.   They give the sources of the material presented, refer to the history of the problem treated, and occasionally indicate the practical importance of the results, their use, and their confirmation by experiment or by other means.   No effort was made to prepare an exhaustive bibliography; for this reason omission of the titles of books or articles does not imply that they are of lesser value than those cited.   Many of the references given presuppose more knowledge of the theory of elasticity and of mathematics than the present text.   Occasionally a special remark calls the attention of the reader to particularly easily readable, important, or difficult papers.

The following abbreviations have been used:

ARC	Aeronautical Research Council (Great Britain)
DVL Jahrbuch	Jahrbuch der Deutschen Versuchsanstalt für Luftfahrt, Oldenbourg, Munich, Germany
J. Aero. Sciences	Journal of the Aeronautical Sciences, Institute of the Aeronautical Sciences, New York, N.Y.
J. Appl. Mech.	Journal of Applied Mechanics, American Society of Mechanical Engineers, New York, N.Y.
J. Roy. Aero. Soc.	The Journal of the Royal Aeronautical Society, London, England
NACA Tech. Memo.	Technical Memorandum of the National Advisory Committee for Aeronautics, Washington, D.C.
NACA Tech. Note	Technical Note of the National Advisory Committee for Aeronautics, Washington, D.C.
NACA Rep.	Technical Report of the National Advisory Committee for Aeronautics, Washington, D.C.
NBS	National Bureau of Standards, Washington, D.C.
Phil. Mag.	The London, Edinburgh, and Dublin Philosphical Magazine and Journal of Science
PIBAL	Polytechnic Institute of Brooklyn Aeronautical Laboratories, Brooklyn, N.Y.
Proceedings ASCE	Proceedings of the American Society of Civil Engineers, New York, N.Y.
Qu. Appl. Math.	Quarterly of Applied Mathematics, Brown University, Providence, R.I.
Rep. and Mem.	Reports and Memoranda of the Aeronautical Research Council (Great Britain)

Transactions ASCE	Transactions of the American Society of Civil Engineers, New York, N.Y.
Transactions ASME	Transactions of the American Society of Mechanical Engineers, New York, N.Y.
WGL Jahrbuch	Jahrbuch der Wissenschaftlichen Gesellschaft für Luftfahrt, Julius Springer, Berlin, Germany
ZAMM	Zeitschrift für Angewandte Mathematik und Mechanik, Verein Deutscher Ingenieure, Berlin, Germany
ZFM	Zeitschrift für Flugtechnik und Motorluftschiffahrt, Oldenbourg, Munich, Germany
ZVDI	Zeitschrift des Vereins Deutscher Ingenieure, Berlin, Germany
McGraw-Hill	McGraw-Hill Book Co., New York, N.Y.
Macmillan	The Macmillan Company, New York, N.Y.
Oldenbourg	R. Oldenbourg, Munich, Germany
Prentice-Hall	Prentice-Hall, New York, N.Y.
Springer	Julius Springer, Berlin, Germany
Teubner	B. G. Teubner, Leipzig, Germany
Van Nostrand	D. Van Nostrand Company, New York, N.Y.
Wiley	John Wiley & Sons, New York, N.Y.

Part 1

1.1–1.6. In a talk given at the annual dinner of the Applied Mechanics Division of the ASME on November 28, 1950, Professor Jesse Ormondroyd of Michigan University made the statement that the principle of virtual displacements was first used by Aristotle in his *Mechanica* in 350 B.C. (see "Applied Mechanics from Aristotle to Rankine," *Mechanical Engineering*, Vol. 73, No. 9, p. 723, September 1951). In reading E. S. Forster's translation of this book in Vol. 6 of the *Opuscula* published under the editorship of W. D. Ross by the Clarendon Press in Oxford, England, in 1913, the author was unable to find anything more than the parallelogram of velocities and the law of the lever; of course, it is not easy to interpret the mechanical concepts of a philosopher who wrote over 2000 years ago, and the definition of the principle can also be given in a narrow or a broad manner.

Professor Ernst Mach of the University of Vienna in his book, *The Science of Mechanics, a Critical and Historical Account of its Development* (first German edition in 1883, first English translation in 1893, fifth English edition in 1942, Open Court Publishing Co., LaSalle, Ill., translation by Thomas J. McCormack), mentions Galileo Galilei as one of the early scientists who recognized the principle (in 1594; see *Discorzi e Dimostrazioni Mathematiche, Intorno a Due Nuove Scienze*, Elsivir, Leyden, Holland, 1638; English translation by H. C. Crew and A. de Salvio entitled *Dialogues Concerning Two New Sciences*, Northwestern University, Evanston, Ill., 1939), but he attributes its first clear and general formulation to John Bernoulli in 1717 (in a letter to Varignon).

The principle is described in all present-day books on mechanics. Lucid presentations can be found in the very readable German book on elasticity entitled *Drang und Zwang*, by August Föppl and Ludwig Föppl (Oldenbourg, 1924, Vol. 1, p. 58) and in the well-known text *Engineering Mechanics* by S. Timoshenko and D. H. Young (McGraw-Hill, Second Edition, 1940, p. 217).

1.7–1.8. The calculation of the displacements of joints of a framework and of the forces in redundant members of frameworks is based on work published by the distinguished English physicist, James Clerk Maxwell in 1864. The paper entitled "On the Calculation of the Equilibrium and the Stiffness of Frames" (*Phil. Mag.*, Series 4, Vol. 27, p. 294) also contains a statement of what is now known as Maxwell's reciprocal theorem. Professor A. S. Niles of Stanford University discusses Maxwell's paper and the further development of its subject matter by E. Betti (1872), Lord Rayleigh (1873), and Otto Mohr (1874), and stresses the importance of the contributions of the last-named author. Niles's article bears the title "Clerk Maxwell and the Theory of Indeterminate Structures" (*Engineering*, Vol. 170, p. 194, Sept. 1, 1950). Interesting historic remarks can also be found in a paper by H. M. Westergaard, "One Hundred Fifty Years Advance in Structural Analysis" (*Transactions ASCE*, Vol. 94, p. 226, Paper 1727, 1930), and in the "Historical Review" section of *An Elementary Treatise on Statically Indeterminate Stresses* by John Ira Parcel and George Alfred Maney, Wiley, Second Edition, 1936, p. 413.

The Maxwell-Mohr, or the dummy load, method is discussed in every modern textbook on structures.

1.9–1.11. These sections were included in this book mainly for the following reasons: Very little is presented in most modern structural texts on space frameworks (except when they can be reduced to plane trusses under symmetric loads); the calculation of warping affords an excellent application of the principle of virtual displacements; the tension coefficient method is certainly worth the space needed for its explanation; and the results obtained in Art. 1.11 are useful in the discussion of Saint-Venant's principle in Art. 4.8.

The tension coefficient method was developed by Sir Richard Southwell in a paper entitled "Primary Stress Determination in Space Frames" (*Engineering*, Vol. 109, p. 165, February 6, 1920). (See also p. 106 of *An Introduction to the Theory of Elasticity for Engineers and Physicists*, by R. V. Southwell, Second Edition, Oxford University Press, 1941). However, it was used earlier by H. F. B. Müller-Breslau in *Die Neueren Methoden der Festigkeitslehre und der Statik der Baukonstruktionen*, Baumgärtner, Leipzig, Germany, 1904, p. 247. Few textbooks discuss tension coefficients; among the exceptions are *Aeroplane Structures*, by A. J. Sutton Pippard and J. Laurence Pritchard (Second Edition, Longmans, Green, and Co., London, 1935, p. 147) whose first edition, published in 1919, was the first textbook on airplane structural analysis; *Airplane Structures* by Alfred S Niles and Joseph S. Newell, Third Edition, Vol. 1, p. 216, Wiley, 1943, whose first edition in 1929 was the first American textbook on airplane structures; and *Strain Energy Methods of Stress Analysis*, by A. J. Sutton Pippard, Longmans, Green, and Co., London, 1928.

The ingenious solution of the torsion problem of the space framework is due to Herbert Wagner, who, however, was apparently unaware of Southwell's work on tension coefficients. Wagner's work is entitled "Ueber räumliche Flugzeug-fachwerke. Die Längsstabkraftmethode," *ZFM*, Vol. 19, No. 15, p. 337, August 14, 1928. It was translated by the NACA as *Tech. Memo. 522*, "The Analysis of Aircraft Structures as Space Frameworks, Method Based on the Forces in the Longitudinal Members." The problem is treated on p. 231 of Niles and Newell's *Airplane Structures* and in David J. Peery's *Aircraft Structures*, McGraw-Hill, 1950, p. 32.

The engine mount ring was analyzed by means of the second minimal principle by Charles E. Mack, Jr., in "Three-Dimensional Analysis of Engine Mount Rings by a Strain Energy Method" in *J. Aero. Sciences*, Vol. 12, No. 1, January 1945. The results of the author's analysis of the forces in a space framework subjected to a warping group were published in the paper "The Applicability of Saint-Venant's Principle to Airplane Structures," *J. Aero. Sciences*, Vol. 12, No. 4, p. 455, October 1945. The derivation of the results appears for the first time in this book.

Space frameworks are treated extensively in August Föppl's *Das Fachwerk im Raum*, Teubner, 1892; in Müller-Breslau's book *Die neueren Methoden*, etc. (see above); in Lebrecht Henneberg's work *Die graphische Statik der starren Systeme*, Teubner, 1911, p. 624; in Benjamin Mayor's *Introduction à la Statique graphique des systèmes de l'espace*, Payot, Lausanne, Switzerland, 1926 (which was preceded by two communications to the Académie des Sciences in Paris, France, printed in the *Comptes Rendus* of December 29, 1902, Vol. 135, p. 1318 under the title "Sur une représentation plane de l'espace et son application à la Statique graphique" and in the issue of January 12, 1903, Vol. 136, p. 85 under the title "Sur la Statique graphique de l'espace"); in "Graphische Statik räumlicher Kräftesysteme" by R. von Mises, *Zeitschrift für Mathematik und Physik*, Teubner, Vol. 64, No. 3, p. 209, 1916; in "Beitrag zur Kinematik des Raumfachwerkes," by W. Prager, *ZAMM*, Vol. 6, No. 5, p. 341, October 1926; and in Friedrich Bleich's *Stahlhochbauten*, Springer, 1932. Short descriptions of space framework theory can be found in *The Theory of Structures*, by Charles M. Spofford, Fourth Edition, McGraw-Hill, 1939, p. 469, and Linton E. Grinter's *Theory of Modern Steel Structures*, Revised Edition, Macmillan, 1949, Vol. 2, p. 14. Of the modern American engineering textbooks S. Timoshenko and D. H. Young's *Theory of Structures* (McGraw-Hill, 1945) devotes considerable discussion to space frameworks (pp. 163–212).

Airplane fuselage frameworks were analyzed by A. J. Sutton Pippard and W. D. Douglas in "Torsional Stresses in the Fuselage of an Airplane," *Rep. and Mem.* 736, June 1921 (see also Pippard's *Strain Energy Methods* etc. cited above) by Aloys van Gries in his *Flugzeugstatik* (Springer, 1921); by Hans Ebner in "Zur Berechnung räumlicher Fachwerke im Flugzeugbau," *DVL Jahrbuch* 1929, p. 371, Report 138; by Edgar Seydel in "Ermittelung der Stabkräfte im Flugzeug-Fachwerkrumpf," *DVL Jahrbuch* 1929, p. 415, Report 139; and by K. Thalau and A. Teichmann in their excellent textbook *Aufgaben aus der Flugzeugstatik*, Springer, 1933. A particular case of the fuselage torsion problem was solved earlier by Hans Reissner in "Neuere Probleme aus der Flugzeugstatik," *ZFM*, Vol. 17, No. 18, p. 384, September 28, 1926.

1.12–1.15. Beam theory was founded by the famous physicists and mathematicians of Basle, Switzerland: James Bernoulli, Daniel Bernoulli, and Leonhard Euler, as well as by the French engineer and scientist, C. L. M. Navier, between 1705 and 1826; however, the problem of the bending of a beam was first formulated by Galileo Galilei in 1638. A discussion of the early history of the problem can be found in the "Historical Introduction" to *A Treatise on the Mathematical Theory of Elasticity* by A. E. H. Love, Fourth Edition, First American Printing, Dover Publications, New York, 1944, p. 1 (First Edition in 1892).

The dummy load method of calculating deflections is amply described in most textbooks on structures, among which the following will be mentioned: Parcel and Maney's *Statically Indeterminate Stresses*, p. 11 (see under 1.7–1.8); J. A. Van den Broek's *Elastic Energy Theory*, Second Edition, Wiley, 1942; *Elementary Structural Analysis*, by J. B. Wilbur and C. H. Norris, McGraw-Hill, 1948, p. 288; and *Airplane Structures* by Niles and Newell, Vol. 1, p. 375 (see under 1.9–1.11).

The three-moment equation was derived by E. Clapeyron in "Calcul d'une poutre élastique reposant librement sur des appuis inégalement espacés," *Comptes rendus des séances de l'Académie des Sciences*, Paris, Vol. 45, p. 1076, December 28, 1857.

1.16. Westergaard and Parcel and Maney (see under 1.7–1.8) attribute the moment-area method to Professor Charles E. Green of the University of Michigan (1874). The method is described in all textbooks on structures.

1.17. The moment-distribution method was described by Professor Hardy Cross, then at the University of Illinois, in a clear and concise paper entitled "Analysis of Continuous Frames by Distributing Fixed End Moments" in the May, 1930, *Proceedings ASCE*, and in the *Transactions ASCE* (Vol. 96, p. 1, 1932). It is discussed in greater detail in *Continuous Frames of Reinforced Concrete*, by Hardy Cross and N. D. Morgan, Wiley, 1932. Similar schemes were used earlier by Manderla (1880) and particularly by Shortridge Hardesty (in Chapter 11 entitled "Secondary Stresses, Temperature Stresses, and Indeterminate Stresses" of J. A. L. Waddell's book *Bridge Engineering*, Vol. 1, p. 178, Wiley, 1916), but only after it had been put into its final form by Hardy Cross was the method generally accepted. This is the reason why his name was given to the moment-distribution method. It is probably the most popular new method in structural analysis and is treated in almost every modern textbook on structures.

An alternative to the Hardy Cross moment-distribution method is the procedure suggested by Dean Linton E. Grinter of the University of Florida under the designation "analysis by balancing end angle changes." The original article was published in the *Proceedings ASCE*, September 1936, p. 995, under the title "Analysis of Continuous Frames by Balancing Angle Changes" and was reprinted in the *Transactions ASCE*, Vol. 102, p. 1020, 1937 (Paper 1976). The method is described on p. 164, Vol. 2 of Grinter's *Theory of Modern Steel Structures* (see under 1.9–1.11). A similar idea was described by C. V. Klouček of Prague, Czechoslovakia, in *Beton und Eisen*, Wilhelm Ernst und Sohn, Berlin, Germany, Vol. 38, No. 24, p. 359, December 20, 1939 ("Das Prinzip der fortgeleiteten Verformung") and was discussed at greater length in the book *Distribution of Deformation* (*a New Method of Structural Analysis*), Prague, Czechoslovakia, 1949. See also "Structural Analysis by Distribution of Deformation," by C. V. Klouček, *Qu. Appl. Math.*, Vol. 9, No. 1, p. 77, April 1951. These methods are related to the moment-distribution method and to Southwell's relaxation method as the complementary energy principle is related to the principle of the minimum of the total potential. They were designated by N. M. Newmark as "continuity restoration" methods, and their significance was discussed in the article "Numerical Methods of Analysis of Bars, Plates, and Elastic Bodies" on p. 138 of the volume *Numerical Methods of Analysis in Engineering* (*Successive Corrections*), edited by L. E. Grinter (Macmillan, 1949).

1.18. Secondary stresses were first analyzed by H. Manderla ("Welche Spannungen entstehen in den Streben eines Fachwerks dadurch, dass die Winkel der Fachwerkdreiecke durch die Belastung eine Änderung erleiden?" Preisarbeit, *Jahresbericht der Technischen Hochschule* München, 1878–79, p. 18) whose work was followed by those of Engesser (1879), Asimont (1879), Winkler (1881), Müller-Breslau (1885), Engesser (*Die Zusatzkräfte und Nebenspannungen eiserner Fachwerkbrücken*, Springer, 1893) and Otto Mohr ("Die Berechnung der Fachwerke mit starren Knotenverbindungen," *Der Civilingenieur*, Organ des Sächsischen Ingenieur und Architektenvereins, Vol. 38, p. 578, 1892, and Vol. 39, p. 69, 1893).

A very complete presentation of the subject can be found in *Die Ermittelung der Nebenspannungen in Eisernen Fachwerkbrücken und das praktische Rechnungsverfahren nach Mohr*, by W. Gehler, Wilhelm Ernst und Sohn, Berlin, 1910. In the American literature much space is devoted to secondary stresses in *The Theory and Practice of Modern Framed Structures*, by J. B. Johnson, C. W. Bryan, and F. E. Turneaure (Vol. 2, p. 423 of the Ninth Edition, rewritten, Wiley, 1911), whose first edition dates back to 1893. An excellent survey of the problem can be found in *Secondary Stresses in Bridge Trusses*, by C. R. Grimm, Wiley, 1908. Of the more recent books a comprehensive treatment of the problem of the secondary stresses is contained in Parcel and Maney's *Statically Indeterminate Stresses* (cited under 1.7–1.8); a shorter presentation may be found in *Theory of Statically Indeterminate Structures* by W. M. Fife and J. B. Wilbur (McGraw-Hill, 1937, p. 202).

1.19. The slope-deflection method was originally developed by H. Manderla (in 1878–79, see under 1.18) and by Otto Mohr (1892–93, see under 1.18) for the purpose of analyzing secondary stresses in plane trusses. It was independently rederived, applied to a broader group of problems, and popularized by George Alfred Maney in "Secondary Stresses and Other Problems in Rigid Frames: A New Method of Solution," *University of Minnesota Studies* in *Engineering* No. 1, Minneapolis, Minn., March 1915. Similarly, Professor A. Ostenfeld in Copenhagen, Denmark, presented the method in 1920 and published a book on it under the title *Die Deformationsmethode*, Springer, 1926. Ostenfeld refers to the publication *Die Methode der Alpha-Gleichungen*, by Ax. Bendixsen, Springer, 1914, as a less complete forerunner of his method.

The slope-deflection method is discussed in all textbooks of structural analysis; an excellent presentation can be found on p. 147 of *Statically Indeterminate Stresses*, by Parcel and Maney (see under 1.7–1.8). A very complete account of the solution of rigid-frame problems by means of this method is given in *Analysis of Rigid Frames*, by A. Amirikian, U. S. Government Printing Office, Washington, 1942.

Part 2

2.1. Today strain energy is a concept known even to beginners in stress analysis; it is discussed in all elementary textbooks on the strength of materials (see, for instance, p. 281 of *Strength of Materials*, by S. Timoshenko, Second Edition, Van Nostrand, 1940). A. E. H. Love, on p. 11 of the "Historical Introduction" to his *Theory of Elasticity* (cited under 1.12–1.15) attributes to George Green the discovery of the existence of a strain-energy function (in 1837).

Of the various formulas derived in this article only those relating to non-uniform torsion are not given in elementary texts. The effect of non-uniform warping on the torsion of an I beam was first investigated by S. Timoshenko in 1905. His first German publication dealing with this subject was entitled "Einige Stabilitätsprobleme der Elasticitätstheorie, Part III, Kipperscheinungen des I-Trägers," *Zeitschrift für Mathematik und Physik*, Vol. 58, p. 337, 1910. The importance of the effect on the buckling of the thin-walled open sections used in the airplane industry was recognized by Herbert Wagner. In 1929 he gave a theory for arbitrary thin-walled sections ("Verdrehung und Knickung von offenen Profilen," in *Festschrift Fünfundzwanzig Jahre Technische Hochschule Danzig*, Kafemann, Danzig, p. 329, 1929, translated as "Torsion and Buckling of Open Sections," *NACA Tech. Memo*. 807, 1937) which was generalized and simplified by Robert Kappus ("Drillknicken zentrisch gedrückter Stäbe mit offenem Profil im elastischen Bereich," *Luftfahrtforschung*, Vol. 14, No. 9, p. 444, 1937, translated as "Twisting Failure of Centrally Loaded Open-Section Columns in the Elastic Range," *NACA Tech. Memo*. 851, 1938).

Further discussion of the problem can be found in "Torsional and Flexural Buckling of Bars of Thin-Walled Open Section under Compression and Bending Loads," by J. N. Goodier, *J. Appl. Mech.*, Vol. 9, No. 3, p. A-103, September 1942; in the appendix to the paper "Stresses in Space-Curved Rings Reinforcing the Edges of Cut-Outs in Monocoque Fuselages," by N. J. Hoff, *J. Roy. Aero. Soc.*, Vol. 47, No. 386, p. 64, February 1943; and in *Airplane Structures*, by Niles and Newell, Vol. 2, p. 316 (cited under 1.9–1.11). A comprehensive treatment of problems involving non-uniform torsion and torsional buckling was given by Timoshenko under the title "Theory of Bending, Torsion, and Buckling of Thin-Walled Members of Open Cross Section," *Journal Franklin Institute*, Vol. 239, 1945, No. 3, p. 201, March; No. 4, p. 249, April; and No. 5, p. 343, May. Warping constants are listed on p. 132 of *Structural Principles and Data*, Royal Aeronautical Society, Fourth Edition, 1952 (the "Structural Analysis" part of the book was written by J. H. Argyris and P. C. Dunne).

2.2. The nomenclature introduced in this article is not the same as that used by other authors. Love discusses the "theorem of minimum energy" on p. 171 of his book cited under 1.12–1.15 and uses the expression "energy," "potential energy," and "potential energy of deformation" for what is called the "total potential" in this article. David Williams uses the term "total potential energy" in his discussion of the various minimal principles and gives many worked-out examples in two easily readable papers prepared for the structural engineer. They are "The Use of the Principle of Minimum Potential Energy in Problems of Static Equilibrium," *Rep. and Mem.* 1827, January 1938, and "The Relations between the Energy Theorems Applicable in Structural Theory," *Phil. Mag.*, Series 7, Vol. 26, No. 177, p. 617, November 1938. Similarly the symbols U and V are used in the literature in various ways. Therefore, the inexperienced reader of scientific articles should be careful to establish from the context what the author means by his particular symbols and how he happens to define the various technical terms.

Chapters on strain-energy methods are contained in all books on the theory of elasticity. By way of example Timoshenko and Goodier's *Theory of Elasticity* might be mentioned (p. 146 cited under 2.2) as well as Southwell's *An Introduction to the Theory of Elasticity* (cited under 1.9–1.11), and the theory of elasticity written primarily for mechanical engineers by C. B. Biezeno of Delft, Holland,

and R. Grammel of Stuttgart, Germany, entitled *Technische Dynamik* (Springer, 1939, reproduced in the United States by Edwards Bros., Ann Arbor, Mich., 1944). A comprehensive presentation of the exact and approximate methods of strain-energy analysis can be found in the chapter on "Variational Methods" in *Mathematical Theory of Elasticity* by I. S. Sokolnikoff, McGraw-Hill, 1946 (p.277), but this book, of course, is intended for the mathematically more sophisticated reader. *Drang und Zwang* by August Föppl and Ludwig Föppl (cited under 1.1–1.6) contains a detailed discussion of the minimal principles and their application to problems in engineering. The book is written for the engineer (p. 58). A more mathematical treatment by E. Trefftz entitled "Mechanik der Kontinua" is contained in Vol. 2 of *Die Differential-und Integralgleichungen der Mechanik und Physik*, by Phillipp Frank and Richard von Mises, Friedr. Vieweg, Braunschweig, Germany, p. 598. Further remarks on the minimal principles of stress analysis can be found under 4.1.

2.4. The theory of the beam column attained great importance during World War I when it was applied to the calculation of the wing spars of biplanes. For this reason solutions of the differential equations were worked out in all the major belligerent countries, and values of the functions involved were tabulated. In England these functions are known as the Berry functions ("The Calculation of Stresses in Aeroplane Wing Spars," by Arthur Berry, *Transactions Royal Aeronautical Society*, No. 1, 1916); in Germany the problem was analyzed by Hans Reissner and E. Schwerin ("Die Festigkeitsberechnung der Flugzeugholme," *Jahrbuch WGL*, Vol. 4, Sonderheft, Springer, 1916); in Russia the work was done by S. Timoshenko; and in the United States tables were prepared by A. S. Niles and J. S. Newell after the war in 1922. Probably the most complete treatment of beam columns can be found on pp. 62-155 of Vol. 2 of Niles and Newell's *Airplane Structures* (cited under 1.9–1.11); see also *Theory of Elastic Stability*, by S. Timoshenko, McGraw-Hill, 1936, p. 1, and *Elasticity* in *Engineering*, by Ernest E. Sechler, Wiley, 1952, p. 359.

The basic equations of beam-column theory were derived and solved long before World War I, but they were not used much, if at all, in structural analysis. A presentation of the theory can be found in Heinrich F. B. Müller-Breslau's *Die graphische Statik der Baukonstruktionen*, Vol. 2, Part 2, p. 623, Second, Enlarged Edition, Alfred Kröner, Leipzig, Germany, 1925 (the first edition of Vol. 1 was published in 1892).

Even earlier E. Winkler included a chapter on beam-column theory in his book entitled *Die Lehre von der Elastizität und Festigkeit* (*Theory of Elasticity and Strength of Materials*, Dominicus, Prague, 1867) in which all the main equations of the theory were derived on pp. 157–184 of Section 5. Winkler did not claim originality for this work, and as the second volume of this book, which was to contain all the references, was probably never published, the author does not know who was the originator of beam-column theory.

In the present article the beam-column differential equation is derived through the variational calculus. An attempt was made to present the theory simply enough for every well-trained engineer to understand it without difficulty. It appears that the average engineer has taken the major hurdle when he gets rid of his feeling that the variational calculus can be understood only by high-brow mathematicians.

To keep the presentation as unencumbered as possible, the analogy between the change in a function due to a variation and the differential of the ordinary calculus was not discussed in detail.   A simple treatment of the variational calculus is available in the book *An Introduction to the Calculus of Variations* by Charles Fox (Oxford University Press, London, 1950).

A graphic procedure for solving the beam-column differential equations was devised by H. G. Howard in "The Graphical and Analytical Determination of Stresses in Single-Span and Continuous Beams under End Compression and Lateral Load with Variation in Shear, Distributed Load, and Moment of Inertia," *Rep. and Mem.* 1233, June, 1928 (on p. 856 of the Technical Reports of the ARC, 1928–29, Vol. 2).   It is discussed on p. 128 of Pippard and Pritchard's *Aeroplane Structures* (see under 1.9–1.11) and in Niles and Newell's *Airplane Structures* (p. 141, see under 1.9–1.11).

For tables of the stiffness coefficients and carry-over factors of bars of constant section see the publications of James, Niles, and Newell, and Lundquist and Kroll listed under 3.8.   The quantities relating to bars whose ends are stiffened by gusset plates are based on the assumption that the bending rigidity of the bar increases according to a hyperbolic law from the beginning of the gusset plate toward the mathematical end point of the bar; at this end point the bar is assumed to be perfectly rigid.   Figures 2.4.9–12 are taken from the paper "Buckling of Rigid-Jointed Plane Trusses," N. J. Hoff, Bruno A. Boley, S. V. Nardo, and Sara Kaufman, *Transactions ASCE*, Vol. 116, p. 958, 1951, Paper 2454.   Comparison with experiments indicates that the assumption is reasonable and slightly conservative.

The notation in beam-column theory varies from author to author. Timoshenko uses u, and Niles and Newell j for the basic parameter.   These symbols are related to k of this article in the following manner:

$$kL = L/j = 2u$$

2.6.   The usefulness of covering a thick layer of inferior wood with one or more layers of high-quality veneer was known to the ancient Egyptians (see p. 1 of *Engineering Laminates* edited by Albert G. H. Dietz, Wiley, 1949).   A scientific study of the mechanical behavior of sandwich structural elements began only when the de Havilland Aircraft Company became interested in using sandwich construction in their airplanes.   The outstanding success of the de Havilland Mosquito fighter bomber, which first flew in 1940, was followed by the publication of many theoretical and experimental studies in this field.   Their results are presented in the author's section entitled "The Strength of Laminates and Sandwich Structural Elements," pp. 6–88 of *Engineering Laminates* (cited above) which also contains an extensive bibliography.

The derivations presented in this article follow the treatment in "Bending and Buckling of Sandwich Beams," by N. J. Hoff and S. E. Mautner, presented at the Sixth International Congress for Applied Mechanics in Paris, France, September 1946 and published in the *J. Aero. Sciences*, Vol. 15, No. 12, p. 707, December 1948.   The same variational approach was used later in the derivation of the equations of equilibrium of sandwich plates in "Bending and Buckling of Rectangular Sandwich Plates," by N. J. Hoff, *NACA Tech. Note* 2225, November 1950.

2.7.   The approximate calculation of deflections, buckling loads, and natural frequencies by the strain-energy method was originated by Lord Rayleigh who in his classical book *The Theory of Sound* (First Edition in 1877; First American Edition, Dover Publications, New York, 1945) presented many examples of its use. Although Lord Rayleigh was perfectly aware of the possibility of representing deflected shapes by series and even made use of this device in some of his work, in the major part of his calculations the deflections were defined by single-term expressions. For this reason it is usual to refer to analyses based on single-term approximations as calculations by the Rayleigh method; the expressions Ritz method, Rayleigh-Ritz method, Timoshenko method, and Rayleigh-Ritz-Timoshenko method are used to indicate strain-energy analyses employing infinite series.

To show that infinite series had been used by Lord Rayleigh, reference is made to two of his publications: "On the Calculation of the Frequency of Vibration of a System in Its Gravest Mode, with an Example from Hydrodynamics," *Phil. Mag.*, Vol. 17, p. 556, 1899; and "On the Calculation of Chladni's Figures for a Square Plate," *Phil. Mag.*, Series 6, Vol. 22, No. 128, p. 225, August 1911.

The rigorous mathematical representation of deflected shapes by infinite series was investigated, the convergence of the series proved, and the series method applied to many problems of practical interest by the Swiss mathematician Walter Ritz. Two of his fundamental papers are: "Über eine neue Methode zur Lösung gewisser Randwertaufgaben," *Göttingener Nachrichten, Math.-Phys. Klasse*, pp. 236–248, 1908; and "Über eine neue Methode zur Lösung gewisser Variationsprobleme der Mathematischen Physik," *Journal für Reine und Angewandte Mathematik* (Crelle, Georg Reimer, Berlin), Vol. 135, No. 1, p. 1, 1909. These two papers can be read without difficulty by engineers who have a sound background knowledge of mathematics.

At the same time when Ritz investigated the mathematical aspects of the Rayleigh method, Timoshenko explored its engineering applications, solved many problems of technical interest, and showed the accuracy obtainable by it. His early Russian papers dated from 1907 and 1908. His work became more generally known when he published some papers in Western European languages. Among these the following may be mentioned: "Sur la stabilité des systèmes élastiques," *Annales des Ponts et Chaussées*, Partie Technique, Series 9 (Part 1 in Vol. 3, p. 496; Part 2 in Vol. 4, p. 73; and Part 3 in Vol. 4, p. 372), 1913; "Einige Stabilitätsprobleme der Elastizitätstheorie," *Zeitschrift für Mathematik und Physik*, Vol. 58, p. 343, 1910; and "The Approximate Solution of Two-Dimensional Problems in Elasticity," *Phil. Mag.*, Sixth Series, Vol. 47, No. 282, p. 1095, June 1924.

Following Professor A. S. Niles' suggestion, in this volume the single-term strain-energy approach will be called the Rayleigh method; the series representation of deflections in the strain-energy analysis of equilibrium problems will be designated as the Rayleigh-Ritz method; and the same procedure followed in stability calculations will be denoted as the Rayleigh-Timoshenko method.

2.8. This introduction to Fourier series resembles the more complete one in *Mathematical Methods in Engineering*, by Th. von Kármán and Maurice A. Biot, McGraw-Hill, 1940, p. 323. There numerical, graphical, and mechanical methods of determining the Fourier coefficients are also treated. Chapters on Fourier series are contained in all books on advanced calculus. (See, for instance, *Advanced Calculus*, by Ivan S. Sokolnikoff, McGraw-Hill, 1939, p. 378). Among more advanced books dealing with this topic are *Fourier's Series and Spherical Harmonics*, by W. E. Byerly, 1893; *Fourier's Series and Integrals*, by H. S. Carslaw, 1930; and *Fourier Series and Boundary Value Problems*, by Ruel V. Churchill, McGraw-Hill, 1941.

2.9. Galerkin's method, first presented in Russian in 1915, is a counterpart of the Rayleigh-Ritz method. With a limited number of terms it gives an approximation, and with an infinite series representing the displacements it yields a rigorous solution. In the western world C. B. Biezeno called attention to this method in 1923 ("Over en Vereenvouding en over en Uitbreiding van de Methode van Ritz," *Zeitschrift Christiaan Huygens*, Vol. 3, p. 69, 1923–24); and in February 1927 H. Hencky commented on it in *ZAMM*, Vol. 7, No. 1, p. 80, under the title "Eine wichtige Vereinfachung der Methode von Ritz zur angenäherten Behandlung von Variationsproblemen." To quote from this article: "The well-known method of Ritz, which has been used extensively in the theory of elasticity, is capable of an elegant as well as practical simplification as was shown by Mr. Galerkin in Leningrad."

In the English literature W. J. Duncan called attention much later to Galerkin's work in the following two papers: "Galerkin's Method in Mechanics and Differential Equations," *Rep. and Mem.* 1798, 1937; and "The Principles of the Galerkin Method, "*Rep. and Mem.* 1848, 1938. The method is described in Biezeno and Grammel's *Technische Dynamik*, p. 137 (see under 2.2); in Sokolnikoff's *Theory of Elasticity*, p. 313 (see under 2.2); in Appendix C entitled "The Galerkin Method" of *NACA Tech. Note* 2556, "Buckling of Rectangular Sandwich Plates Subjected to Edgewise Compression with Loaded Edges Simply Supported and Unloaded Edges Clamped," by Kuo Tai Yen, V. L. Salerno, and N. J. Hoff, January 1952; and in "A Simplified Method of Elastic Stability Analysis for Thin Cylindrical Shells," Part II, "Modified Equilibrium Equation," by S. B. Batdorf, *NACA Tech Note* 1342, June 1947, p. 7.

Part 3

3.1. According to Mach (*The Science of Mechanics*, cited under 1.1–1.6) the minimal property of stable equilibrium (in contrast to the stationary property of equilibrium) was first pointed out by Maupertuis in a communication to the Paris Academy in 1740. The beginning of the theory of elastic stability is usually given as 1744 when Leonhard Euler published his *Methodus inveniendi lineas curvas maximi minimive proprietate gaudentes* in Lausanne, Switzerland. In the appendix (Additamentum) to this treatise entitled "De curvis elasticis" the differential equation of the elastic rod was derived and solved for the case of axial compressive loading. This work of Euler also marks the beginning of the calcus of variations because the buckled shape of the rod was obtained from considerations of the minimum of the total potential. An English translation by W. A. Oldfather, C. A. Ellis, and D. M. Brown is available under the title *Leonard Euler's Elastic Curves* (1933).

Euler's stability analysis of the elastic rod was generalized by G. H. Bryan to apply to elastic bodies of arbitrary shape. In his paper, "On the Stability of Elastic Systems" (*Proceedings Cambridge Philosophical Society*, Vol. 6, p. 199, 1889, presented on February 27, 1888), he found from an order of magnitude consideration that only thin bodies, such as thin rods, plates, and shells, were likely to buckle. One of his conclusions turned out to be incorrect; he thought that extensional deformations had to be absent at buckling, and thus a thin spherical shell could not buckle when subjected to an external hydrostatic pressure.

Stability is discussed in all textbooks on strength of materials and the theory of elasticity. Very readable comprehensive treatises for the engineer are *Theory of Elastic Stability*, by S. Timoshenko (cited under 2.4), 1936, which probably did more than any other book to popularize the rigorous treatment of stability problems among engineers; and *Buckling Strength of Metal Structures*, by Friedrich Bleich, McGraw-Hill, 1952, whose author had given an excellent account of stability analyses in his outstanding earlier book *Theorie und Berechnung der Eisernen Brücken*, Springer, 1924 (pp. 99–239).

A few other articles dealing with the general theory of elastic stability are: "On the General Theory of Elastic Stability," by R. V. Southwell, *Philosphical Transactions Royal Society London*, Series A, Vol. 213, p. 187, 1914; "Buckling of Elastic Structures," by H. M. Westergaard, *Transactions ASCE*, Vol. 85, p. 576, Paper 1490, 1922; "The General Variational Principle of the Theory of Structural Stability," by W. Prager, *Qu. Appl. Math.*, Vol. 4, No. 4, p. 378, January 1947 (uses tensor notation); and "Some Observations on Elastic Stability," by J. N. Goodier, *Proceedings First U. S. National Congress of Applied Mechanics*, ASME, 1952, p. 193 (easy to read).

The numerical treatment of buckling load problems is discussed, among other books in *Eigenwertprobleme und ihre numerische Behandlung*, by Lothar Collatz, Becker and Erler, Leipzig, Germany, 1945, reprinted in the U.S.A. by the Chelsea Publishing Co., New York, 1948; *Numerical Methods in Engineering*, by Mario G. Salvadori, Prentice-Hall, 1952 (short, easy to read); *Applied Elasticity*, by Chi-Teh Wang, McGraw-Hill, 1953; *Relaxation Methods in Engineering Science*, by R. V. Southwell, Clarendon Press, Oxford, 1940; *Relaxation Methods in Theoretical Physics*, by R. V. Southwell, Clarendon Press, Oxford, 1946; and *An Introduction to Relaxation Methods*, by F. S. Shaw, Dover Publications, New York, 1953.

The approach to stability problems by equations of finite differences is described in *Die gewöhnlichen und partiellen Differenzengleichungen der Baustatik*, by Fr. Bleich and E. Melan, Springer, 1927. A graphic procedure was given by Luigi Vianello in "Graphische Untersuchung der Knickfestigkeit gerader Stäbe" in *ZVDI*, Vol. 42, No. 52, p. 1436, December 24, 1898, and an equivalent numerical procedure by Dana Young in "Inelastic Buckling of Variable Section Columns," *J. Appl. Mech.*, Vol. 12, No. 3, p. *A*-165, September 1945.

In aeronautical engineering circles the use of the expression "critical buckling load" has become widespread in the last few years. This is an unnecessarily complex expression because either one of the adjectives "critical" and "buckling" qualifies the noun "load" sufficiently. When a choice has to be made from among a number of buckling loads corresponding to different deflected shapes, the "smallest buckling load" is of particular interest in engineering; the smallest as well as the larger ones are all critical loads.

3.2. The behavior of initially curved columns is treated on p. 31 of the *Theory of Elastic Stability* by Timoshenko (cited under 2.4). A great deal of useful information on columns, including the secant formula for eccentric load application (p. 85), is contained in *Aeroplane Structures* by Pippard and Pritchard (cited under 1.9–1.11). For lateral loads see the above two references and p. 62 of Vol. 2 of *Airplane Structures* by Niles and Newell (cited under 1.9–1.11).

The Southwell plot was first described by R. V. Southwell in the paper, "On the Analysis of Experimental Observations in Problems of Elastic Stability," *Proceedings Royal Society London*, Ser. A, Vol. 135, p. 601, 1932. Modifications of and extensions to Southwell's method can be found in the following papers:

"Some Tests on the Stability of Thin Strip Material under Shearing Forces in the Plane of the Strip," by H. J. Gough and H. L. Cox, *Proceedings Royal Society London*, Series A, Vol. 137, p. 145, 1932; "On the Application of Southwell's method for the Analysis of Buckling Tests," by L. H. Donnell, *Stephen Timoshenko 50th Anniversary Volume*, p. 27, McGraw-Hill, 1938; "Experimental Study of Deformation and Effective Width in Axially Loaded Sheet-Stringer Panels," by W. Ramberg, A. E. McPherson, and S. Levy, *NACA Tech. Note* 684, January 1939; and "Heterostatic Loading and Critical Astatic Loads," by L. B. Tuckerman, NBS Research Paper RP 1163, *Journal of Research NBS*, Vol. 22, p. 1, January 1939.

A book dealing with the determination of buckling loads by vibration testing was written by Professor C. Massonnet under the title, *Les Relations entre les modes normaux de vibration et la stabilité des systèmes élastiques*, Goemaere, Brussels, Belgium, 1940.

3.3. This article contains a short presentation of the theory of the buckling of a sandwich column. For details see "Bending and Buckling of Sandwich Beams" by N. J. Hoff and S. E. Mautner which was cited under 2.6. Other references to the literature are also given under 2.6.

It is of interest to note that Eq. 3.3.15 is Engesser's formula for a column which is weak in shear; it was first published in the paper "Die Knickfestigkeit gerader Stäbe, *Zentralblatt der Bauverwaltung*, Vol. 11, No. 49, p. 483, December 5, 1891. A formula similar to Eq. 3.3.13, derived by P. P. Bijlaard by his method of "split rigidities," was given in the *J. Aero. Sciences*, Vol. 16, No. 9, p. 573, September 1949 under the title "Stability of Sandwich Plates." It was derived earlier in the publications of the Netherlands Academy of Sciences in 1946 and 1947.

The formulas derived in this article were proved experimentally. A description of the experiments can be found in the paper by Hoff and Mautner cited under 2.6.

The literature of the buckling of sandwich plates is quite extensive. It is not given here because its understanding requires a knowledge of the theory of elasticity and is, therefore, beyond the scope of this textbook.

3.4. A complete dynamic investigation of the stability of elastic systems and a consideration of large, not infinitesimal, disturbances is not necessary in most cases of practical importance. Exceptions are systems that are intrinsically non-linear, such as curved shells. The study of such systems, and of the dynamic stability, is still in its beginning and offers great mathematical difficulties.

In non-linear systems details of the loading process have considerable influence on the buckling load. The effect of the properties of the testing machine on the buckling load was discussed by Th. von Kármán in his dissertation "Untersuchungen über Knickfestigkeit," *Forschungsarbeiten, Verein Deutscher Ingenieure*, No. 81, Berlin, 1910; by Th. von Kármán and H. S. Tsien in "The Buckling of Thin Cylindrical Shells under Axial Compression" in the *J. Aero. Sciences*, Vol. 8, No. 8, p. 303, June 1941; and particularly by H. S. Tsien in "Buckling of a Column with Non-Linear Lateral Supports," *J. Aero. Sciences*, Vol. 9, No. 4, p. 119, February 1942, and in "A Theory for the Buckling of Thin Shells," *J. Aero. Sciences*, Vol. 9, No. 10, p. 373, August 1942.

The relationship between the classical instability load, the buckling load observed in tests of various kinds, and the maximum load-carrying capacity of columns in slow and in rapid loading processes was investigated by means of the dynamic criteria of stability and described by the author and his collaborators at the Polytechnic Institute of Brooklyn in a number of papers: "Dynamic Criteria of Buckling," by N. J Hoff, *Research, Engineering Structures Supplement*, Butterworths Scientific Publications, London, and Academic Press, New York, 1949, p. 121; "The Process of Buckling of Elastic Columns," by N. J. Hoff, *PIBAL Report* 163, December 1949; "The Dynamics of the Buckling of Elastic Columns," presented by N. J. Hoff at the annual summer meeting of the applied mechanics division of the ASME at Purdue University, Lafayette, Ind., in June 1950 and published in the *J. Appl. Mech.*, Vol. 18, No. 1, p. 68, March 1951; "An Experimental Investigation of the Process of Buckling of Columns," by N. J. Hoff, S. V. Nardo, and Burton Erickson, *PIBAL Report* 170, August 1950, printed in the *Proceedings Society for Experimental Stress Analysis*, Vol. 9, No. 1, 1951, p. 201. See also three papers presented at the First U. S. National Congress of Applied Mechanics in Chicago, Ill., in June 1951 and published in the *Proceedings First U. S. National Congress of Applied Mechanics*, ASME, 1952: "The Maximum Load Supported by an Elastic Column in a Rapid Compression Test," by N. J. Hoff, S. V. Nardo, and Burton Erickson, p. 419; "The Behavior of a Simply Supported Column under Constant or Varying End Load with Transverse Displacement of One Point of Support," by V. L. Salerno, Frances Bauer, and James Sheng, p. 425; and "Numerical Analysis of the Process of Buckling of Elastic and Inelastic Columns," by J. P. Chawla, p. 435.

The need for a dynamic criterion of buckling for non-conservative systems was pointed out by Professor H. Ziegler of the Polytechnic Institute of Zürich, Switzerland, in "Die Stabilitätskriterien der Elastomechanik," *Ingenieur-Archiv*, Vol. 20, No. 1, p. 49, 1952.

The strain-energy method of buckling load calculation was originated by Lord Rayleigh. A rather comprehensive collection of his work is the latest edition of *The Theory of Sound* (First Edition in 1877, see remarks under 2.7). One of the early investigators of shell stability by strain-energy analysis was G. H. Bryan ("On the Stability of a Plane Plate under Thrusts in Its Own Plane with Applications to the 'Buckling' of the Sides of a Ship," *Proceedings London Mathematical Society*, Hodgson, London, Vol. 22, No. 399, p. 54, presented on December 11, 1890). Walter Ritz, S. Timoshenko, and R. V. Southwell pioneered in applying series and strain-energy considerations to the investigation of structural elements of importance in engineering. References to their work can be found under 2.7 and 3.1.

A short and readable presentation of the strain-energy method is *Rayleigh's Principle and Its Application to Engineering*, by G. Temple and W. G. Bickley, Oxford University Press, 1933. A discussion of the various ramifications of the energy method as applied to buckling load calculations was given by Hans Reissner in "Energiekriterium der Knicksicherheit" (*ZAMM*, Vol. 5, No. 6, p. 475, December 1925).

3.5. Buckling loads of columns with variable moment of inertia are treated by Fr. Bleich in *Buckling Strength of Metal Structures* (see under 3.1) on p. 186; by Timoshenko in *Theory of Elastic Stability* (see under 2.4) on p. 128; by Pippard and Pritchard in *Aeroplane Structures* (see under 1.9–1.11) on p. 95; and by others.

3.6. A different derivation of the same results was given by N. J. Hoff in the

J. Roy. Aero. Soc., Vol. 40, No. 309, p. 663, September 1936, under the title "Elastically Encastred Struts." The solution was obtained from the differential equation, and not by the strain-energy method.   Figures 3.6.4 to 3.6.7 are taken from this paper.

The same problem was treated independently and in a different manner but with equivalent results by W. Prager, "Elastic Stability of Plane Frameworks," *J. Aero. Sciences*, Vol. 3, No. 11, p. 388, September 1936. See also Alfred Teichmann "Einspannwirkung bei Knickstäben in Flugzeug-Fachwerken," *ZFM*, Vol. 21, No. 10, p. 249, May 28, 1930, and *DVL Jahrbuch* 1930, Rep. 183, p. 221, translated as "Effects of the End Fixation of Airplane Struts," *NACA Tech. Memo.* 582, September 1930; Alfred Teichmann "Das räumliche Knicken einiger Stabverbindungen des Flugzeugbaus," *ZFM*, Vol. 22, No. 17, p. 525, September 14, 1931, and *DVL Jahrbuch* 1931, Rep. 224, p. 230, translated as "Spatial Buckling of Various Types of Airplane Strut Systems," *NACA Tech. Memo.* 647, November 1931; Henri Rivière "Calcul des poutres comprimées encastrées élastiquement à leurs extrémités—Flambage des barres de treillis assemblées rigidement," *Bulletin L'Aérotechnique* 200, p. 7, 1936 (Gauthier-Villars, Paris); and K. Borkmann "Kurventafeln für den Stabilitätsnachweis ebener Stabgruppen," *Luftfahrtforschung*, Vol. 13, No. 1, p. 1, January 20, 1936 and *Luftfahrtforschung*, Vol. 14, No. 2, p. 86, February 20, 1937, translated as "Charts for Checking the Stability of Compression Members in Trusses," *NACA Tech. Memo.* 800, July 1936, and "Charts for Checking the Stability of Plane Systems of Rods," *NACA Tech. Memo.* 837, September 1937. The effect of end restraint was also discussed by Paul E. Sandorff in "Notes on Columns" in the *J. Aero. Sciences*, Vol. 11, No. 1, p. 1, January 1944.

It is worth noting that the strain-energy solution given here is not approximate, but rigorous, because the three terms assumed to represent the deflections are capable of defining exactly the deflected shape at buckling.

3.7. Torsional buckling was first analyzed in the literature by Herbert Wagner in *Verdrehung und Knickung von offenen Profilen* (cited under 2.1). He discovered the phenomenon while working on the structural problems of the all-metal Rohrbach airplanes. The paper was translated by the NACA under the title "Torsion and Buckling of Open Sections," *Tech. Memo.* 807, 1936. An error in the derivation was corrected by Robert Kappus in "Drillknicken zentrisch gedrückter Stäbe mit offenem Profil im elastischen Bereich," *Luftfahrtforschung*, Vol. 14, p. 44, 1937, translated as *Tech. Memo.* 851, 1938.

A paper based on Wagner's work in which the importance of torsion in general column theory was pointed out is "A Theory for Primary Failure of Straight Centrally Loaded Columns," by E. E. Lundquist and C. M. Fligg, *NACA Report* 582, 1937. The theory was further clarified and simplified by J. N. Goodier in two papers: "The Buckling of Compressed Bars by Torsion and Flexure," *Cornell University Engineering Experiment Station Bulletin* 27, December 1941, and "Torsional and Flexural Buckling of Bars of Thin-Walled Open Section under Compressive and Bending Loads," *J. Appl. Mech.*, Vol. 64, p. A-103, September 1942. Experimental papers on the subject are: "Verdrehung und Knickung von offenen Profilen," by H. Wagner and W. Pretschner, *Luftfahrtforschung*, Vol. 11, No. 6, p. 174, December 5, 1934 (translated as *NACA Tech. Memo.* 784, January 1936, and "Experimental Study of Torsional Column Failure," by A. S. Niles, *NACA Tech. Note* 733, 1939.

The strain-energy derivation given here follows the lines of the paper "A

Strain-Energy Derivation of the Torsional-Flexural Buckling Loads of Straight Columns of Thin-Walled Open Sections," by N. J. Hoff, *Qu. Appl. Math.*, Vol. 1, No. 4, p. 341, January 1944. The physical explanation of the phenomenon is taken from the author's discussion of Goodier's paper in the *J. Appl. Mech.*, Vol. 10, No. 2, p. A-110, June 1943.

The treatment of the problem in this article is rigorous because the infinite series are capable of representing the deflected shape exactly. Of course, the results show that each value of n corresponds to a different buckling load, and thus the deflected shape at buckling is described by single-term expressions in u, v, and β.

Torsional buckling is treated in great detail by Niles and Newell in *Airplane Structures*, Vol. 2, p. 316 (see under 1.9–1.11), and by Bleich in "Buckling Strength of Metal Structures," p. 104 (see under 3.1). Other references of interest in this connection are "Torsional Instability of Struts," by A. G. Pugsley, *Aircraft Engineering*, Vol. 4, No. 43, p. 229, September 1932, and "Primary Instability of Open-Section Stringers Attached to Sheet," by S. Levy and W. D. Kroll, *J. Aero. Sciences*, Vol. 15, No. 10, p. 580, October 1948.

3.8. A method involving the use of the so-called "four-moment equations" was presented in 1919 by the great structural engineer of Vienna, Friedrich Bleich, for the analysis of the secondary stresses and the stability of plane frameworks ("Die Knickfestigkeit elastischer Stabverbindungen," *Der Eisenbau*, Engelmann, Leipzig, Vol. 10, No. 2, p. 27, 1919). The problem of the stability of a plane truss as a unit was treated by the eminent mathematician R. von Mises in the paper "Über die Stabilitätsprobleme der Elastizitätstheorie," *ZAMM*, Vol. 3, No. 6, p. 406, December 1923, and by the same author jointly with J. Ratzersdorfer in "Die Knicksicherheit von Fachwerken," *ZAMM*, Vol. 5, No. 3, p. 218, June 1925. A short book on the buckling of rigid frames entitled *Knickfestigkeit der Stabverbindungen* by H. Zimmermann, appeared in 1925 (Wilhelm Ernst and Sohn, Berlin, Germany).

The calculation of the stiffness-coefficient and carry-over factors of bars subjected to axial loads by B. W. James in a thesis prepared under Professor A. S. Niles's direction at Stanford University and entitled "Principal Effects of Axial Load on Moment-Distribution Analysis in Rigid Structures" in 1933, and publication of the thesis by the NACA as *Tech. Note* 534 in 1935, established the foundations of a stability check by moment distribution. Subsequently E. E. Lundquist presented two stability criteria based on moment distribution in the papers, "Stability of Structural Members under Axial Load," *NACA Tech. Note* 617, 1937; "Method of Estimating the Critical Buckling Load for Structural Members," *NACA Tech. Note* 717, 1939; and "Principles of Moment Distribution Applied to Stability of Structural Members," *Proceedings Fifth International Congress of Applied Mechanics*, Wiley, 1938, p. 145. More detailed tables of the stiffness coefficients and carry-over factors were published by E. E. Lundquist and Wilhelmina D. Kroll under the title "Tables of Stiffness and Carry-Over Factor for Structural Members under Axial Load," *NACA Tech. Note* 652, 1938.

The proof of the convergence criterion was given by me in a paper read before the second annual summer meeting of the Institute of the Aeronautical Sciences in Pasadena, Calif., on June 24, 1940, under the title "Instability of Aircraft Frameworks." This was followed by the publications "Stable and Unstable Equilibrium of Plane Frameworks," *J. Aero. Sciences*, Vol. 8, No. 3, p. 115,

January 1941, and "Stress Analysis of Aircraft Frameworks," *J. Roy. Aero. Soc.*, Vol. 45, No. 367, p. 241, July 1941.

In "The Proportioning of Aircraft Frameworks" (*J. Aero. Sciences*, Vol. 8, No. 8, p. 319, June 1941) a three-step procedure was recommended and illustrated by a numerical example. In the first step, the sizes of the members are chosen on the basis of rough assumptions regarding the end fixation. In the second, the framework is broken up into groups of bars separated by fictitious ideal pin joints, and the stability of each group is investigated as described in Art. 3.6. After the necessary changes are made in the sizes of the members, a final check of the stability is obtained through the convergence criterion of the moment-distribution process.

Experiments carried out to prove the validity of the theory are described in "Buckling of Rigid-Jointed Plane Trusses," by N. J. Hoff, Bruno A. Boley, S. V. Nardo, and Sara Kaufman, *Transactions ASCE*, Vol. 116, p. 958, 1951, Paper 2454.

The convergence criterion is fully treated in *Airplane Structures*, by Niles and Newell, Vol. 2, p. 303 (cited under 1.9–1.11), and in *Buckling Strength of Metal Structures*, by Friedrich Bleich, on p. 193 (cited under 3.1). It is applied to the frameworks and rigid frames of civil engineering structures in "Buckling of Trusses and Rigid Frames," by George Winter, P. T. Hsu, Benjamin Koo, and M. H. Loh, *Cornell University Engineering Experiment Station Bulletin* 36, April 1948.

A paper dealing with a generalization of the theory of the stability of assemblies of bars, valid for space frameworks whose members buckle spatially, is "Die Stabilität räumlicher Stabverbindungen," by Friedrich Bleich and Hans Bleich, *Zeitschrift des Oesterreichischen Ingenieur und Architekten-Vereins*, No. 37–38, p. 345, September 21, 1928.

3.9. The Euler formula, after its derivation by Leonhard Euler, the famous mathematician of Basle, Switzerland, in 1744, slowly fell into disrepute because it did not agree with experimental evidence. Considère and Engesser were the first to point out that the Euler formula was valid only when the buckling stress calculated from it was smaller than the elastic limit; it had been used incorrectly by engineers to predict the buckling of short columns. In a lucid paper presented on September 14, 1889, before the Congrès International des Procédés de Construction in Paris, Considère proved theoretically that the correct value of the modulus to be used in the Euler formula had to be between the tangent modulus and Young's modulus; his experiments showed that the buckling stress depended only on the slenderness ratio of the column for any given material. The paper was published in the proceedings of the congress under the title "Resistance des pièces comprimées" on p. 371 of the *Annexe* (Librairie Polytechnique, Paris, France, 1891). Engesser suggested a modified Euler formula in which Young's modulus was replaced by the tangent modulus ("Über die Knickfestigkeit gerader Stäbe," *Zeitschrift des Architekten und Ingenieurvereins zu Hannover*, Vol. 35, No. 4, p. 455, 1889, Jänecke, Hannover, Germany). For this suggestion he was attacked sharply by Professor F. Yasinski of St. Petersburg, Russia. In his answer Engesser acknowledged his mistake and proceeded to a logical and rigorous formulation of the inelastic buckling problem. He was the first to show that the reduced modulus had to be used in the Euler formula in order to obtain the stability limit of a column in the classical sense ("Über Knickfragen," *Schweizerische Bauzeitung*, Vol. 26, No. 4, p. 24, July 1895).

In 1910 Th. von Kármán presented a more detailed analysis ("Untersuchungen über Knickfestigkeit," cited under 3.4) and proved his conclusions by a series of careful experiments. An independent analysis of the same subject was presented by R. V. Southwell ("The Strength of Struts," *Engineering*, Vol. 94, p. 249, August 23, 1912). In the United States the treatise "Strength of Steel Columns—Theory of Steel Columns Which Are Stressed Beyond the Proportional Limit and Which Are Eccentrically Loaded or Initially Curved," by H. M. Westergaard and W. R. Osgood, *Transactions ASME*, Vol. 49–50, Part 1, Paper APM-50-9, 1927–28, and William R. Osgood's paper "Column Curves and Stress-Strain Diagrams," Research Report 492, *Journal of Research NBS*, Vol. 9, p. 571, October 1932, in which the reduced modulus was designated as the double modulus, contributed to the general acceptance of the modified Euler formula by engineers.

Buckling above the elastic limit is discussed in many textbooks and handbooks, for instance in Timoshenko's *Theory of Elastic Stability* (see under 2.4, p. 156), and in Bleich's *Buckling Strength of Metal Structures* (p. 8, see under 3.1). The idea of buckling during the loading process, without strain reversal, was proposed by F. R. Shanley ("The Column Paradox," *J. Aero. Sciences*, Vol. 13, No. 12, p. 678, December 1946, and "Inelastic Column Theory," *ibid.*, Vol. 14, No. 5, p. 261, May 1947).

A dynamic investigation of the buckling process was undertaken by me in the papers cited under 3.4. The effect of initial stresses on the buckling load was discussed by W. R. Osgood at the First U. S. National Congress of Applied Mechanics in Chicago, June 1951 ("The Effect of Residual Stresses on Column Strength," published in the *Proceedings First U. S. National Congress of Applied Mechanics*, ASME, 1952, p. 415) and by Ching Huan Yang, Lynn S. Beedle, and Bruce G. Johnston ("Residual Stress and the Yield Strength of Steel Beams," *Welding Journal*, Welding Research Supplement, Vol. 31, No. 4, p. 205-s, April 1952). A numerical method for the calculation of the buckling load of a variable section column was presented by Dana Young for the case when the elastic limit of the material is exceeded ("Inelastic Buckling of Variable Section Columns," *J. Appl. Mech.*, Vol. 12, No. 3, p. A-165, September 1945). Vianello's graphic procedure (cited under 3.1) is also suitable for the analysis of this problem.

The thin-walled columns used in airplanes can buckle locally below or above the elastic limit of the material. This problem is discussed at length in books on airplane structures, for instance in *Airplane Structural Analysis and Design*, by Ernest E. Sechler and Louis G. Dunn, Wiley, 1942; *Aircraft Structural Mechanics*, by Franz R. Steinbacher and George Gerard, Pitman Publishing Corp., New York, 1952; and *Structural Principles and Data*, Part 2, "Structural Analysis," by J. H. Argyris and P. C. Dunne, published under the authority of the Council of the Royal Aeronautical Society by Sir Isaac Pitman and Sons, London, 1952 (Fourth Edition).

Part 4

4.1. The complementary energy principle was first stated by Fr. Engesser, the ingenious professor of mechanics at the University of Karlsruhe, in his paper "Über statisch unbestimmte Träger bei beliebigem Formänderungsgesetze und über den Satz von der kleinsten Ergänzungsarbeit," published in the *Zeitschrift des Architekten- und Ingenieur-Vereins zu Hannover*, Vol. 35, 1889, column 733. It was a generalization of Castigliano's theorems which were proposed by

Alberto Castigliano in his thesis for the diploma of an engineer at the Polytechnic Institute of Torino, Italy, in 1873, and was described later in the *Atti della Academia delle Scienze* under the title "Nuova teoria intorno dell'equilibrio dei sistemi elastici" (1875) and in the book *Théorie de l'équilibre des systèmes élastiques*, published by Negro in Torino, Italy, in 1879. A translation of this book by E. S. Andrews is available under the title *Elastic Stresses in Structures* (Scott, Greenwood, and Son, London, 1919). Castigliano in turn stated that the least-work principle was originally enunciated by General L. F. Menabrea in a memoire presented to the Academy of Sciences in Turin, Italy, in 1857. However, the proof was not rigorous, and the principle was not generally accepted until Castigliano's work became known.

The complementary-energy principle has not been used very widely in mechanics. Its advantages were pointed out by H. M. Westergaard, the late dean of the graduate school of engineering at Harvard University, Cambridge, Mass., in an article entitled "On the Method of Complementary Energy," *Transactions ASCE*, Vol. 107, p. 765, 1942. In solving vibration problems Westergaard used the principle in a generalized form in which the minimization was carried out through a variation of both the stresses and the displacements. This possibility was fully exploited later by Eric Reissner, professor of mathematics at the Massachusetts Institute of Technology, Cambridge, Mass., who derived a new minimal principle in the paper, "On a Variational Theorem in Elasticity," *Journal of Mathematics and Physics*, Vol. 29, No. 2, p. 90, July 1950. While in the first minimal principle arbitrary continuous deformations are assumed and the equilibrium conditions are enforced through a minimization with respect to displacements, and in the second minimal principle the minimization with respect to stress is used to select the true state of equilibrium from among static equilibrium states established with no regard to continuity of deformations, in Reissner's general minimal principle both displacements and stresses are assumed arbitrarily, and equilibrium as well as continuity is enforced by means of minimization.

Another method in which neither the equilibrium nor the continuity conditions are satisfied rigorously was presented in an easily understandable manner by W. Prager in "The Extremum Principles of the Mathematical Theory of Elasticity and Their Use in Stress Analysis," *University of Washington Engineering Experiment Station Bulletin* 119, Seattle, Wash., 1951. See also W. Prager and J. L. Synge "Approximations in Elasticity Based on the Concept of Function Space," *Qu. Appl. Math.*, Vol. 5, No. 3, p. 241, October 1947 (uses tensor notation).

Today Castigliano's theorems are commonly used in engineering and are described in detail in many textbooks. A great deal of space is devoted to it in the classical work *Graphische Statik der Baukonstruktionen* by the famous German civil engineer H. Müller-Breslau (see under 2.4), in Pippard's *Strain-Energy Methods* etc. cited under 1.9–1.11, and in Sir Richard Southwell's textbook, *An Introduction to the Theory of Elasticity* (see under 1.9–1.11).

The nomenclature used in this article was devised to emphasize the similarities between the first and the second minimal principles.

4.2. Castigliano's second theorem was derived in the publications mentioned under 4.1. A problem of a three-bar truss beyond the elastic limit was solved by Westergaard in the paper cited under 4.1. However, he made use of a stress-strain diagram consisting of two straight lines. The representation of the stress-strain relationship by Eq. 4.2.12 was suggested by W. R. Osgood ("Plastic Bending," *J. Aero. Sciences*, Vol. 11, No. 3, p. 213, July 1944).

4.3. Castigliano's and Müller-Breslau's publications mentioned under 4.1 contain examples of rigid-frame analysis by means of Castigliano's theorem. Many textbooks on structures discuss rigid-frame analysis. The following works may be mentioned in this connection: *Strain-Energy Methods of Stress Analysis*, by Pippard (cited under 1.9–1.11); *The Theory of Structures*, by C. M. Spofford, McGraw-Hill, Fourth Edition, p. 423, 1939; *Airplane Structural Analysis and Design*, by E. E. Sechler and L. G. Dunn, Wiley, p. 147, 1942; *Theory of Structures*, by Timoshenko and Young (cited under 1.9–1.11); *Elementary Structural Analysis*, by Wilbur and Norris (cited under 1.12–1.15); and *Fundamentals of Aircraft Structures* by Millard V. Barton, Prentice-Hall, p. 99, 1948.

Müller-Breslau showed that an elastic center of a rigid frame can be defined so that the equations for the three redundant quantities become independent of one another if they are assumed to be acting at the elastic center. This method of frame analysis is not discussed in the present textbook because no real saving of work results from it. It is used extensively by E. F. Bruhn in *Analysis and Design of Airplane Structures*, Tri-State Offset Co., Cincinnati, Ohio. The method of replacing a complete section by several joints was described by me in "Stress Analysis of Rings for Monocoque Fuselages," *J. Aero. Sciences*, Vol. 9, No. 7, p. 245, May 1942. The basic idea, as well as much advice regarding the choice of the redundant quantities, can be found in *Aufgaben aus der Flugzeugstatik*, by Thalau and Teichmann (cited under 1.9–1.11).

References to the Hardy Cross moment-distribution method can be found under 1.18. The geometry and the distribution factors of the building frame shown in Fig. 4.3.5 were taken from "The Approximate Design of Small Rigid-Framed Structures," by F. S. Shaw, *Aeronautical Research Laboratories Report* SM 153, Melbourne, Australia, June 1950. The paper contains simplified formulas developed with the aid of the moment-distribution method.

Building frames are analyzed by means of the moment-distribution method on p. 415 of Wilbur and Norris' textbook cited under 1.12–1.15; in Grinter's *Theory of Modern Steel Structures* cited under 1.9–1.11; on p. 36 of Vol. 2 of Niles and Newell's book cited under 1.9–1.11; and in *Statically Indeterminate Structures*, by L. C. Maugh, Wiley, 1946.

4.4. The reciprocal theorem was first published by Maxwell in 1864 (see under 1.7–1.8). It is treated in all books on structures. For Castigliano's work see the references under 4.1.

4.5. The contents of this article were presented by me in a greatly abbreviated form, under the title "Complementary Energy Analysis of the Failing Load of a Clamped Beam," in the *J. Appl. Mech.*, Vol. 19, No. 4, p. 563, December 1952.

4.6. The two original enthusiastic promoters of the idea of limit analysis and design in the English-speaking world are Professor J. F. Baker of Cambridge University, England (formerly of Bristol University, England) and Professor J. A. Van den Broek of the University of Michigan, Ann Arbor, U.S.A. The former published, together with J. W. Roderick, two interim reports on an investigation into the behavior of welded rigid-frame structures in the British periodical *Transactions Institution of Welding*: "An Experimental Investigation of the Strength of Seven Portal Frames" in October 1938 (Vol. 1, No. 4, p. 206), and "Further Tests on Beams and Portals" in April 1940 (Vol. 3, No. 2, p. 83). Since that time Professor Baker and his group have been most successful in expanding and popularizing their ideas. Two comparatively recent summary reports containing references to the literature are "A Review of Recent Investigations into the Behavior of Steel Frames in the Plastic Range," by J. F. Baker, *Journal Institution of Civil Engineers*, Vol. 31, No. 3, p. 188, January 1949, and "The Design of Steel Frames," by J. F. Baker, *Structural Engineer*, Vol. 27, p. 397, October 1949.

Van den Broek's first publication on this topic dates back to October 1939. It was entitled *Theory of Limit Design* and was printed in Vol. 44 of the *Journal Western Society of Engineers*. Two more publications bearing the same title followed. One was an article in Vol. 105, p. 638, of the *Transactions ASCE* (1940) and the second a book published by Wiley in 1948.

The idea of limit design, however, is much older. The first publication dealing with it appeared in the Hungarian periodical *Betonszemle* in 1914 and was written by Gábor Kazinczy under the title "Kisérletek befalazott tartókkal" ("Experiments with Clamped Girders"; Vol. 2, No. 4, p. 68, April 1; Vol. 2, No. 5, p. 83, May 1; and No. 6, p. 101, June 1). On p. 102 Kazinczy explained the behavior of the girders at collapse by the appearance of three yield hinges which transformed the redundant structure into a mechanism. At a congress in Vienna he proposed the establishment of safety factors on the basis of collapse loads and probability calculations (unpublished but discussed on p. 656 of the *Final Report Third Congress International Association for Bridge and Structural Engineering*, Liège, Belgium, 1948). Kazinczy's first publication in a Western European language is *Bemessung von statisch unbestimmten Konstruktionen unter Berücksichtigung der bleibenden Formänderungen*, International Congress for Metallic Structures, Liège, Belgium, September 1930. This was followed by "Statisch unbestimmte Tragwerke unter Berücksichtigung der Plastizität," *Der Stahlbau*, 1931, No. 5, p. 58, and by other papers on the subject as well as by a book, *Az anyagok képlékenységének jelentősége a tartószerkezetek teherbirása szempontjából* ("The Significance of Plasticity to the Collapse Loads of Structures"), Royal Hungarian University Press, Budapest, Hungary, 1942. Data on Kazinczy and a translation of his first article on limit design were published in the *Welding Journal Research Supplement*, Vol. 19, No. 1, p. 14-s, January 1954.

In 1917 Professor N, C. Kist of the Polytechnic Institute of Delft, Holland, chose limit design as the topic of his inaugural lecture. He also discussed the subject in an article entitled "Die Zähigkeit des Materials als Grundlage für die Berechnung von Brücken, Hochbauten und ähnlichen Konstruktionen aus Flusseisen" in the German periodical *Der Eisenbau*, Vol. 2, p. 425, in 1920. Other contributions were made by M. Grüning (1926), H. Maier-Leibnitz (1928), Fr. Bleich (1932), Hans Bleich (1932), and others.

The most recent scientific development in this field took place in the graduate division of applied mathematics of Brown University (Providence, R. I.) under the direction of Professor W. Prager. Starting out from principles derived by H. J. Greenberg of Brown University (now at Carnegie Institute of Technology, Pittsburgh, Pa.) and S. M. Feinberg (Moscow, Russia), a method of analysis was devised by H. J. Greenberg and W. Prager, and published in the *Proceedings ASCE* (Separate 59, February 1951) under the title "Limit Design of Beams and

Frames" (see also *Transactions ASCE*, Vol. 117, p. 447, Paper 2501, 1952). A related paper is "On Plastic-Rigid Solutions and Limit-Design Theorems for Elastic-Plastic Bodies," by D. C. Drucker, H. J. Greenberg, E. H. Lee, and W. Prager, *Proceedings First National Congress of Applied Mechanics*, American Society of Mechanical Engineers, New York, N.Y., 1952, p. 533. Further advances were made by Professor P. S. Symonds of Brown University and B. G. Neal of Cambridge University, Cambridge, England. The presentation of the material in Art. 4.6 is based on their joint papers, "The Rapid Calculation of the Plastic Collapse Load for a Framed Structure" (*Institution of Civil Engineers Structural Paper* 29, November 1951) and "Recent Progress in the Plastic Methods of Structural Analysis" (*Journal Franklin Institute*, Vol. 252, No. 5, p. 383, November 1951).

Many small-scale tests were carried out in Cambridge by Baker's group. Full-size structures have been tested recently at Lehigh University (Bethlehem, Pa.) under Professor Bruce G. Johnston's direction. A sample of the results is "Progress Report 4, Part I—Test Results and Requirements for Connections," by A. A. Topractsoglou, L. S. Beedle, and B. G. Johnston, *Welding Journal*, p. 359s, July 1951.

The magnitude of the fully plastic moment was investigated in detail by J. W. Roderick and I. H. Phillips in "Carrying Capacity of Simply Supported Mild Steel Beams," *Research*, Engineering Structures Supplement, Butterworths Scientific Publications, London, and Academic Press, New York, p. 9, 1949.

Limit analysis is discussed in the very readable textbook *Theory of Perfectly Plastic Solids*, by William Prager and Philip G. Hodge, Jr., Wiley, 1951.

The shapes of most of the structures and the relative values of their yield moments (or fully plastic moments) are patterned on the small office buildings described by F. S. Shaw (see note under Art. 4.3).

4.7. The development of monocoque construction was sketched by the author in "A Short History of the Development of Airplane Structures," *American Scientist*, Vol. 34, No. 2, p. 212, April 1946, and Vol. 34, No. 3, p. 370, July 1946. Other historic and technical data can be found in "Thin-Walled Monocoques," by me in *Proceedings Aeronautical Conference London*, Royal Aeronautical Society, London, p. 313, 1948.

The theory given in Art. 4.7 was first presented by me at the Chicago meeting of the applied mechanics division of the ASME in June 1944 under the title "Stresses in a Reinforced Monocoque Cylinder under Concentrated Symmetric Transverse Loads." It was published in the *J. Appl. Mech.*, Vol. 11, No. 4, p. A-235, December 1944. An independent but less accurate analysis was given by J. E. Wignot, Harry Combs, and A. F. Ensrud of the Lockheed Airplane Corporation in *NACA Tech. Note* 929 during the same year under the title "Analysis of Circular Shell-Supported Frames." Good agreement between my theory and experiment was found at the Langley Memorial Aeronautical Laboratory of the NACA, as described in *Wartime Report* L-66, December 1945, entitled "The Effect of Concentrated Loads on Flexible Rings in Circular Shells," by P. Kuhn, J. E. Duberg, and G. E. Griffith.

The theory was further developed at the NACA in the following two publications: "Stress Analysis by Recurrence Formula of Reinforced Circular Cylinders under Lateral Loads," by John E. Duberg and Joseph Kempner, *NACA Tech. Note* 1219, 1947, and "Charts for Stress Analysis of Reinforced Circular Cylinders under Lateral Loads," by Joseph Kempner and John E. Duberg,

NACA Tech. Note 1310, 1947. Another paper presenting numerical results is due to L. Beskin; its title is "Local Stress Distribution in Cylindrical Shells." It was published in the *J. Appl. Mech.*, Vol. 13, No. 2, p. A-137, June 1946. Two more papers of interest are "The Stresses in a Circular Fuselage," by W. J. Goodey, *J. Roy. Aero. Soc.*, Vol. 50, No. 431, p. 833, November 1946, and "Sul Calcolo delle Strutture a Guscio," by P. Cicala, *L'Aerotecnica*, Vol. 26, No. 3, p. 138, September 1946.

Theoretical as well as experimental work on the subject was continued at the Polytechnic Institute of Brooklyn and resulted in a series of articles in the *Reissner Anniversary Volume*, published by J. W. Edwards, Ann Arbor, Mich., in 1949 under the common title "Concentrated Load Effects in Reinforced Monocoque Structures." The individual papers are "Basic Theory," by N. J. Hoff, p. 277; "The Applicability of Saint-Venant's Principle to Reinforced Monocoque Structures," by V. L. Salerno, p. 290; "The Effect of Bending of Stringers in a Reinforced Monocoque with an Arbitrary Number of Fields," by Harold Liebowitz, p. 300; "Consideration of the Eccentricity and the Shearing and Extensional Deformations of the Rings," by Bruno A. Boley, p. 313; and "Experimental Investigation," by Sebastian V. Nardo, p. 321. Final design charts were presented in the paper, "Shear Stress Concentration and Moment Reduction Factors for Reinforced Monocoque Cylinders Subjected to Concentrated Radial Loads," by N. J. Hoff, Vito L. Salerno, and Bruno A. Boley, *J. Aero. Sciences*, Vol. 16, No. 5, p. 277, May 1949. The curves of Fig. 4.7.5 were taken from this publication. They show that the elementary analysis of the problem leads to grossly inaccurate results. The curves of Figs. 4.7.1 and 4.7.3 were taken from V. L. Salerno's doctoral thesis "Theoretical and Photoelastic Investigation of the Concentrated Load Problem in Reinforced Monocoque Cylinders" submitted to the Polytechnic Institute of Brooklyn in 1947. Another doctoral thesis submitted to the Polytechnic Institute of Brooklyn is due to Robert S. Levy. Its title is "Effect of Bending Rigidity of Stringers upon Stress Distribution in Reinforced Monocoque Cylinder under Concentrated Transverse Loads." It was published in the *J. Appl. Mech.*, Vol. 15, No. 1, p. 30, March 1948.

The main result of all these theoretical and experimental investigations is that the theory presented in Art. 4.7 and its extension to cylinders reinforced by several elastic rings is accurate enough for practical purposes except when it predicts shear stress-concentration factors greatly exceeding 10 or even 20. In such cases all the secondary effects neglected in the treatment in Art. 4.7, such as the strain energy of shear and extension in the rings and the strain energy of bending in the stringers, attain some local importance and contribute to a reduction in the peak stresses.

4.8. A strain-energy analysis similar to that contained in Art. 4.8 was first given by the author in a discussion of the paper, "The Elastic Sphere under Concentrated Loads," by E. Sternberg and F. Rosenthal, *J. Appl. Mech.*, Vol. 20, No. 2, p. 305, June 1953. It was based on the following papers by Professor J. N. Goodier of Stanford University, Calif.: "A General Proof of Saint-Venant's Principle," *Phil. Mag.*, Series 7, Vol. 23, p. 607, April 1937; "Supplementary Note on A General Proof of Saint-Venant's Principle," *Phil. Mag.*, Series 7, Vol. 24, p. 325, August 1937; and "An Extension of Saint-Venant's Principle, with Applications," *Journal of Applied Physics*, Vol. 13, No. 3, p. 167, March 1942. Another paper written by the author on the subject is "The

Applicability of Saint-Venant's Principle to Airplane Structures," *J. Aero. Sciences*, Vol. 12, No. 4, p. 455, October 1945.

The quotation is from p. 132 of Love's *Theory of Elasticity* cited under 1.12–1.15. A proof of the principle for the case of a concentrated force acting perpendicularly to the plane surface of a semi-infinite half-space was given by J. Boussinesq in *Applications des potentiels à l'étude de l'équilibre et du mouvement des solides élastiques*, Paris, 1885. In a discussion of the problem Professor R. von Mises of Harvard University showed ("On Saint-Venant's Principle," *Bulletin American Mathematical Society*, Vol. 51, p. 555, 1945) that an extension of the principle to finite bodies is, in general, not permissible.

The problem of the strip of flat sheet under equal and opposite concentrated loads was solved by Fr. Bleich in *Der Bauingenieur*, Vol. 4, p. 255, 1923 ("Der gerade Stab mit Rechteckquerschnitt als ebenes Problem"). The calculations are also presented on p. 344 of Vol. 1 of *Drang und Zwang*, by A. Föppl and L. Föppl (cited under 1.1–1.6), and the results are quoted in *Theory of Elasticity*, by Timoshenko and Goodier (cited under 2.2) on p. 52. The solution of the framework problem was given by the author in his paper of 1945 cited above.

The validity of Saint-Venant's principle for space frameworks was investigated experimentally by Professor Pippard at Imperial College, London. Two papers dealing with this subject are "On an Experimental Verification of Castigliano's Principle of Least Work and of a Theorem Relating to the Torsion of a Tubular Framework," by A. J. S. Pippard and J. F. Baker, *Phil. Mag.*, July 1925, and "On an Experimental Investigation of the Applicability of Saint-Venant's Principle to the Case of Frameworks Having Redundant Bracing Members," by A. J. S. Pippard and P. H. W. Clifford, *Phil. Mag.*, January 1926. The experiments are also described in Pippard's *Strain-Energy Methods* etc. (see under 1.9–1.11).

Additional References. Since this bibliography was completed, a few more interesting books and articles have been published on topics discussed in this volume. Foremost among them is the *History of Strength of Materials* by Stephen P. Timoshenko (McGraw-Hill, 1953) which contains a great deal of information not only on the development of structural theory but also about the persons who contributed to it. A new edition of the outstanding *Technische Dynamik* by C. B. Biezeno and R. Grammel has been prepared in English under the title *Engineering Dynamics*. It is being published in four volumes during 1954–55 by Blackie & Son, London, England. The fourth edition of Vol. 1 of *Airplane Structures* by Niles and Newell was published in 1954 by John Wiley & Sons, and *Statically Indeterminate Stresses* by Parcel and Maney was replaced in 1955 by *Analysis of Statically Indeterminate Structures* by Parcel and Moorman (Wiley).

To the literature of numerical analysis has been added *Relaxation Methods* by D. N. de G. Allen, a student and co-worker of Sir Richard Southwell (McGraw-Hill, 1954). Information on the dynamic aspects of the buckling process has been collected in the Forty-First Wilbur Wright Memorial Lecture entitled "Buckling and Stability" by N. J. Hoff, *J. Roy. Aero. Soc.*, Vol. 58, No. 1, p. 3, January 1954. The complementary-energy theorems have been re-examined by Henry L. Langhaar in "The Principle of Complementary Energy in Nonlinear Elasticity Theory," *Journal Franklin Institute*, Vol. 256, No. 3, p. 255, September 1953 (uses tensor notation). Saint-Venant's principle has been further clarified by E. Sternberg in "On Saint-Venant's Principle," *Qu. Appl. Math.*, Vol. 11, No. 4, p. 393, January 1954.

Appendix 2

Problems

The first two digits of the number of each problem refer to the chapters of the book. Numerical values of the solutions were generally obtained with slide-rule accuracy only.

1.5.1. Find the mechanical advantage of the pulley system shown in the figure.

Ans. $P = W/6$.

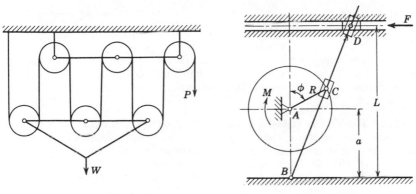

PROB. 1.5.1   PROB. 1.5.2

1.5.2. The quick-return mechanism shown in the figure consists of a crank AC of length R and a straight driving rod BCD. As the crank rotates around point A at a uniform speed $d\phi/dt$, its end point C slides in a slot along rod BCD. Consequently the rod oscillates about point B and causes point D to move in the horizontal slot with a variable velocity. Find the moment M necessary for static equilibrium when the force F and the angle ϕ are given.

Ans. $M = FL(R/a)[(R/a) + \cos\phi][1 + (R/a)\cos\phi]^{-2}$.

1.7.1. The inclined force W is acting on joint B of the statically determinate framework shown in the figure. The members of the framework are steel tubes of $1\frac{1}{4}$ in. over-all diameter and 0.049 in. wall thickness. The cross-sectional area of such a tube is 0.1849 sq in., and the modulus of elasticity of the material can be taken as 30×10^6 psi. Suitable supports prevent buckling. What is the vertical component of the displacement of joint B?

Ans. 0.0615 in.

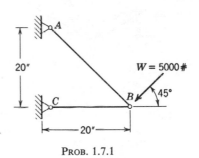

PROB. 1.7.1

471

1.7.2. Find the displacement component of point C of the statically determinate framework in the 45° direction indicated in the figure if the cross-sectional area of each bar is 0.1 sq in. and Young's modulus for the material is 30×10^6 psi.

Ans. $\delta = 0.03414$ in.

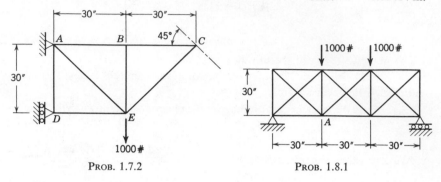

PROB. 1.7.2   PROB. 1.8.1

1.8.1. The redundant pin-jointed ideal framework shown in the figure has bars of 0.1 sq in. cross-sectional area. What is the vertical downward displacement of joint A if the modulus is 30×10^6 psi? (It is assumed that the bars do not buckle). *Ans.* $\delta = 0.0296$ in.

1.8.2. The load of 10,000 lb is carried by seven bars of equal cross section. The cross-sectional area of each bar is 0.1 sq in., and the modulus is 30×10^6 psi. The angles subtended by the bars with the vertical are 60°, 45°, 30°, 0°, −30°, −45°, and −60°. What is the ratio of the load in bar 3 to that in bar 1? What is the vertical displacement of the joint to which the load is applied?

Ans. The ratio is 1/2, and the displacement is $\delta = 0.0308$ in.

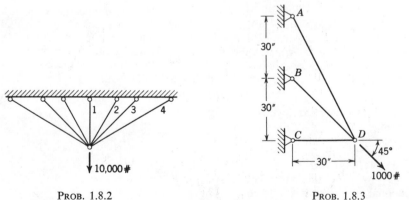

PROB. 1.8.2   PROB. 1.8.3

1.8.3. The inclined load of 1000 lb is applied to joint D as shown in the figure. If the cross-sectional area of each bar is 0.1 sq in. and Young's modulus for the material is 30×10^6 psi, what are the forces in the bars?

Ans. In bar AD the force is 379 lb, in bar BD it is 522 lb, and in bar CD it is 170 lb, all in tension.

1.8.4. Find the vertical downward displacement of point D of the redundant framework shown in the figure if A is the cross-sectional area of each bar and E is Young's modulus of the material. *Ans.* $\delta = \sqrt{2}\, WL/EA$.

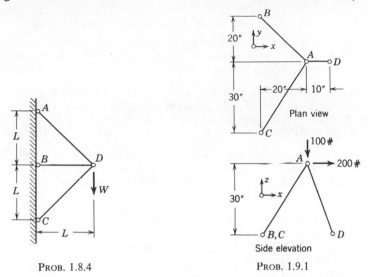

Plan view

Side elevation

PROB. 1.8.4   PROB. 1.9.1

1.9.1. A load is applied to joint A of a statically determinate space framework. It has an x component of 200 lb and a z component of -100 lb. The load is transmitted to points B, C, and D of a rigid plane by means of three bars. Find the forces in the bars.

Ans. In bar AB the force is 137 lb tension, in bar AC it is 104 lb tension, and in bar AD it is 282 lb compression.

1.9.2. An elevated platform is connected with the ground by n rods symmetrically arranged as shown in the figure. A moment M is applied to the

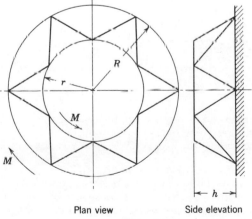

Plan view   Side elevation

PROB. 1.9.2

platform and is balanced by an equal and opposite moment at the ground. What are the forces in the connecting bars if the plane of the platform and that of the ground are permitted to warp?

Ans. The forces alternate between $+F$ and $-F$ with

$$F = M[h^2 + R^2 + r^2 - Rr \cos (2\pi/n)]^{1/2}/[nRr \sin (2\pi/n)].$$

1.12.1. A simply supported beam of length L is subjected to a lateral load W at a distance $(L/5)$ from one of the supports. What is the deflection of the beam at a distance $(L/4)$ from the same support? *Ans.* $\delta = 0.0099 \, WL^3/EI.$

1.12.2. A beam of length L is simply supported. It has a moment of inertia I from $x = 0$ to $x = L/3$ and a moment of inertia $2I$ from $x = L/3$ to $x = L$. A vertical downward load W is applied to the beam at $x = 2L/3$. What is the deflection under the load? *Ans.* $\delta = (13/1458)(WL^3/EI).$

1.12.3. A beam of length L is simply supported. It has a bending rigidity EI from $x = 0$ to $x = L/2$ and a bending rigidity $2EI$ from $x = L/2$ to $x = L$. A vertical downward load W is applied to the beam at its middle. What is the deflection under the load? *Ans.* $\delta = WL^3/64EI.$

1.12.4. A cantilever beam of constant bending rigidity is under the action of a vertical downward load W applied at a distance $L/3$ from the free end. The total length of the beam is L. What is the deflection of a section of the beam at a distance $L/4$ from the fixed end? *Ans.* $\delta = (7/384)WL^3/EI$

1.13.1. A beam whose two ends are rigidly fixed is also supported at two intermediate points. What is the bending-moment distribution if the bending rigidity is constant?

Ans. At the rigid wall to the left $M = -(12/40)Pa$; under the load in the left span $M = (11/40)Pa$; over the intermediate support to the left $M = -(6/40)Pa$; at the midpoint of the central span $M = 0$; and in the right-hand half of the beam the moments follow from the antisymmetry of the structure. Between any two values given the bending moment varies linearly.

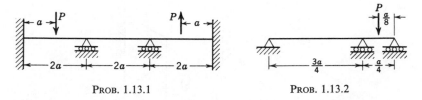

PROB. 1.13.1   PROB. 1.13.2

1.13.2. Calculate the bending moments in the redundant beam shown in the figure, assuming that the bending rigidity is constant.

Ans. The moment under the load is $(29/512)Pa$, and at the middle support it is $-(6/512)Pa$.

1.13.3. A cantilever beam is loaded with a uniformly distributed load of 10 lb per in. of its length. The cross section is a solid square of 1 in. side length, the length of the beam is 30 in., and the material is steel with $E = 29 \times 10^6$ psi. A support is provided 0.2 in. below the free end of the beam; that is, there is a gap of 0.2 in. between the beam and the support before the load is applied.   Calculate the bending moments.

   Partial ans.   The maximum moment occurs at the fixed end, and it is $M = -2742$ in.-lb.

1.13.4.   A beam of constant moment of inertia is subjected to a uniformly distributed load w.   The two ends of the beam are rigidly fixed, and a support is provided in the middle a distance Δ below the beam. Find the end moments.

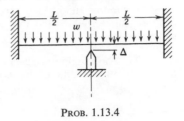

   Ans.   The end moment is $-wL^2/12$, provided $w \leq w_\Delta$; and it is $-(wL^2/48) - (24EI\Delta/L^2)$, provided $w \geq w_\Delta$; and $w_\Delta = 384EI\Delta/L^4$.

PROB. 1.13.4

1.15.1.   A continuous beam on four supports is provided with a frictionless ideal hinge at its midpoint.   The bending rigidity EI of the beam is constant. Find the deflection δ of the hinge when a load W is applied to the middle of the first span.   *Ans.* $\delta = -WL^3/64EI$ (upward).

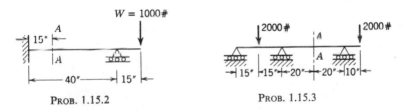

PROB. 1.15.1

1.15.2.   Find the deflection δ of the statically indeterminate beam at section AA under the action of the load W when the cross section is a solid rectangle measuring 2 in. by 1 in. (with 2 in. the vertical dimension) and when the material is steel with $E = 30 \times 10^6$ psi,   *Ans.* $\delta = -0.0263$ in. (upward).

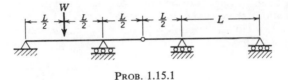

PROB. 1.15.2   PROB. 1.15.3

1.15.3.   The continuous beam on three supports has a constant section.   It is a solid rectangle 2 in. deep and 1 in. wide.   The material is steel with $E = 30 \times 10^6$ psi.   Find the deflection of section AA.   *Ans.* $\delta = -0.10$ in. (upward).

1.15.4.   What is the deflection of the middle of the beam shown in Prob. 1.13.1, and what is the deflection under the load?

   Ans.   At the middle of the beam the deflection is zero because of the antisymmetry of the system;   under the load to the left it is $\delta = (13/240)(Pa^3/EI)$ (downward).

1.15.5. Calculate the deflection of the midpoint of the beam shown in Prob. 1.13.2; that is, the deflection of the beam at a distance $a/2$ from the end support. *Ans.* $\delta = (5/12,288)(Pa^3/EI)$ (upward).

1.15.6. A uniformly distributed load w is acting on a beam of length L and constant bending rigidity EI, one end of which is rigidly fixed while the other is simply supported. Find the deflection at a distance $L/3$ from the simply supported end. *Ans.* $\delta = (5/972)(wL^4/EI)$.

1.15.7. The continuous beam has a constant bending rigidity. Find the deflection in the middle of the second span.

Ans. $\delta = (3/896)(WL^3/EI)$ (upward).

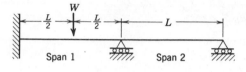

PROB. 1.15.7

1.15.8. A small carriage is resting on a cantilever beam. In the calculation of the deformations the weight of the carriage is disregarded. What is the angle of rotation of the rigid carriage if $L = 20$ in. $I = 2$ in.^{4} $a = 6$ in. $E = 29 \times 10^6$ psi and $W = 1200$ lb? *Ans.* $\phi = 0.011$ rad $= 0.633°$.

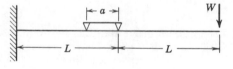

PROB. 1.15.8

1.15.9. The continuous beam on three supports has a constant bending rigidity EI. Find the deflection of the beam under the load W.

Ans. $\delta = (11/96)(WL^3/EI)$.

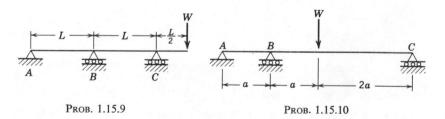

PROB. 1.15.9   PROB. 1.15.10

1.15.10. The continuous beam on three supports has a constant bending rigidity EI. What is the rotation of the beam at support C?

Ans. $\phi = (17/72)(Wa^2/EI)$.

1.17.1. The end portions of a beam between $x = 0$ and $x = s$ as well as between $x = L - s$ and $x = L$ are assumed to be perfectly rigid. The middle portion extending from $x = s$ to $x = L - s$ has a finite bending rigidity EI, Determine the stiffness coefficient S and the carry-over factor C for the beam.

Ans.

$$S = \frac{[1 - (s/L)]^3 - (s/L)^3}{[1 - 2(s/L)]^4} \frac{4EI}{L}$$

$$C = \frac{1}{2}\left\{2 - \frac{[1 - 2(s/L)]^3}{[1 - (s/L)]^3 - (s/L)^3}\right\}$$

1.19.1. The rigid frame shown in the figure is subjected to uniformly distributed loads of intensity w over span BC. Find the bending moments.

Ans. The moments are shown in part (*b*) of the figure.

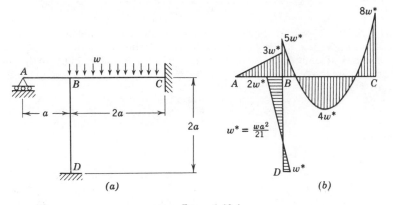

PROB. 1.19.1

2.7.1. A vertical downward load P is acting at the midpoint of a horizontal beam of length L whose two ends are rigidly fixed. The bending rigidity of the beam is EI_0 when $0 \leq x < (L/3)$; it is $4EI_0$ when $(L/3) < x < (2L/3)$; and it is again EI_0 when $(2L/3) < x \leq L$. Find the midpoint deflection by the Rayleigh-Ritz method.

Ans. The exact answer is $y_{max} = (19/36)y^*$, where y^* is the deflection of the midpoint of a uniform beam of bending rigidity $4EI_0$ whose two ends are simply supported: $y^* = (1/48)(PL^3/4EI_0)$.

2.7.2. A vertical downward load P is acting at the midpoint of a simply supported horizontal beam of length L. The bending rigidity of the beam is EI_0 when $0 \leq x < (L/3)$; it is $4EI_0$ when $(L/3) \leq x \leq (2L/3)$; and it is again EI_0 when $(2L/3) < x \leq L$. Find the midpoint deflection by the Rayleigh-Ritz method.

Ans. The exact solution is $y_{max} = (17/9)y^*$, where y^* is the deflection of the midpoint of a simply supported uniform beam of bending rigidity $4EI_0$: $y^* = (1/48)(PL^3/4EI_0)$.

2.7.3. The variation in the bending rigidity of a simply supported beam is given by $EI = EI_0(2x/L)$, provided $x \leq (L/2)$; in the other half of the beam, the bending rigidity varies symmetrically with respect to the center of the beam. Find the deflection of the midpoint of the beam under the action of the load Q by means of the Rayleigh-Ritz method. The exact answer is $y_{\max} = 0.03125 QL^3/EI_0$.

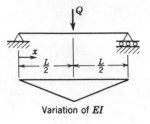

Variation of EI

PROB. 2.7.3

2.7.4. A simply supported beam-column of length L and uniform moment of inertia EI is under the action of a lateral load Q applied to its middle and two compressive axial loads P applied to its ends. Find the deflection of the midpoint by means of the Rayleigh-Ritz method.

Ans.

$$y_{\max} = \frac{2QL^3}{\pi^4 EI} \sum_{n=1,3,5,\cdots}^{\infty} (-1)^{(n-1)/2} \frac{1}{n^4} \left(\frac{1}{1 - (P/n^2 P_E)} \right)$$

where

$$P_E = \pi^2 EI/L^2$$

2.7.5. The bending rigidity of the middle half of a beam-column is $3EI_0$, while that of the end quarters is EI_0. Find the maximum deflection of the beam under the simultaneous action of a lateral load Q in the middle and axial compressive loads P at the ends by means of the Rayleigh method.

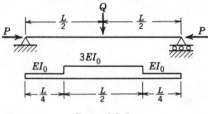

PROB. 2.7.5

The answer is, if a single-sine function is chosen to represent the deflected shape:

$$y_{\max} = \frac{2QL^3}{\pi^4 EI_0} \frac{1}{(2/\pi)(\pi + 1) - (P/P_E)}$$

where

$$P_E = \pi^2 EI_0/L^2$$

3.1.1. Determine the buckling load of a column of constant bending rigidity EI when one end of the column is rigidly fixed while the other end is free to rotate but is not permitted to deflect laterally.

Ans. $P_{cr} = 2.04 P_E$ with $P_E = \pi^2 EI/L^2$.

3.1.2. A rod of constant bending rigidity EI and of length $3L$ is simply supported at its two ends and at a distance L from one end, and axial compressive forces P are applied to its two ends. What is the critical value of the load P at which the rod buckles?

Ans. $P_{cr} = 1.5 P_E$ where $P_E = \pi^2 EI/(2L)^2$.

3.1.3. A rod of length $2L$ is rigidly fixed at one end and simply supported at the middle and at the second end. Its constant bending rigidity is EI. What must be the value of the compressive axial loads P applied to the ends of the rod to cause buckling? *Ans.* $P_{cr} = 1.28 P_E$ where $P_E = \pi^2 EI/L^2$.

3.5.1. A column of length L and of a variable moment of inertia is pinned at both ends. The bending rigidity varies according to the law $EI = EI_0(x/L)$. The two ends of the column are at $x = 0$ and at $x = L$ respectively. What is the value of the axial compressive force P at which the column buckles?

Ans. A single-sine-term approximation to the buckled shape yields $P_{cr\,1} = \pi^2 E(I_0/2)/L^2$ while a two-term approximation gives $P_{cr\,2} = 0.84 P_{cr\,1}$.

3.5.2. A column extending from $x = 0$ to $x = L$ has a stepwise constant moment of inertia. From $x = 0$ to $x = L/2$ the bending rigidity is EI_1 and from $x = L/2$ to $x = L$ it is EI_2. The two ends of the column are pinned. What is the value of the axial compressive load P under which the column buckles when $I_2 = 3I_1$?

Ans. A single-sine-term approximation yields $P_{cr\,1} = \pi^2 E(2I_1)/L^2$ while a two-term approximation gives $P_{cr\,2} = 0.88 P_{cr\,1}$.

3.5.3. A column of variable cross section is fixed at its base and free at its upper end. The cross sections are solid squares and the side length is 1 in. at the bottom and $\frac{3}{4}$ in. at the top. The side length varies linearly with height. The total length of the column is 20 in. What is the buckling load if the material is steel with $E = 29 \times 10^6$ psi?

Ans. $P_{cr} = 10,500$ lb (this is the rigorous solution).

3.5.4. A column whose ends are pinned is subjected to an axial compressive load P. The length of the column is L, and the moment of inertia of the cross section is given by $I = I_0[1 - 3(x/L)^2]$ with x measured from the middle of the column. Find the buckling load.

3.5.5. A column whose ends are pinned is subjected to an axial compressive load P. The length of the column is L, and the moment of inertia of the cross section is given by $I = I_0 \sin(\pi x/L)$, with x measured from one end of the column. Find the buckling load.

3.5.6. A column whose two ends are pinned is subjected to an axial compressive load P. The bending rigidity is constant. The middle of the column is restrained from lateral displacement by a spring whose constant is K (that is, a

lateral displacement of 1 in. of the middle of the column causes a force of K lb
to appear in the spring).   Calculate the buckling load.

Ans.   A single-sine-term approximation yields $P_{cr} = P_E + (2/\pi^2)KL$ where
L is the length of the column.

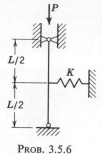

PROB. 3.5.6

3.8.1.   Determine the buckling load of the rod described in Prob. 3.1.1 by
the convergence criterion.   As the sum of the stiffness at the middle support must
vanish, the critical load can be obtained directly with the aid of the tables on
p. 122 of Niles and Newell's textbook, without recourse to a step-by-step
moment-distribution process.

3.8.2.   Determine the buckling load of the rod described in Prob. 3.1.3.   See
hint in Prob. 3.8.1.

4.2.1.   Calculate the bending moments in the statically indeterminate frame
by means of Castigliano's theorem.

Ans.   At the point of application of the load $M = [3/(\pi + 1)](Pr/2)$ and
in the straight portion of the frame $M = -[(\pi - 2)/(\pi + 1)](Pr/2)$ with the
positive sign indicating tension on the outside of the frame.

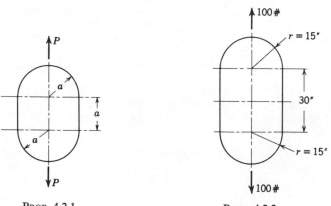

PROB. 4.2.1   PROB. 4.2.2

4.2.2.   Calculate the bending moments in the statically indeterminate frame
by means of Castigliano's theorem.

Ans. At the point of application of the load $M = 584$ in.-lb. and in the straight portions of the frame $M = -166$ in.-lb. with the positive sign indicating tension on the outside of the frame.

4.2.3. Calculate the bending moments in a beam whose two ends are rigidly fixed when the load is zero at one fixed end A and w_{max} at the other fixed end B and varies linearly between the two ends.

$$\text{*Ans.*} \quad M_A = -(2/60)w_{max} L^2 \text{ and } M_B = -(3/60)w_{max} L^2.$$

4.2.4. Calculate the end moments in a beam whose two ends A and B are rigidly fixed when a lateral load P is applied to the beam at a distance $L/4$ from end A. The length of the beam is L, and its constant bending rigidity is EI.

$$\text{*Ans.*} \quad M_A = -(9/64)PL \text{ and } M_B = -(3/64)PL.$$

4.2.5. Calculate the bending moments in a beam with one end simply supported and the other end rigidly fixed when a lateral load P is applied to the middle of the beam. The length of the beam is L, and its constant bending rigidity is EI.

Ans. The moment under the load is $(5/32)PL$, while at the fixed end it is $-(6/32)PL$.

4.2.6. The force P applied at point A to the circular ring of constant bending rigidity is balanced by two forces of magnitude $P/2$, each acting on the ring at points C and E respectively. Find the bending moments.

Partial ans. · At A the bending moment is $0.28Pa$.

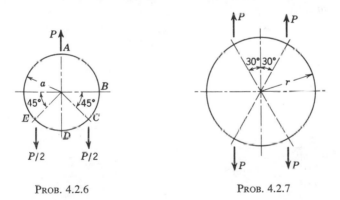

PROB. 4.2.6   PROB. 4.2.7

4.2.7. The circular ring is subjected to four forces P as shown in the figure. Find the bending moments.

Partial ans. The maximum bending moment occurs at the ends of the horizontal diameter. Its value is $0.282Pr$.

4.2.8. The statically indeterminate frame of constant bending rigidity is subjected to four forces P as shown in the figure. Find the bending moments.

Partial ans. The value of the horizontal shear force in each straight vertical portion of the frame is given by $V = (3\pi + 12R)P/(4R^3 + 6\pi R^2 + 24R + 3\pi)$ where $R = b/a$. With the shear force known the bending moments follow from statics.

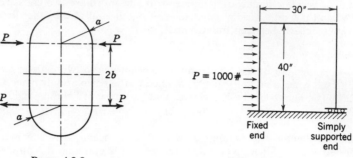

PROB. 4.2.8   PROB. 4.3.1

4.3.1. The singly redundant frame is subjected to a distributed lateral load of 1000 lb total. Find the bending moments.

           Partial ans. The maximum moment is 14,666 in.-lb.

4.3.2. Find the bending moments in the rigid frame of constant bending rigidity shown in the figure.

     Ans. At the three points A, B, and C the bending moments are, respectively, 0, $-0.606\,Wa$, and $0.808\,Wa$, with the positive sign indicating tension on the outside of the structure. Between any two of these points the moments vary linearly.

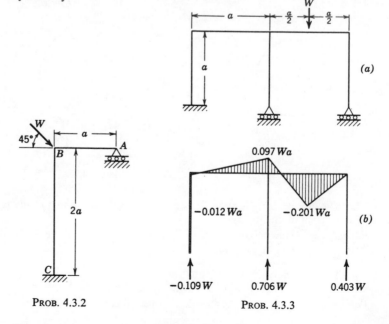

PROB. 4.3.2   PROB. 4.3.3

4.3.3. One column of the rigid frame is rigidly fixed to the ground, while the lower ends of the remaining two columns are free to move laterally and to rotate. Calculate the bending moments caused by the load W.

   Ans. The moment diagram is shown and the maximum moments as well as the reactions are indicated in part (*b*) of the figure. Moments are considered positive if they cause tension on the outside of the frame. The two simply supported columns are not subjected to bending moments.

4.4.1. A vertical downward load P is acting at the midpoint of a simply supported horizontal beam of length L. The bending rigidity of the beam is EI_0 when $0 \leq x < (L/3)$; it is $4EI_0$ when $(L/3) \leq x \leq (2L/3)$; and it is again EI_0 when $(2L/3) < x \leq L$. What is the deflection of the beam at $x = (L/2)$?

   Ans. $y_{max} = (17/9)y^*$, where y^* is the deflection of the midpoint of a simply supported uniform beam of bending rigidity $4EI_0$: $y^* = (1/48) \times (PL^3/4EI_0)$.

4.4.2. A vertical downward load P is acting at the midpoint of a horizontal beam of length L whose two ends are rigidly fixed. The bending rigidity of the beam is EI_0 when $0 \leq x < (L/3)$; it is $4EI_0$ when $(L/3) \leq x \leq (2L/3)$; and it is again EI_0 when $(2L/3) < x \leq L$. What is the deflection of the midpoint of the beam?

   Ans. $y_{max} = (19/36)y^*$, where y^* is the deflection of the midpoint of a simply supported uniform beam of bending rigidity $4EI_0$: $y^* = (1/48) \times (PL^3/4EI_0)$.

4.4.3. A vertical force P is applied to the crown of a rigid frame whose center line is a semicircle. The two ends of the frame are rigidly fixed to a non-yielding foundation. Find the bending-moment distribution.

   Ans. $M = (-0.610 + 0.458 \cos \phi + 0.5 \sin \phi)Pr$.

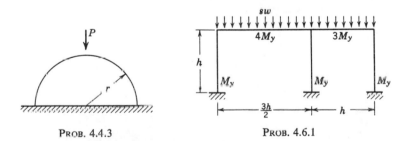

PROB. 4.4.3   PROB. 4.6.1

4.4.4. Calculate the deflection of the midpoint of the beam of Prob. 4.2.3.

   Ans. $\delta = (1/768)(wL^4/EI)$.

4.6.1. The one-story building frame is subjected to a uniform vertical downward load of intensity sw per unit length of the beam. The strength of the elements of the structure is characterized by the yield moment, which is M_y for the columns, $4M_y$ for the longer beam, and $3M_y$ for the shorter beam. Find the value of the safety factor s at which the building frame collapses.

   Ans. $s = 22.8(M_y/h^2w)$.

4.6.2. The two-story building frame is subjected to a uniformly distributed lateral load of intensity sw per unit height and to a uniformly distributed vertical downward load of intensity $3sw$ per unit length of the beams of the second floor. The strength of the elements of the structure is characterized by their yield moments. The value of the yield moment is M_y for the beams and columns of the second story, $2M_y$ for the columns of the first story, and $3M_y$ for the beams of the second floor. Find the value of the safety factor s at which the structure collapses.

Ans. $s = 4M_y/h^2w$.

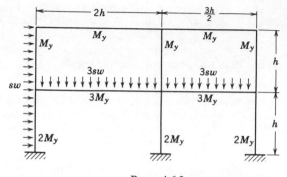

PROB. 4.6.2

A U T H O R   I N D E X

S U B J E C T   I N D E X

Aileron, 81
Antisymmetry, 351
  in loading of rigid frame, 118
  in space framework, 49
Approximate representation of functions, 201
  by polynomials, 254
  by trigonometric series, 202
  with least square error, 202

Beam, analysis of, by Castigliano's theorem, 342
  by Hardy Cross method, 95
  continuous, 77
    on elastic supports, 79, 81
    with support displacements, 79, 81
  convention for sign of moments, 98
  deflections of, 69, 70
  elastic, 63
  rigid, 14
  rotation of end section of, 71
  ultimate load of clamped, 380
Beam column, 28, 151
Bent, simple, 357
Bernoulli-Navier assumptions, 63
Buckling, by torsion, 291
  by torsion and flexure, 280, 291
  during loading process, 325
Buckling load, 224
  calculation of, by Rayleigh's method, 254, 256
  charts, 275
  effect of initial stress on, 328
  effect of residual stress on, 328
  minimization of, 274

Buckling load, of column of variable cross section, 256, 259, 265
  of column with elastic end fixation, 266, 276, 278, 279
  of elastic column, 229
  of sandwich column, 245, 247, 248
  of truss, 294
Building frame, see Frame

Carry-over factor, 99
  for compression bar with gusset plates, 168
  for tension bar with gusset plates, 170
  of beam column, 163, 164
    by Galerkin's method, 211, 215
Castigliano's theorem, first, 377
  second, 342
  use of, in frame analysis, 357
Choice of redundants in space framework, 51
Circular frame, 349, 353, 422
Collapse mechanisms, 400, 405
Collapse of structure, 380
  see also Limit-load analysis
Column, analysis of, by Rayleigh's method, 251, 256
  curves, empirical, 327
  deflections of, see Deflections
  effective length of, with elastic end fixation, 276–279
  initially crooked, 230
  with eccentric loads, 238
  with elastic end fixation, 266
  with variable cross section, 256
Complementary energy, 332